Advances in Processing, Simulation and Characterization of Alloys

Advances in Processing, Simulation and Characterization of Alloys

Guest Editors

Marianthi Bouzouni
Shouxun Ji

MDPI

Basel • Beijing • Wuhan • Barcelona • Belgrade • Novi Sad • Cluj • Manchester

Guest Editors
Marianthi Bouzouni
Department of Physical Metallurgy and Forming
Hellenic Research Centre for Metals S.A.—ELKEME S.A
Oinofyta
Greece

Shouxun Ji
Brunel Centre for Advanced Solidification Technology (BCAST)
Brunel University of London
Uxbridge
UK

Editorial Office
MDPI AG
Grosspeteranlage 5
4052 Basel, Switzerland

This is a reprint of the Special Issue, published open access by the journal *Crystals* (ISSN 2073-4352), freely accessible at: https://www.mdpi.com/journal/crystals/special_issues/2NZL13BXH4.

For citation purposes, cite each article independently as indicated on the article page online and as indicated below:

Lastname, A.A.; Lastname, B.B. Article Title. *Journal Name* **Year**, *Volume Number*, Page Range.

ISBN 978-3-7258-6618-2 (Hbk)
ISBN 978-3-7258-6619-9 (PDF)
https://doi.org/10.3390/books978-3-7258-6619-9

Contents

About the Editors

Marianthi Bouzouni

Marianthi Bouzouni is a senior metallurgical engineer at the Hellenic Research Centre for Metals S.A. located in Greece. Since May 2015, she has been working in the Departments of Physical Metallurgy and Forming and Process Metallurgy, providing scientific services in the frame of scientific projects particularly in alloy design of steel, aluminium and copper alloys and process improvement especially with the utilization of computational tools such as the CALPHAD method for thermodynamic and kinetics simulation as well as phase-field modelling. She received her PhD. from the National Technical University of Athens in 2021 in the field of physical metallurgy and computational materials science. The title of her PhD. thesis was "Simulation of phase transformations and microstructural evolution of low alloyed steels during ultra-fast heat treatment". She continues to actively collaborate with the National Technical University of Athens through scientific projects. Within the frame of PhD. studies, she has supervised 17 diploma theses. She has published 26 research papers in scientific journals and delivered 24 presentations at international and national conferences. Her work has been cited 219 times and she has h-index of 8.

Shouxun Ji

Shouxun Ji is currently a Professor at Brunel University of London. He has been focusing on the development of lightweight metallic materials and structures, in particular aluminium alloys and magnesium alloys. He also works on the additive manufacturing of alloys. He received the award of national innovation in 2017 from CMF UK for advancing the casting materials and the development of aluminium alloys, the award of 'Person of the Year 2022' from the International Magnesium Science and Technology Society for his achievement in magnesium research and development, and the 2023 award of excellence in the commercial cast product category from the International Magnesium Association (IMA) for high-temperature magnesium alloys in a small engine cylinder. Prof. Ji has published over 200 scientific papers in peer-reviewed journals with an h-index of 43, and has filed about 30 patents worldwide. He is an Editor or Editorial Board Member of 5 international scientific journals including *Journal of Magnesium Alloys, International Journal of Metallic Materials, Crystals, Advanced Manufacturing*, and *Frontier in Materials*. Since 2020, he has continuously been on the list of top 2% scientists, compiled by Elsevier and Stanford University.

crystals

MDPI

Article

Microanalysis-Based Simulation of Heterogeneous Dispersoid Distribution in an Al Alloy After the Homogenization Stage

Nicolás García Arango [1,*], Roman Schuster [1,2], Rainer Abart [2] and Erwin Povoden-Karadeniz [1,3]

1 Christian Doppler Laboratory for Interfaces and Precipitation Engineering CDL-IPE, Institute of Materials Science and Technology, TU Wien, Getreidemarkt 9, 1060 Vienna, Austria; erwin.povoden-karadeniz@tuwien.ac.at (E.P.-K.)

2 Department of Lithospheric Research, University of Vienna, Josef-Holaubek-Platz 2, 1090 Vienna, Austria

3 Institute of Materials Science and Technology, TU Wien, Getreidemarkt 9, 1060 Vienna, Austria

* Correspondence: nicolas.arango@tuwien.ac.at

Abstract

We simulate the dispersoid distribution within the Al matrix grains of an aluminum 6082 alloy by combining finite difference cell modeling with mean-field precipitation simulations. The results demonstrate that the initial as-cast microstructure and the heating rate during the ramp-up to the isothermal homogenization temperature are the most important factors governing the dispersoid particle distribution. The simulation results are validated by Electron Probe Microanalysis (EPMA) and Optical Microscopy on experimental run products. The results indicate that dispersoids can only achieve uniform distribution throughout the grain when the heating rate to the homogenization temperature is sufficiently slow.

Keywords: AA6082; dispersoids; phase transformations; simulation; thermokinetics

1. Introduction

During the fabrication process of a wide variety of aluminum products, a fine and homogeneous distribution of small precipitates, referred to as dispersoids, is desired. The fine distribution of dispersoids has a significant, positive effect on the mechanical properties of the material, such as their role in preventing recrystallization during cold rolling and extrusion at elevated temperatures [1–4]. During the different stages of the processing of the 6xxx series Al alloys containing Mn, cubic α-$Al_{15}(Fe,Mn)_3Si_2$ dispersoids act as preferential nucleation sites for both metastable (β'-$Mg_3Si_{1.67}$, B'-$Al_4Mg_8Si_7$) and stable (β-Mg_2Si) phases, involved in the hardening response of the material [5–7].

Dispersoids typically precipitate during the heating ramp to the homogenization temperature. The 6xxx series Al alloys usually contain small amounts of dispersoid-forming elements, such as Mn, Fe, and Cr. The supersaturation of these elements in the matrix after the casting leads to the formation of highly stable intermetallic phases inside the grains. The crystal structures of the dispersoids depend on the nominal alloy composition [8] and particularly on the Mn:Fe ratio [9]. In the Al 6082 alloy, which is relatively high-alloyed by Mn, cubic α-$Al_{15}(Fe,Mn)_3Si_2$ dispersoids are predominantly stabilized [10–12].

A detailed analysis of the microstructure [13] revealed that dispersoids do not nucleate evenly throughout the grain. Two distinctive zones depleted of dispersoids are observed. One is located near the grain boundaries, which is known as the Dispersoid Free Zone (DFZ). A second area can be found in the center of the grains, sometimes referred to as the Coarse Dispersoid Zone (CDZ). In the CDZ, dispersoids are typically larger than in other domains of the grains and occur with a significantly lower number density. Both the CDZ and the DFZ depend on the heat treatment applied. Different heating rates or temperature profiles lead to different dispersoid distributions, as reported by the addition of a holding step to the heating process [13,14]. This paper aims to explain the thermodynamic reasons for the observed formation of the CDZ in the center of the grains, including experimental results and a computational analysis of the problem.

2. Experimental

EN-AW 6082 ingots with the nominal composition shown in Table 1 were prepared by the company partner, Neuman Aluminium, via direct chill casting. The samples were cut from the central region of the as-cast material. Using a Bähr 805 A/D dilatometer (TA Instruments, New Castle, DE , USA), four different heating rates from room temperature to 560 °C, which is a typical homogenization temperature for the processing of Al 6082 alloys, were applied: 0.01 K/s, 0.1 K/s, 1 K/s, and 10 K/s, with a holding time of 1 h at 560 °C, followed by rapid quenching. The heat treatments were performed under vacuum. To observe the intermetallic phases and their distribution across the grains, samples were polished and etched with a solution of 10% NaOH and 90% distilled water. The etching was performed for 75 s.

Table 1. Chemical composition of the Al 6082 alloy in mass percent.

	Si	Fe	Mg	Mn	Cu + Cr + Zn + Ti	Al
AA6082	1.2	0.25	0.65	0.57	0.1	Balance

Electron probe microanalysis (EPMA) was performed on samples in their as-cast state and after reaching 300, 400, and 560 °C when heated at 0.01 K/s. These measurements were performed using a CAMECA SX Five Field Emission Gun Electron Microprobe (CAMECA, Gennevilliers, France) at the EPMA laboratory of the Core Facility Electron Beam Microanalysis, Faculty of Earth Sciences, Geography and Astronomy at the University of Vienna (AT). The instrument was operated at an acceleration voltage of 18 keV, a beam current of 62 nA, and a working distance of 10 mm.

3. Modeling

To perform a computational study with high predictive capacity on the dispersoid nucleation and their heterogeneous distribution inside the Al matrix grains, the underlying modeling of the simulation must take into account the relevant physical mechanisms at all stages of the process, from the initial as-cast condition, where microsegregation is of the utmost importance, to the fundamental laws for nucleation (Classical Nucleation Theory) and diffusion-mediated precipitate growth, including the thermodynamic CALPHAD description of all relevant phases present in the material [15].

Section 3.1 explains the definition of the simulation domain and how it is discretized to account for the initial chemical heterogeneity, with elements being unevenly distributed within the grains. Sections 3.2–3.4 discuss the thermodynamic essence of the nucleation process, starting with the Classical Nucleation Theory (CNT), followed by the treatment of the nucleation sites and the particular nucleation sequence in these alloys. The chapter concludes with the treatment of chemical gradients inside the grain. It is shown how CNT can provide a reasonable explanation on why α-$Al_{15}(Fe,Mn)_3Si_2$ dispersoids can only distribute evenly throughout the grain when the heating rate to the homogenization temperature is sufficiently slow.

3.1. Domain Discretization

The typical microstructure after solidification is characterized by the relatively large primary phases located near the grain or dendrite boundaries of the Al matrix, and a non-uniform distribution of solute elements in the Al matrix grains, commonly known as microsegregation [16]. Microsegregation results from thermodynamically controlled minor element partitioning between the liquid and the Al-fcc matrix phase and its temperature dependence during the solidification stage.

The segregated as-cast state of an alloy can be approximated by the Scheil–Gulliver hypothesis [17–19], according to which the solidification of the material is treated as a sequential concatenation of concentric layers (or shells) of newly formed solid. Each layer has the equilibrium concentration of solute elements given by the phase diagrams, with no diffusion allowed between the successive solid shells. This results in the widely described layered nature of the as-cast material [20]. Contrary to the solid portion, the Scheil–Gulliver hypothesis considers diffusion-mediated elemental redistribution to be fast enough in the liquid to assume instantaneous chemical homogenization and uniform composition of the liquid, which includes the solidification front. The Scheil–Gulliver hypothesis has been extensively employed to describe the inhomogeneous compositional nature of a wide variety of materials in their as-cast state [21–25].

The Scheil–Gulliver calculation can be used to obtain the theoretical element distribution in mass % in the Al-fcc matrix grains as a function of the liquid fraction, F_{liquid}. It is possible to estimate the element distribution within the grains by relating the amount of liquid remaining (F_{liquid}) in mol. % and the measured mean grain diameter once solidification is completed, as shown in Equation (1). The x coordinate has its origin at the hypothetical center of the Al matrix grain and its maximum (when the phase fraction of liquid approaches 0) at half of the average grain diameter $\left(\frac{d}{2}\right)$, where the grain boundary is located. For the case under study, the mean grain diameter after solidification is approximately 120 μm.

$$x = \left(1 - F_{liquid}\right) \cdot \frac{d}{2}. \tag{1}$$

Using the MatCalc software, version 6.05 [26], along with its thermodynamic and diffusion databases, mc_al.tdb [27] and mc_al.ddb [28] respectively, the primary phases formed during casting and their relative amounts are calculated with the Scheil–Gulliver module for the composition presented in Table 1. Results are shown in Figure 1a. The main primary intermetallic phase is the α-$Al_{15}(Fe,Mn)_3Si_2$ phase. In lower amounts, the presence of silicon in its diamond crystal structure, Mg_2Si, and β-$Al_{4.5}FeSi$ is also predicted in the as-cast microstructure. Similar results were obtained by [11,29].

As can be interpreted from the element distribution calculation shown in Figure 1b, there are significant Mg and Si concentration differences between the center of the grain or dendrite arm, representing the location of the first Al-fcc crystallization from the liquid, and the regions located closer to the boundary, which correspond to the stage when the solid fraction approaches 1, i.e., where the latest portion of the solid is formed. As explained in the following paragraphs, the concentration gradients of the solute elements within the Al-fcc matrix grains (microsegregation) are responsible for the heterogeneous distribution of precipitates formed during homogenization.

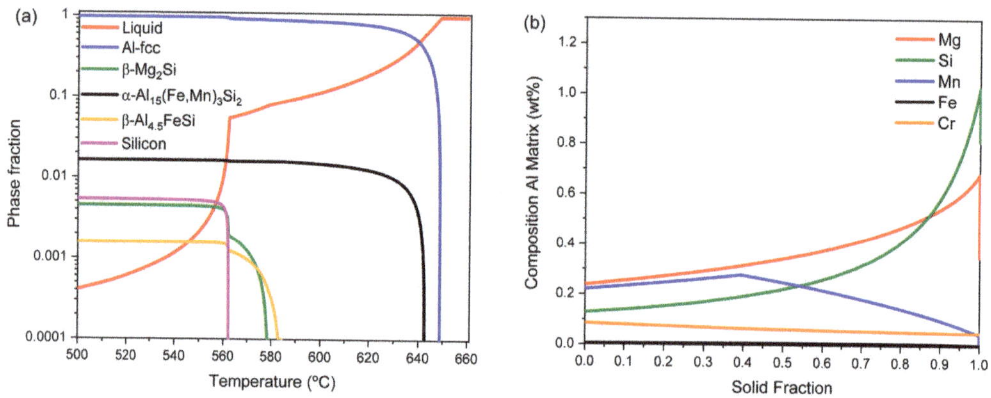

Figure 1. Phase fraction as a function of temperature resulting from the Scheil–Gulliver calculation (**a**). Element distribution in the Al-fcc matrix as a function of fraction solidified (**b**).

To establish a reasonable computational study for the highly heterogeneous element distribution present within the microstructure, the simulation domain is treated as a linear 1D geometrical setup, divided into 10 regions ("cells"). The simulation cells cover all domains from the center to the border of the grain, as schematized in Figure 2a. To account for the initial as-cast state, Mg and Si concentrations are assigned to each simulation cell following the Scheil–Gulliver calculation, shown as continuous dashed lines in Figure 2b. This simulation is carried out using the "Cell Simulation" module available in the MatCalc software, which has been introduced in [30]. A finite differences formulation solving Fick's second law of diffusion is used to couple the transient multi-element diffusion between the discretized subdomains. The formulation is shown for one element in Equation (2).

$$\left[1+\frac{D_{i,i+1}\Delta t}{(\Delta x)^2}+\frac{D_{i,i-1}\Delta t}{(\Delta x)^2}\right]c_i^{t+}-\frac{\Delta t}{(\Delta x)^2}\left[D_{i,i+1}c_{i+1}^{t+}+D_{i,i-1}c_{i-1}^{t+}\right]=c_i^{t}. \quad (2)$$

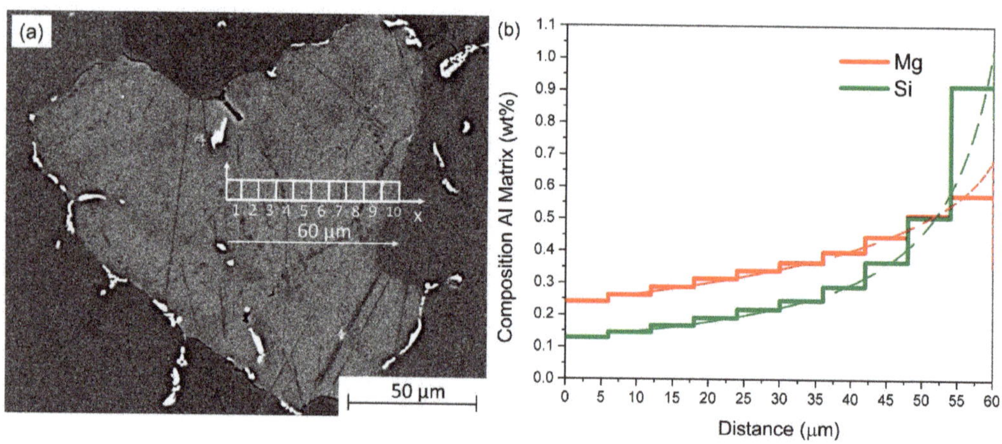

Figure 2. As-cast state of the material. (**a**) Discretization of a grain into subdomains along the horizontal x-axis, extending from the hypothetical grain center (left) to its boundary (right). (**b**) Initial Mg and Si concentrations of the discretized cells are shown as solid lines, while the calculated Scheil–Gulliver concentration profiles are depicted with dashed lines. The horizontal x-axis represents the distance from the grain center to its boundary.

With c_i^t being the element content of cell i at time t, and c_i^{t+} the concentration of the element at t + Δt. $D_{i,i+1}$ and $D_{i,i-1}$ are the mean diffusion coefficients of cell i + 1 and cell i − 1, respectively. Δx represents the width of one cell. A system containing n cells and j elements results in j linear equation systems, each with n equations and n variables.

3.2. Classical Nucleation Theory

The nucleation of a new crystalline phase within a solid can be addressed by means of the CNT [31]. In CNT, the exact location and time at which an individual particle forms cannot be determined, as the process is of stochastic nature. However, the rate of new stable nuclei in a volume under a given temperature and chemical condition can be calculated, giving rise to the fundamental parameter in precipitation kinetics, the nucleation rate, J (nuclei/s·m^3) [31]. The equation describing the nucleation rate reads as follows:

$$J = N_0 \cdot Z \cdot \beta^* \cdot e^{-G_c^*/K_B T} \cdot e^{-\frac{\tau}{t}}. \quad (3)$$

The pre-exponential terms N_0, Z, and β^* represent the number of available nucleation sites, the Zeldovich factor, and the atomic attachment rate, respectively. G_c^*, in J/atom, is the Gibbs energy of a cluster with critical size, which may be understood as the transition point between stable and unstable nuclei, k_B is the Boltzmann constant, T is the absolute temperature, τ is the incubation time, and t is the time of the precipitation event. The effect of G_c^* is particularly relevant for the nucleation rate, as a small variation in G_c^* strongly affects the probability of nucleation events. To obtain an expression for G_c^*, the energy required to form a nucleus, ΔG_{nucl} (J), can be split into a negative contribution, stemming from the driving force for the formation of a favorable phase, ΔG_{vol}, and a positive contribution, resulting from the creation of a new interface between the nucleating and parent phase, ΔG_{surf}. For a spherical nucleus of radius ρ, ΔG_{nucl} is then given by the following:

$$\begin{aligned} \Delta G_{nucl} = \Delta G_{vol} &+ \Delta G_{surf} = \tfrac{4}{3}\pi\rho^3 \Delta g_{vol} + 4\pi\rho^2 \Delta g_{surf} \\ &= \tfrac{4}{3}\pi\rho^3\left(-\tfrac{d_{chem}}{\upsilon^\alpha} + \Delta G_{elas}\right) + 4\pi\rho^2\gamma, \end{aligned} \quad (4)$$

where γ (J/m^2) represents the interfacial energy, ΔG_{elas} (J/m^3) is the elastic energy related to the misfit in crystal structure between precipitate and matrix, υ^α is the molar volume of the matrix phase (m^3/mol), and d_{chem} is the chemical driving force for nucleation (J/mol). This value can be directly calculated from the thermodynamic database.

In CNT, the Gibbs energy of the critical cluster, G_c^*, is calculated as the maximum of ΔG_{nucl} in the ΔG_{nucl} vs. ρ curve by the following:

$$\frac{\partial \Delta G_{nucl}}{\partial \rho} = 0, \quad (5)$$

which results in the expression of the Gibbs energy of the critical cluster:

$$G_c^* = \frac{16\pi}{3} \frac{\gamma^3}{\left(-\frac{d_{chem}}{\upsilon^\alpha} + \Delta G_{elas}\right)^2}. \quad (6)$$

This states that the larger the chemical driving force for the nucleation of a phase, the lower the energy barrier required to form a stable nucleus.

3.3. Nucleation Sequence and Nucleation Sites

To perform a computational analysis with predictive capacity on the spatial distribution of α-$Al_{15}(Fe,Mn)_3Si_2$ dispersoids, the simulation scheme must include the microsegregation and the sequence of the nucleation events. The complete sequence of precipitates forming in the Al-Mg-Si system has been extensively studied [32–35]. For the current analysis, metastable phases, such as Guinier–Preston zones and β″-Mg_5Si_6, nucleating and dissolving at temperatures lower than the appearance of dispersoids, are not considered, as they do not appear to contribute to their nucleation. Within the alloy composition range considered in this study, and based on Transmission Electron Microscopy images, the metastable β′-$Mg_3Si_{1.67}$ and B′-$Al_4Mg_8Si_7$ phases are reported to nucleate during the heating stage of the homogenization treatment at a temperature of ~250 °C [5,6,10]. These precipitates play an essential role in the microstructure evolution since they act as preferential nucleation sites for dispersoids at temperatures of ~300–350 °C [5,8,36–38]. It should be noted that for a similar composition range, Lodgard et al. [10] reported the presence of a second intermediate phase nucleating in between β′-$Mg_3Si_{1.67}$ and the dispersoids, the "u-phase". Due to the lack of unambiguous evidence, this additional transformation is not included in the present simulation, and β′ will be the term used to group and address the metastable phases leading to dispersoid nucleation.

A simplified schematic illustration of the dispersoid nucleation sequence, as inferred from previous microanalytical findings [5,8,10], is shown in Figure 3. Initially, as the temperature rises to T_1, metastable β′ particles nucleate within the Al matrix grains. As the temperature increases further to the range at which dispersoids nucleate (T_2), the dispersoids precipitate preferentially at the interfaces between the β′ and the Al matrix. Once the solvus of the metastable β′ particles is exceeded (T_3), they dissolve while the dispersoids remain.

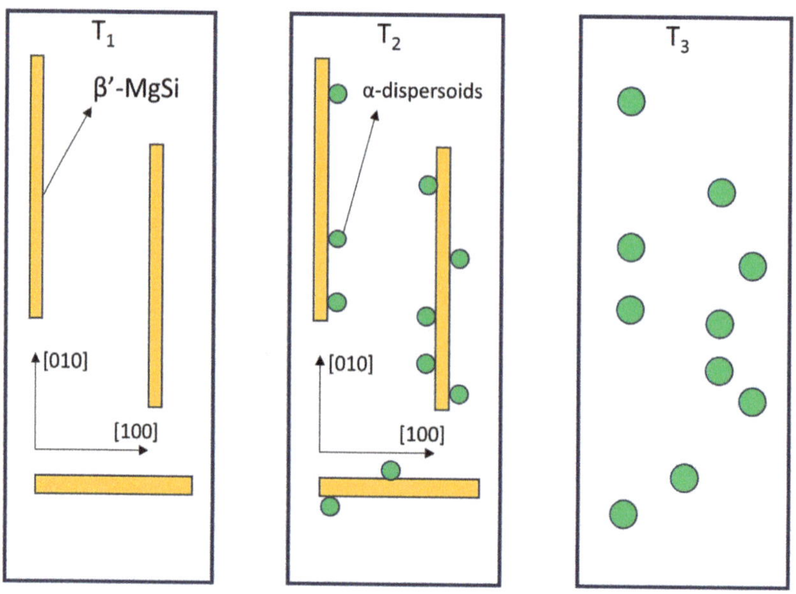

Figure 3. Schematic illustration of the sequence leading to the heterogeneous nucleation of α-$Al_{15}(Fe,Mn)_3Si_2$ dispersoids in 6xxx series Al alloys. $T_1 < T_2 < T_3$. The indicated crystallographic directions refer to the lattice of the Al matrix phase.

The metastable MgSi-rich phases nucleating during the heating to homogenization temperature have a platelet morphology, as schematized in Figure 3, and are of a semi-coherent nature with the Al matrix, yielding crystallographic coherency in the <100> direction [33]. The nucleation of these phases occurs inside the grain, probably at dislocation cores or other defects in the matrix phase.

The simulation aims to link the nucleation events of metastable MgSi-rich phases and favored dispersoid formation via the "excess" availability of nucleation sites due to MgSi pre-nuclei. To achieve such a scenario in a physically sound simulation setup, the β' phase nucleation in the bulk is initially computed, following CNT. Since these metastable precipitates are known to distribute evenly within the grain [8,10], the number of their possible nucleation sites, N, can be considered to be constant throughout the volume of the grain. To simplify the nucleation nature of β', a homogeneous nucleation is assumed. Therefore, spatial variations in the nucleation rate of the β' phase can only be explained by the value of $G_c{}^*$, which depends on the chemical driving force ($G_c{}^* \propto 1/d_{chem}{}^2$); see Equation (6).

Subsequently, in the simulation, α-$Al_{15}(Fe,Mn)_3Si_2$ dispersoids are set to nucleate on the surface of the metastable β' phase. A mean-field model connecting the available nucleation sites for a precipitating phase to the presence of a precursor phase—a pre-nucleus—is included in MatCalc in the section of "nucleation on precipitate surface", obtaining the parameter $N = N_\alpha$ in the formula of the nucleation rate (Equation (7)) as the number of atoms at the surface of the parent precipitate. The calculation of N_α in MatCalc has been addressed in our research group [39] as the ratio between the interfacial area of precursor phase $A_{int/\beta'}$ and the mean area of a single atom A_{at}.

$$N_\alpha = \frac{A_{int/\beta'}}{A_{at}}. \tag{7}$$

It should be noted that N_α depends on time and temperature, but also on the spatial coordinate $N_\alpha = N_\alpha(x)$, as the precursor phase will populate the inner and outer regions of the grain differently due to the differences in the respective chemical driving forces. At temperatures above the solvus of β', and as it dissolves, dispersoid nucleation will stop, even at large driving forces, due to the absence of available nucleation sites ($N_\alpha \rightarrow 0$).

3.4. Analysis of the Interrelation Between Chemical Gradients and Precipitation

The impact of microsegregation on the nucleation of precipitates can be understood by examining the differences in the evaluated chemical driving forces between solute-enriched and solute-depleted zones within the grain (see Figure 4a,b). After solidification, areas of the microstructure closer to the grain boundaries contain Mg and Si atoms at higher concentrations than the more internal regions of the grains. The associated higher solute supersaturation translates into larger driving forces for the nucleation of metastable phases. The calculated gradient in the chemical driving force for the precipitation of α-$Al_{15}(Fe,Mn)_3Si_2$ dispersoids and β' at 200 and 300 °C is shown in Figures 4a and 4b, respectively.

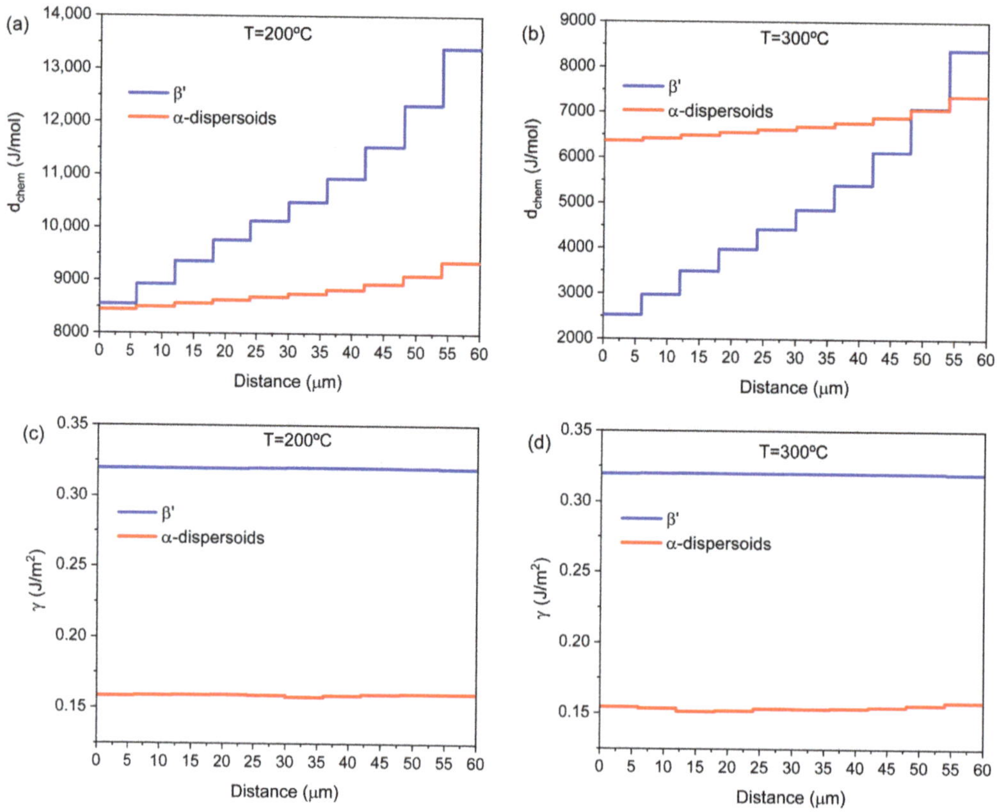

Figure 4. Calculated chemical driving forces for precipitation of β′-MgSi and α-$Al_{15}(Fe,Mn)_3Si_2$ dispersoids in the discretized domain at T = 200 °C (**a**) and T = 300 °C (**b**). Calculated interfacial energies with the Al-fcc matrix of β′-MgSi and α-$Al_{15}(Fe,Mn)_3Si_2$ dispersoids at T = 200 °C (**c**) and T = 300 °C (**d**). The *x*-axis has its origin in the theoretical center of the grain and extends to its boundary.

The elastic strain energy (ΔG_{elas}) always acts against the nucleation event [31]. For nucleation to occur, the driving force for the nuclei formation must be larger than the elastic strain energy ($d_{chem}/\upsilon^{\alpha} > \Delta G_{ela}$). As a simplification, ΔG_{elas} is ignored in the analysis. Furthermore, it can be assumed that interfacial energies are largely independent of element distribution in the grain, as shown in Figure 4c,d, where interfacial energies for β′-MgSi and α-dispersoids are calculated at 200 and 300 °C. As a consequence, the ratio of critical nucleation energy between areas closer to the boundary, $G_c{}^*{}_{bound}$, and the center of the grain, $G_c{}^*{}_{center}$, can be approximated from Equation (6) as follows:

$$\frac{G^*_{c_{center}}}{G^*_{c_{bound}}} \sim \left(\frac{d_{chem\,bound}}{d_{chem\,center}}\right)^2. \tag{8}$$

As shown in Figure 4a,b, upon temperature evolution, differences in Mg and Si concentration between the outer and inner regions of the grain translate into differences in β′ nucleation driving forces between $\left.\frac{d_{chem\,bound}}{d_{chem\,center}}\right|_{T=200^\circ C} \sim 1.5$ and $\left.\frac{d_{chem\,bound}}{d_{chem\,center}}\right|_{T=300^\circ C} \sim 3$. Following Equation (8), the ratios in critical nucleation energy can be calculated as a first approximation by the following: $\frac{G^*_{c_{center}}}{G^*_{c_{bound}}} \sim 2.25 - 9$.

Recalling the nucleation rate expression (Equation (5)) and assuming all factors multiplying the exponential term (N_0, Z, and β^*) remain constant throughout the grain, the ratio between the nucleation rate in the region close to the grain boundary and at the central area of the grain can be expressed as follows:

$$\frac{J_{bound}}{J_{center}} \sim \left(\frac{e^{-Gc^*_{bound}/K_B T}}{e^{-Gc_{center}^*/K_B T}} \right) = e^{-\frac{Gc^*_{bound}}{K_B T} \cdot \left(\frac{Gc^*_{center}}{Gc^*_{bound}} - 1 \right)}. \tag{9}$$

With a characteristic value of $G_c{}^* = 5 \cdot 10^{-20}$ at 200 °C, the previous ratio in nucleation probability of metastable β' between the internal and external domains of a grain can be estimated as approximately 4–5 orders of magnitude higher in the outer regions of the grain as compared to the inner zone, where the CDZ typically appears.

Mn is the slowest diffusing element of the microstructure. Therefore, its availability is the limiting factor for the nucleation and growth of α-dispersoids. Mn microsegregation resulting from the solidification of the material does not produce such sharp concentration gradients within the grain as in the case of Mg and Si, as seen in Figure 5. EPMA results indicate that Mn is evenly distributed in the dendrite arms, having a pronounced peak at the boundaries, where the large primary Mn-rich intermetallic particles are located. For simplicity, the Mn concentration is subsequently treated as a constant averaged value from the Scheil–Gulliver calculation.

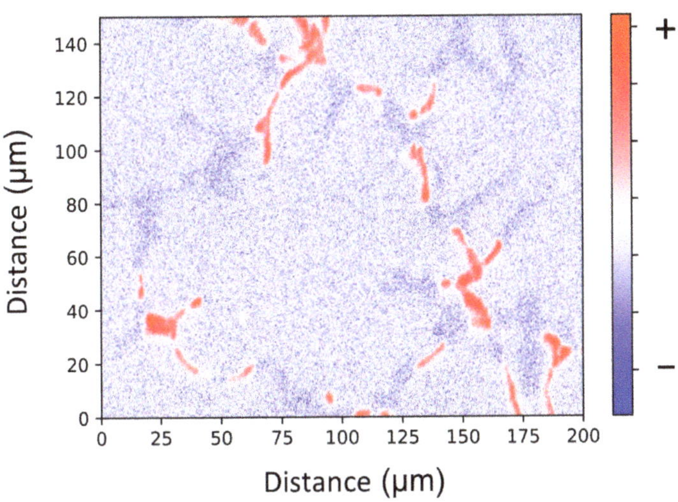

Figure 5. EPMA analysis of the Mn distribution in the as-cast state.

Compared to metastable MgSi phases, the differences in driving force for the precipitation of dispersoids between the different regions within the grain are insignificant (Mn is not strongly segregated). Consequently, their associated nucleation barrier is similar between the central and outer areas of the grain. Added to the nucleation site availability, the ratio of dispersoid nucleation rate can be expressed as follows:

$$\frac{J_{bound}}{J_{center}} \sim \frac{N_{bound}}{N_{center}} \cdot \left(\frac{e^{-Gc^*_{bound}/K_B T}}{e^{-Gc_{center}^*/K_B T}} \right) \sim \frac{N_{bound}}{N_{center}}. \tag{10}$$

This results in a linear dependence of the nucleation rate ratio on the available nucleation sites.

4. Results and Discussion

4.1. Formation of Coarse Dispersoid Zones

The prediction of the effect of the heating rate on dispersoid nucleation is demonstrated in Figure 6, where etched samples are shown after heating to 560 °C at four different rates (0.01, 0.1, 1, and 10 K/s) and held at 560 °C for 1 h. For the slowest heating rate, 0.01 K/s, small regions of DFZ appear near the grain boundaries, where primary phases are prominent. At the same time, α-$Al_{15}(Fe,Mn)_3Si_2$ dispersoids distribute evenly throughout the grain, and the CDZ is absent (Figure 6a,b). On the contrary, CDZ develops at faster heating rates and becomes more prominent at higher rates. In Figure 6c,d, at a heating rate of 0.1 K/s, the heterogeneous distribution of dispersoids is seen. Marked in red, a CDZ appears in the inner section of the grains. At 1 K/s, shown in Figure 6e,f, the area of the CDZ within the grain significantly increases, where there is an even more prominent effect at the highest heating rate (10 K/s), displayed in Figure 6g,h. The optical microscopy results suggest that a larger CDZ results from faster ramps to homogenization temperatures.

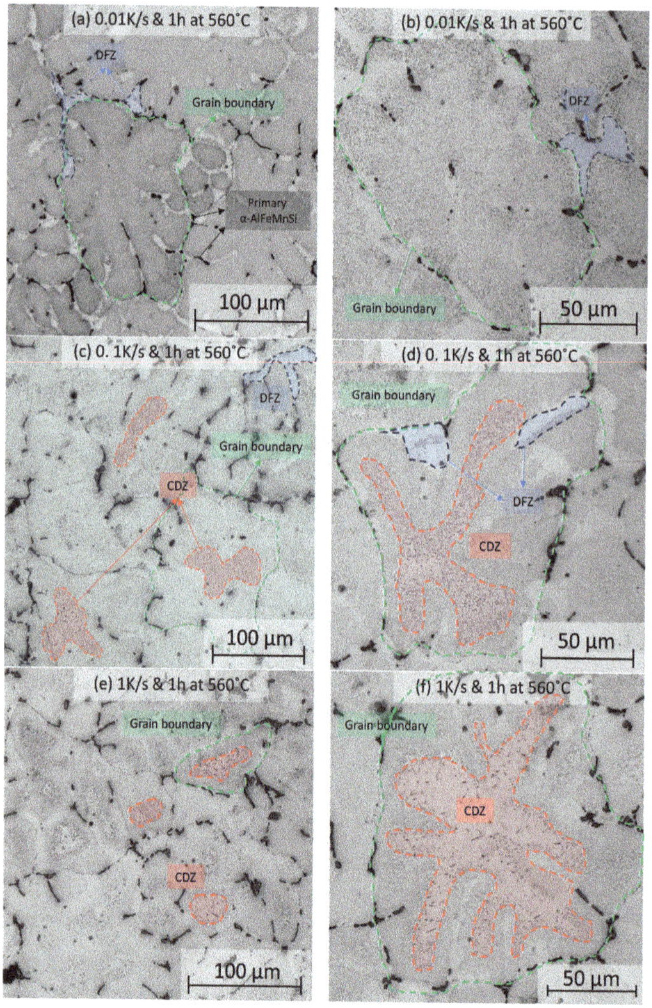

Figure 6. *Cont.*

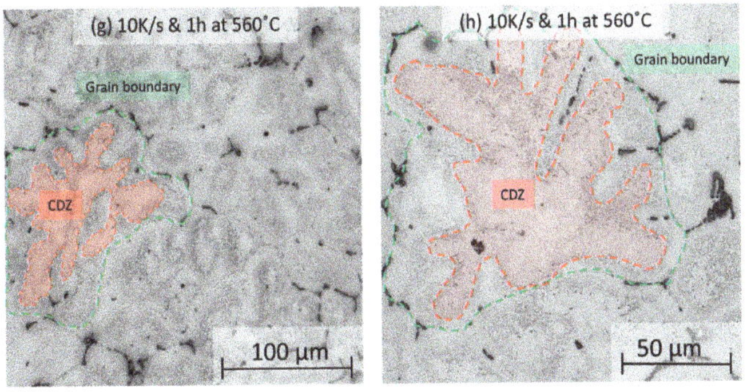

Figure 6. Optical microscopy images showing the dispersoid distribution in samples that were heated with four different heating rates. No CDZ is observed at a heating rate of 0.01 K/s (**a**,**b**). CDZ appears at the grain centers after heating at 0.1 K/s (**c**,**d**), 1 K/s (**e**,**f**), and 10 K/s (**g**,**h**) heating rates and keeping the samples at 560 °C for 1 h.

4.2. Concentration Gradients

We hypothesize that the heterogeneous dispersoid distribution within the grain is directly related to the initial microsegregation. The variation in dispersoid distribution at different heating rates could be linked to changes in the extent of element diffusion under the various heating conditions. By this reasoning, concentration gradients would be effectively removed by diffusion at sufficiently low heating rates. A more even element distribution would then translate into a uniform nucleation of the initial β' precursor phase and, subsequently, of the α-$Al_{15}(Fe,Mn)_3Si_2$ dispersoids, hindering the formation of CDZs.

To test this conjecture, the current discretized setup is used to simulate the diffusion effect on concentration gradients. Initially, a diffusion calculation is performed in MatCalc from room temperature until the homogenization temperature of 560 °C at the slowest heating rate of 0.01 K/s. The results of element distribution at selected temperatures are shown in Figure 7.

As shown in Figure 7, only at temperatures higher than 400 °C, even for the slowest heating rate studied, is the diffusion of Mg and Si atoms from the more concentrated areas to the less concentrated ones appreciable. Diffusion results at higher heating rates are not included, as faster heating only results in lower overall element redistribution. It should be noted that the Mn evolution can be neglected due to its low segregation tendency and its slow diffusion. An estimation of the average distance an atom can travel at a given temperature is given by the relation:

$$l \approx \sqrt{2Dt}, \tag{11}$$

where D is the diffusivity of the given element in the matrix, and t is the time. The diffusivity can be calculated as follows:

$$D = D_o e^{-Q_R/RT}, \tag{12}$$

where D_0 is the frequency factor, with unit m^2/s, and Q_R is the activation energy, expressed in J/mol. Accepted values of D_0 and Q_R for both Mg and Si can be retrieved from published diffusion databases. In MatCalc [28], these values are $D_{0\text{-}Mg} = 7.5 \times 10^{-6}$ m^2/s, $Q_{R\text{-}Mg}$ = 115,000 J/mol, $D_{0\text{-}Si} = 3 \times 10^{-5}$ m^2/s, and $Q_{R\text{-}Si}$ = 123,900 J/mol.

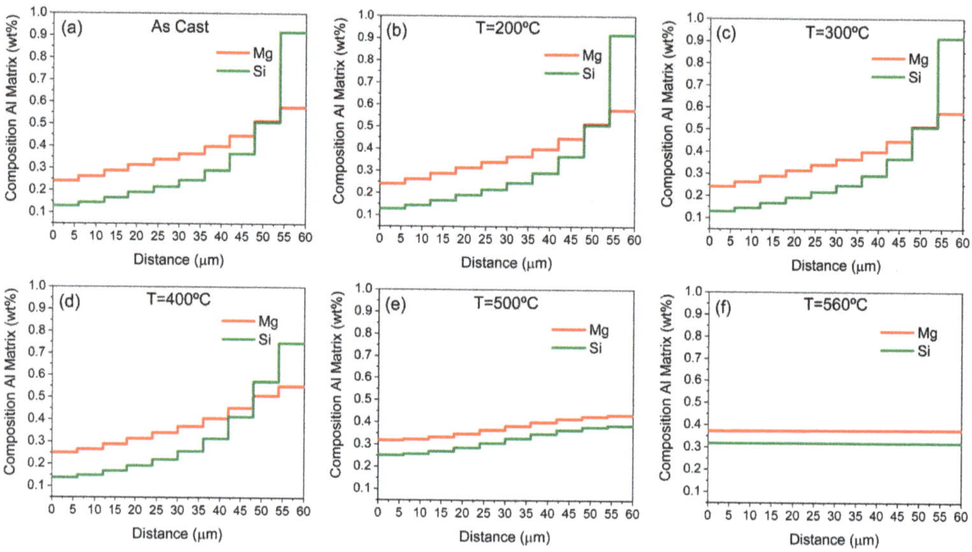

Figure 7. Diffusion calculation on the discretized domain at different states during heating to the homogenization temperature at a rate of 0.01 K/s, **(a)** at the as-cast state, **(b)** at 200 °C, **(c)** at 300 °C, **(d)** at 400 °C, **(e)** at 500 °C, **(f)** at 560 °C.

By heating the material at 350 °C for 1 h, the average length that Mg and Si atoms can move within the grain, according to Equation 11, is around 2.7 μm and 3.2 μm, respectively, a distance more than an order of magnitude lower than the mean grain radius of ~60 μm. This suggests that during the heating stage of the homogenization treatment, at temperatures where metastable phases nucleate and regardless of the heating rate, the atomic mobility is insufficient to level out the concentration gradients present in the grain. This can only happen at the prolonged holding stages at elevated temperatures: the average distance Mg and Si atoms can move, calculated after 1 h at 560 °C, results in values of 49.8 μm and 64.5 μm, respectively, in the same order of magnitude as the average grain radius.

EPMA results confirm that segregation of Mg and Si is still present at 300 and 400 °C during the heating stage of the homogenization treatment, as shown in Figure 8c–f, and it is only at higher temperatures that elements distribute evenly in the grain, as seen in Figure 8g,h. Due to its low mobility in the Al matrix, changes in Mn distribution from its initial state, shown in Figure 5, are not observed.

Chemical gradients translate into driving force gradients for the precipitation inside the grain. Due to the sparse mobility at temperatures where metastable phases nucleate, it has already been shown that the original compositional gradients are largely retained despite the large chemical potential gradients acting against them, and therefore, the evolution of the microstructure follows, regardless of the heating rate, the kinetic environment established by the initial element segregation.

The formation of the large areas of depleted and coarse dispersoid zones within the grain cannot be understood as a diffusion effect but rather as a situation where differences in the chemical driving forces for the nucleation of precursor β′ phase, and therefore in critical nucleation energy G_c*, scale exponentially in the nucleation rate formula that describes the probability of a single nucleation event. This results in several orders of magnitude of difference in probable nucleation events between areas near the boundary and the center of the grain, linearly determining the available nucleation sites for subsequent dispersoid nucleation.

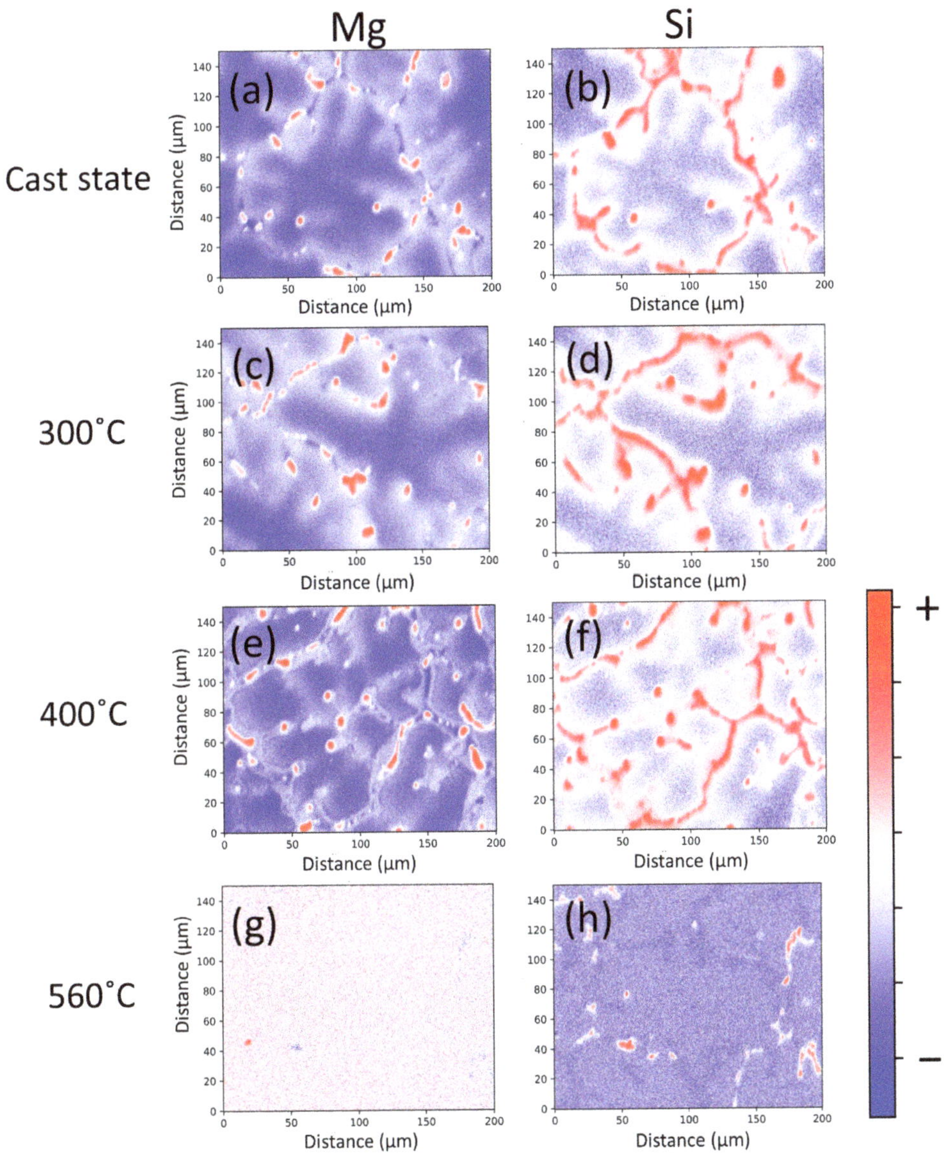

Figure 8. EPMA results showing the distribution of Mg and Si atoms within the grain at various stages of the homogenization heating ramp. (**a**,**b**) correspond to the as-cast state. (**c**,**d**) associate with the grain at 300 °C; (**e**,**f**) at 400 °C; and (**g**,**h**) at 560 °C.

Lower heating rates or isothermal steps at sufficiently high temperatures allow more time for nucleation, leveling out the effect of nucleation rate differences between zones depleted in Mg and Si, where the nucleation of metastable β′ is less likely, and zones with higher Mg and Si concentrations, which are more prone to the nucleation of metastable β′. At sufficiently low heating rates, metastable β′ particles may thus nucleate even in larger numbers and also in the less favorable zones for nucleation. At the same time, initial β′ nuclei formed near the boundaries of the grain grow (and later dissolve) upon heating. Assuming a temperature range for preferred β′ nucleation, ΔT, alloys heated at 0.01 K/s would have 10^3 more seconds to undergo particle generation compared to a material heated at 10 K/s, which would be enough time to compensate for the calculated differences in nucleation rates within the grain.

4.3. Simulation

The simulation results are shown in Figure 9. After one hour at 560 °C, differences in dispersoid precipitation for the four heating rates studied can be expressed in terms of the Mn content of the Al matrix, predicted phase fraction, calculated number densities, and remaining driving force for α-$Al_{15}(Fe,Mn)_3Si_2$ nucleation. The evolution of Mn concentration in the Al matrix depends indirectly on the heating rate and the associated dispersoid evolution. Supersaturated Mn from the casting process is consumed by the formation but mainly by the growth of dispersoids, as their growth potential will be higher at lower heating rates. Mn supersaturation is entirely eliminated throughout the entire grain at 0.01 K/s heating rate, while Mn remains supersaturated in the inner regions of the grain for the faster heating rates, as shown in Figure 9a, the latter being associated with incomplete dispersoid nucleation.

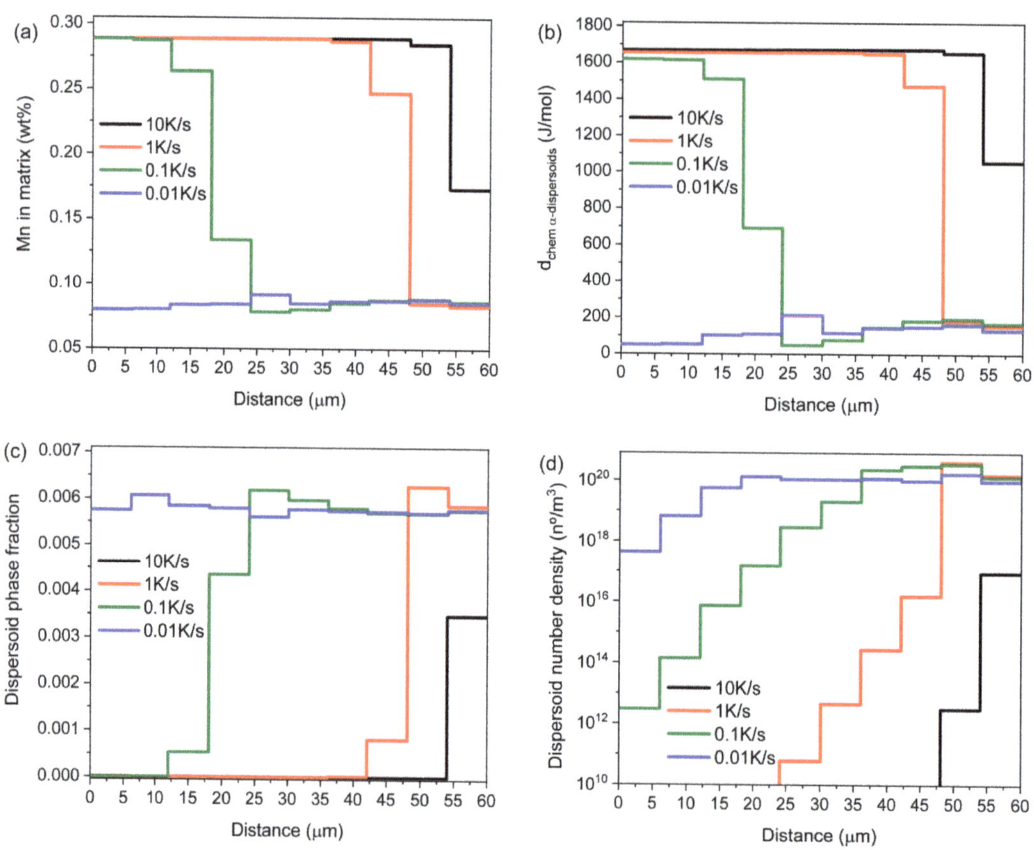

Figure 9. Simulation results showing (**a**) Mn dissolved in the Al matrix as solid solution, (**b**) driving force for α-$Al_{15}(Fe,Mn)_3Si_2$ nucleation, (**c**) α-$Al_{15}(Fe,Mn)_3Si_2$ dispersoids phase fraction, (**d**) and α-$Al_{15}(Fe,Mn)_3Si_2$ number densities as function of the spatial coordinate of the grain for four different heating rates.

The nucleation of dispersoids is expected to be completed when, after the heat treatment, the remaining driving force for their formation is close to zero, i.e., it is too low to enable any further nucleation event in all grain regions. It is seen in Figure 9b that this is only achieved at the lowest heating rate, while large driving forces for the nucleation of dispersoid remain present in the inner regions of the grain, reaching broader areas at higher heating rates, as exemplified in the 10 K/s heating instance.

Figure 9c,d show the calculated dispersoid phase fraction and number density after the heat treatments. For the lowest heating rate, the variation in the dispersoid fraction is small throughout the grain, which is supported by the experimental evidence shown in Figure 6a,b. Number densities calculated for a heating rate of 0.01 K/s vary from 10^{18} n°/m^3 in the central domain to 10^{20} n°/m^3 in the outer regions. Compared to published data by [40], for an Al 6082 alloy, density values after a homogenization treatment at a slightly higher temperature of 570 °C and 2 h were approximately 10^{12} n°/m^2. Assuming the characteristic dimension of dispersoids to be $l_c \approx 100$ nm, the measured number densities were in the range of 10^{19} n°/m^3. At higher heating rates, the dispersoid number density and phase fraction in the inner regions of the grains tend to zero, demonstrating the formation of the experimentally observed CDZ (Figure 6c–h).

5. Summary

The heating rate effect on the evolution of the spatial distribution of α-$Al_{15}(Fe,Mn)_3Si_2$ dispersoid precipitation during the heating of an Al 6082 alloy from the as-cast state at room temperature to the homogenization temperature was studied, combining experiments and thermokinetic modeling. The experiments show that after relatively rapid heating at heating rates of 0.1 K/s to 10 K/s, dispersoid particles are unevenly distributed within the Al matrix grains, typically showing a Coarse Dispersoid Zone (CDZ) in the internal regions of the grains. Only after comparatively slow heating at 0.01 K/s is a nearly uniform distribution of dispersoid particles is obtained. Both experimental and computational results point to the initial concentration gradients of solute Mg and Si within the Al matrix grains (microsegregation) attained during the casting as the main cause for the heterogeneous dispersoid distribution after rapid heating. Element diffusion is insufficient to eliminate the initial solute concentration differences during heating to the temperatures where dispersoid nucleation occurs. The resulting differences in the driving forces for the nucleation of the intermediate β' phase between the inner and outer areas of the grain lead to the observed heterogeneous dispersoid distribution. Only at the low heating rate of 0.01 K/s is there enough time available in the relevant temperature range to allow for the sufficient nucleation of the intermediate β' phase, which largely compensates for the chemically induced nucleation rate differences and results in a more homogeneous dispersoid distribution throughout the Al matrix grains.

Author Contributions: Conceptualization, N.G.A.; methodology, N.G.A., R.S., R.A. and E.P.-K.; software, N.G.A.; investigation, N.G.A., R.S., R.A. and E.P.-K.; writing—original draft preparation, N.G.A.; writing—review and editing, N.G.A., R.S., R.A. and E.P.-K.; visualization, N.G.A.; supervision, E.P.-K. All authors have read and agreed to the published version of the manuscript.

Funding: This research was funded by Christian Doppler Forschungsgesellschaft in the framework of the CD-Laboratory of Interfaces and Precipitation Engineering (CDL-IPE).

Data Availability Statement: Data will be made available on request.

Acknowledgments: The financial support from the Austrian Federal Ministry for Digital and Economic Affairs and the National Foundation for Research, Technology, and Development is gratefully acknowledged. The authors acknowledge TU Wien Bibliothek for financial support through its Open Access Funding Programme. The authors gratefully acknowledge the provision of sample material by Neuman Aluminium.

Conflicts of Interest: The authors declare no conflict of interest. The funders had no role in the design of the study, in the collection, analysis, or interpretation of data, in the writing of the manuscript, or in the decision to publish the results.

References

1. Zhang, G.W.; Nagaumi, H.; Han, Y.; Xu, Y.; Parish, C.M.; Zhai, T.G. Effects of Mn and Cr additions on the recrystallization behavior of Al-Mg-Si-Cu alloys. *Mater. Sci. Forum* **2017**, *877*, 172–179. [CrossRef]
2. Dorward, R.C.; Bouvier, C. A rationalization of factors affecting strength, ductility and toughness of AA6061-type Al-Mg-Si-(Cu) alloys. *Mater. Sci. Eng. A* **1998**, *254*, 33–44. [CrossRef]
3. Arnoldt, A.R.; Schiffl, A.; Höppel, H.W.; Österreicher, J.A. Influence of different homogenization heat treatments on the microstructure and hot flow stress of the aluminum alloy AA6082. *Mater. Charact.* **2022**, *191*, 112129. [CrossRef]
4. Polmear, I.; St. John, D.; Nie, J.F.; Qian, M. *Light Alloys: Metallurgy of the Light Metals*, 5th ed.; Butterworth-Heinemann: Oxford, UK, 2017; pp. 1–525.
5. García Arango, N.; Kahlenberg, R.; Schuster, R.; Miesenberger, B.; Maawad, E.; Wojcik, T.; Povoden-Karadeniz, E. In-situ Analysis of EN-AW 6082 during Homogenization. *Soc. Sci. Res. Netw.* **2025**. [CrossRef]
6. Milkereit, B.; Starink, M.J.; Rometsch, P.A.; Schick, C.; Kessler, O. Review of the quench sensitivity of aluminium alloys: Analysis of the kinetics and nature of quench-induced precipitation. *Materials* **2019**, *12*, 4083. [CrossRef] [PubMed]
7. Kahlenberg, R.; Wojcik, T.; Falkinger, G.; Krejci, A.L.; Milkereit, B.; Kozeschnik, E. On the precipitation mechanisms of β-Mg_2Si during continuous heating of AA6061. *Acta Mater.* **2023**, *261*, 119345. [CrossRef]
8. Farh, H.; Djemmal, K.; Guemini, R.; Serradj, F. Nucleation of dispersoids study in some Al-Mg-Si alloys. *Ann. Chim. Sci. Matér.* **2010**, *35*, 283–289. [CrossRef]
9. Becker, H.; Bergh, T.; Vullum, P.E.; Leineweber, A.; Li, Y. Effect of Mn and cooling rates on α-, β- and δ-Al–Fe–Si intermetallic phase formation in a secondary Al–Si alloy. *Materialia* **2019**, *5*, 100198. [CrossRef]
10. Lodgaard, L.; Ryum, N. Precipitation of dispersoids containing Mn and/or Cr in Al-Mg-Si alloys. *Mater. Sci. Eng. A* **2000**, *283*, 144–152. [CrossRef]
11. Mrówka-Nowotnik, G.; Sieniawski, J.; Wierzbinska, M. Intermetallic phase particles in 6082 aluminium alloy. *Arch. Mater. Sci. Eng.* **2007**, *28*, 69–76.
12. Davignon, G.; Serneels, A.; Verlinden, B.; Delaey, L. An isothermal section at 550 °C in the Al-rich corner of the Al-Fe-Mn-Si system. *Metall. Mater. Trans. A* **1996**, *27*, 3357–3361. [CrossRef]
13. Lodgaard, L.; Ryum, N. Distribution of Mn- and Cr-containing dispersoids in Al-Mg-Si-alloys. *Mater. Sci. Forum* **2000**, *331–337*, 945–950. [CrossRef]
14. Li, Z.; Qin, J.; Zhang, H.; Wang, X.; Zhang, B.; Nagaumi, H. Improved distribution and uniformity of α-Al(Mn,Cr)Si dispersoids in Al-Mg-Si-Cu-Mn (6xxx) alloys by two-step homogenization. *Metall. Mater. Trans. A* **2021**, *52*, 3204–3220. [CrossRef]
15. Kozeschnik, E. *Modeling Solid-State Precipitation*; Momentum Press: New York, NY, USA, 2013.
16. Ohno, M.; Yamashita, M.; Matsuura, K. Importance of microstructural evolution on prediction accuracy of microsegregation in Al-Cu and Fe-Mn alloys. *Int. J. Heat Mass Transf.* **2019**, *133*, 1120–1128. [CrossRef]
17. Scheil, E. Bemerkungen zur Schichtkristallbildung. *Int. J. Mater. Res.* **1942**, *34*, 70–72. [CrossRef]
18. Gulliver, G.H. The quantitative effect of rapid cooling upon the constitution of binary alloys. *J. Inst. Met.* **1913**, *9*, 120–157.
19. Glicksman, M.E. *Principles of Solidification: An Introduction to Modern Casting and Crystal Growth Concepts*; Springer: New York, NY, USA, 2011.
20. Schaffnit, P.; Stallybrass, C.; Konrad, J.; Stein, F.; Weinberg, M. A Scheil-Gulliver model dedicated to the solidification of steel. *Calphad* **2015**, *48*, 184–188. [CrossRef]
21. Samaras, S.N.; Haidemenopoulos, G.N. Modelling of microsegregation and homogenization of 6061 extrudable Al-alloy. *J. Mater. Process. Technol.* **2007**, *194*, 63–73. [CrossRef]
22. Basak, C.B.; Krishnan, M. Applicability of Scheil-Gulliver solidification model in real alloy: A case study with Cu-9wt%Ni-6wt%Sn alloy. *Philos. Mag. Lett.* **2015**, *95*, 376–383. [CrossRef]
23. Sarafoglou, P.I.; Serafeim, A.; Fanikos, I.A.; Aristeidakis, J.S.; Haidemenopoulos, G.N. Modeling of microsegregation and homogenization of 6xxx Al-alloys including precipitation and strengthening during homogenization cooling. *Materials* **2019**, *12*, 1421. [CrossRef]

24. Cai, M.; Robson, J.D.; Lorimer, G.W. Simulation and control of dispersoids and dispersoid-free zones during homogenizing an AlMgSi alloy. *Scr. Mater.* **2007**, *57*, 603–606. [CrossRef]
25. Malakhov, D.V.; Panahi, D.; Gallerneault, M. On the formation of intermetallics in rapidly solidifying Al-Fe-Si alloys. *Calphad* **2010**, *34*, 159–166. [CrossRef]
26. Kozeschnik, E.; MatCalc. Solid State and Precipitation Kinetics Simulation Software. Available online: http://www.matcalc.at (accessed on 10 June 2025).
27. Institute of Materials Science and Technology, TU Wien. Open-License Multi-Component MatCalc Thermodynamic Database mc_al Version 2.036. *Vienna.* Available online: https://www.matcalc.at/index.php/databases/open-databases (accessed on 10 June 2025).
28. Institute of Materials Science and Technology, TU Wien. Open-License Multi-Component MatCalc Diffusion Mobilities Database mc_al Version 2.008. *Vienna.* Available online: https://www.matcalc.at/index.php/databases/open-databases (accessed on 10 June 2025).
29. Sarafoglou, P.I.; Haidemenopoulos, G.N. Phase fraction mapping in the ascast microstructure of extrudable 6xxx aluminum alloys. *Int. J. Mater. Res.* **2014**, *105*, 1202–1209. [CrossRef]
30. Retzl, P.; Zamberger, S.; Kozeschnik, E. Computational analysis of austenite film thickness and C-redistribution in carbide-free bainite. *Mater. Res. Express* **2021**, *8*, 7. [CrossRef]
31. Christian, J.W. The Classical Theory of Nucleation. In *The Theory of Transformations in Metals and Alloys*; Springer: Pergamon, Turkey, 2002; Chapter 10; pp. 422–479.
32. Feufel, H.; Gödecke, T.; Lukas, H.L.; Sommer, F. Investigation of the Al-Mg-Si system by experiments and thermodynamic calculations. *J. Alloys Compd.* **1997**, *247*, 31–42. [CrossRef]
33. Matsuda, K.; Sakaguchi, Y.; Miyata, Y.; Sato, T.; Kamio, A.; Ikeno, S. Precipitation sequence of various kinds of metastable phases in Al-1.0mass% Mg2Si-0.4mass% Si alloy. *J. Mater. Sci.* **2000**, *35*, 179–189. [CrossRef]
34. Marioara, C.D.; Andersen, S.J.; Zandbergen, H.W.; Holmestad, R. The Influence of Alloy Composition on Precipitates of the Al-Mg-Si System. *Metall. Mater. Trans. A* **2005**, *36*, 691–702.
35. Povoden-Karadeniz, E.; Lang, P.; Warczok, P.; Falahati, A.; Jun, W.; Kozeschnik, E. CALPHAD modeling of metastable phases in the Al-Mg-Si system. *CALPHAD* **2013**, *43*, 94–104. [CrossRef]
36. Hu, R.; Ogura, T.; Tezuka, H.; Sato, T.; Liu, Q. Dispersoid formation and recrystallization behavior in an Al-Mg-Si-Mn alloy. *J. Mater. Sci. Technol.* **2010**, *26*, 237–243. [CrossRef]
37. Li, Y.J.; Arnberg, L. Quantitative study on the precipitation behavior of dispersoids in DC-cast AA3003 alloy during heating and homogenization. *Acta Mater.* **2003**, *51*, 3415–3428. [CrossRef]
38. Kemsies, R.H.; Milkereit, B.; Wenner, S.; Holmestad, R.; Kessler, O. In situ DSC investigation into the kinetics and microstructure of dispersoid formation in Al-Mn-Fe-Si(-Mg) alloys. *Mater. Des.* **2018**, *146*, 96–107. [CrossRef]
39. Sobotka, E.; Kreyca, J.; Fuchs, N.; Wojcik, T.; Kozeschnik, E.; Povoden-Karadeniz, E. The Role of MX Carbonitrides for the Particle-Stimulated Nucleation of Ferrite in Microalloyed Steel. *Metall. Mater. Trans. A* **2023**, *54*, 2903–2923. [CrossRef]
40. Strobel, K.; Easton, M.A.; Sweet, L.; Couper, M.J.; Nie, J.F. Relating quench sensitivity to microstructure in 6000 series aluminium alloys. *Mater. Trans.* **2011**, *52*, 914–919. [CrossRef]

crystals

MDPI

Article

Modeling Fatigue Data of Complex Metallic Alloys Using a Generalized Student's *t*-Birnbaum–Saunders Family of Lifetime Models: A Comparative Study with Applications

Farouq Mohammad A. Alam *, Fouad Khalawi and Abdulkader Monier Daghistani

Department of Statistics, Faculty of Science, King Abdulaziz University, Jeddah 21589, Saudi Arabia; fakhlwi@kau.edu.sa (F.K.); amadaghistani@kau.edu.sa (A.M.D.)

* Correspondence: fmalam@kau.edu.sa

Abstract

The mechanical reliability of metallic alloys under cyclic loading is crucial for optimizing their microstructure–property relationships. Understanding the statistical behavior of fatigue failure data is essential for designing alloys that endure extreme environmental conditions. This study introduces a generalization of the Student's *t*-Birnbaum–Saunders distribution to improve the modeling of fatigue life data, which often exhibit heavy tails and are common in advanced alloy systems. Seven different estimation methods are employed to estimate and compare the parameters of the proposed distribution, providing a comprehensive statistical framework for fatigue failure analysis. The goodness-of-fit of the proposed model and its sub-models, along with the joint relative efficiency of parameter estimates, is assessed using real fatigue data within the maximum likelihood framework. Additionally, the robustness of estimation methods is examined through Monte Carlo simulations across various sample sizes and parameter configurations. The results highlight the effectiveness of the generalized Student's *t*-Birnbaum–Saunders distribution in capturing the stochastic nature of fatigue failure in metallic alloys, offering valuable insights for materials design and predictive reliability modeling. These findings align with advancements in computational modeling and simulation, contributing to developing alloys with tailored mechanical properties.

Keywords: generalized Birnbaum–Saunders distribution; point estimation; data analysis; optimization

MSC: 62F10; 62F30; 62F40; 60E05

1. Introduction

The mechanical reliability of metals and alloys under repeated stress (e.g., cyclic loading) is of paramount importance in structural engineering, aerospace, and materials science. Fatigue failure, defined as progressive structural damage resulting from the combined action of one or more stress factors (e.g., loading), plays a crucial role in determining the lifespan of metallic components. As industrial researchers continue to develop advanced metallic alloys with tailored properties, there is a pressing need for statistical models that accurately capture the stochastic nature of fatigue life data. In practice, well-known lifetime models (e.g., exponential and Weibull distributions) often fail to accommodate the heavy-tailed behavior observed in such data, particularly for complex or novel alloys.

In recent years, various studies have addressed the modeling of fatigue failure. For example, ref. [1] proposed a fatigue crack growth model by integrating finite element modeling with an approximate Bayesian computation framework to predict rail defects, specifically transverse defects, in data from a U.S. Class I railroad. Ref. [2] introduced a new empirical method for estimating the cumulative distribution function of fatigue life under given loads by statistically transforming fatigue data across different stress or strain levels, thereby improving distribution estimation. Ref. [3] proposed a probability-based basis for predicting the fatigue life of notched components, accounting for size effects. The underlying life distribution of pultruded fiber-reinforced compounds was studied by [4], who explored its statistical characteristics. A comprehensive review on fatigue modeling using neural networks was conducted by [5], summarizing past progress and outlining future research directions. Ref. [6] examined the extreme value distribution of maximum fatigue indicators in large-scale engineering components using an upscaling method extrapolated from simulations involving statistical volume elements, each containing approximately 264 grains. Lastly, ref. [7] fitted a set of 950 NASA-conducted gear surface fatigue test results to a three-parameter Weibull distribution, while ref. [8] recently discussed advances in probabilistic modeling of low-cycle fatigue.

In practice, objects made of materials and alloys are subjected to various loading parameters, such as cyclic loading, which causes fatigue in the long run. The latter phenomenon describes the beginning and spread of cracks in materials and alloys due to loading parameters. This phenomenon is considered in the design of structural components in many engineering disciplines. Fatigue modeling and prediction are essential to ensure the reliability and safety of elements considered in cyclic loading conditions, especially in applications like civil infrastructure and automotive. From a statistical perspective, fatigue can be considered as a random variable from a lifetime model, which is typically a probability distribution with non-negative support and at least one parameter that can be linked to a stress factor. Various lifetime probability distributions and statistical inferential methods have been applied to model the statistical variability in fatigue data. For example, the applicability of the normal distribution versus the three-parameter Weibull distribution was investigated and compared by [9] according to the results of six series of 18 to 30 similar fatigue tests on three specimens. A random fatigue-limit model was considered for data of specimens in four-point out-of-plane bending tests of carbon eight-harness satin/epoxy laminate by [10], while ref. [11] analyzed nickel base superalloy fatigue data with runouts via a model with nonconstant standard deviation and a fatigue limit parameter. In this connection, an article by [12] investigated some issues when applying Markov chain Monte Carlo and Laplacian approximations estimators to the experimental design problem in the random fatigue-limit model with application to laminate panel data. In [13], a competitive Bayesian treatment of stress-life data was drawn from a collection of records of fatigue experiments that were performed on 75S-T6 aluminum alloys, assuming at least one model was provided. Recently, ref. [14] proposed the Bayesian model averaging method, which offers a general framework for developing probabilistic fatigue models with superior robustness and precision in their predictions. A direct application of statistical distribution theory in fatigue analysis was a probabilistic model called the Birnbaum–Saunders (BS) fatigue life distribution [15,16]. The latter distribution was proposed to model fatigue failure under cyclic loading based on a physical mechanism in which failure results from the progressive growth of initial micro-cracks. Specifically, crack growth is assumed to be normally distributed, arising from the accumulation of minor damage due to repeated stress cycles, with failure occurring once a critical crack length is reached. On the one hand, the BS model offers a probabilistic framework based on a stochastic representation of damage accumulation. On the other hand, physics-based fatigue models (e.g., the Coffin–Manson model) rely on material-specific parameters such as strain energy amplitude. Overall, the BS model provides a data-driven approach for lifetime prediction under repetitive stress, particularly when detailed material or loading information is unavailable.

Assume that X is a continuous random variable denoting the time-to-failure due to fatigue. Hence, X is considered to follow the unimodal two-parameter BS distribution if its cumulative distribution function (CDF) is defined as:

$$F(x;\alpha,\beta) = \Phi\left(\alpha^{-1}\left[\sqrt{\frac{x}{\beta}} - \sqrt{\frac{\beta}{x}}\right]\right), \quad x > 0, \tag{1}$$

such that $\Phi(\cdot)$ is the CDF of the standard normal distribution, while $\alpha > 0$ is a shape parameter, and $\beta > 0$ is a scale parameter. The study by [17] offered a broader derivation of the BS distribution based on a biological framework, reinforcing the physical justification for its application by easing the initial assumptions made in [15,16]. Since its introduction, the BS distribution has received considerable attention from many researchers due to its close relation to the normal distribution and other desirable properties. According to a thorough review by [18], over two hundred research articles and one dedicated monograph have been published detailing the properties, advancements, and various extensions and generalizations of this distribution. Recently, ref. [19] discussed frequentist and Bayesian estimation methods for the two-parameter conventional BS distribution and the prediction of missing data (i.e., failure times) under the Type-II censoring plan. The latter lifetime distribution is used to analyze the time to fatigue of a sample of a specific type of aluminum coupon. The coupons are cut in a corresponding direction to the rolling, and they were oscillated at specific levels of cycles per second and maximum stress per cycle. The study by [20] considered a goodness-of-fit test for the Birnbaum–Saunders distribution based on the probability plot and provided an application to the strength of glass fiber of length 15 cm. The BS distribution does not accurately model the heavy-tailed pattern that may occur in fatigue data of complex alloys. Researchers have considered modifying or generalizing the BS distribution to compensate for this limitation. For instance, refs. [21,22] developed the generalized Birnbaum–Saunders distribution (GBS) by substituting the normal kernel with elliptically symmetric kernels (e.g., Student's t, Cauchy, etc.) to enhance the model's flexibility. Afterward, ref. [23] considered Student's t BS survival regression models with heavy-tailed errors, presented some inferential results, and performed diagnostics analysis, while ref. [24] introduced the BS distributions based on scale mixtures of normal models and illustrated the results by analyzing fatigue failure data of aluminum specimens of type 6061-T6. Recently, another approach considered by [25] to generalize this model is by changing the square root in the CDF by another shape parameter $\theta > 0$. Recent research about the BS distribution and its generalizations has been considered by researchers, including, but not limited to, the mentioned contributions. A study by [26] introduced a new type of BS model as an alternative to the conventional model to fit fatigue data, while ref. [27] used a new bivariate BS distribution to establish a regression of generalized linear models. One of the desirable properties of the BS distribution is that its scale parameter represents the median lifetime, unlike the scale parameter of the exponential distribution, which represents the mean lifetime. Consequently, some researchers focused on studying the median of the BS distribution. For instance, ref. [28] developed a test to assess the impact of two interacting factors on the median for BS distributed response via the integrated likelihood ratio test framework. To address a serious challenge in lifetime data analysis, ref. [29] considered the model misspecification problem between the log-normal and BS distributions. A study by [30] assumed a varying-stress accelerated life test for a generalized BS model to establish an extension for this model. The study outlined the aspects of this highly flexible distribution, obtained the classical maximum likelihood estimators (MLEs), developed a novel goodness-of-fit procedure, and proposed a new inference approach using Bayesian theory. Under the latter framework, ref. [31] considered estimating the parameters of the BS distribution in the presence of right-censored data. A study by [32] recently employed the BS distribution to model metallic materials fatigue life under cyclic loading. The study compared this model to the normal distribution. The considered models were used to analyze three sets of data of unnotched specimens of 75S-T6 aluminum alloys and carbon coating with several types of loading. Recently, ref. [33] investigated the asymptotic aspects of the method of moments estimators for a newly parametrized BS distribution.

Estimating the model parameters is a topic of significant interest to researchers and has been extensively studied in the statistical literature. Since there are many estimation methods, researchers conducted Monte Carlo comparative simulation studies to compare their performance in terms of various statistical and computational aspects. Such studies are imperative for evaluating under different settings, providing robust comparisons between estimation accuracy and efficiency. Several studies demonstrate this, notably those listed in [34–43], among others.

This study proposes a robust statistical framework based on the generalized Student's t-Birnbaum–Saunders (GTBS) distribution as a suitable model for fatigue failure. The aim of this study is twofold. First, a flexible generalization for the BS distribution is proposed to model fatigue failure. Second, since estimating the distributional properties of a probability distribution (e.g., survivability, hazard rate, etc.) necessitates obtaining appropriate estimations for the model parameters using appropriate estimation procedures, the model parameters are being estimated via seven frequentist parametric estimation procedures. For this study, the considered frequentist parametric estimation methods are the maximum likelihood estimation (MLE), the least-squares estimation (LSE), the weighted least-squares estimation (WLSE), the maximum product of spacings estimation (MPSE), the Cramér–von Mises estimation (CVME), the Anderson–Darling estimation (ADE), and the right-tailed Anderson–Darling estimation (RADE) methods. It is important to mention that it is assumed that the obtained estimators from these methods exist and are unique. It is essential to note that the GTBS model and its sub-models do not directly incorporate microstructural variables (e.g., grain size or phase distribution). However, its formulation inherently captures the statistical characteristics of fatigue life data, often resulting from such underlying heterogeneity. In complex metallic alloys, variations in microstructural features frequently lead to substantial scatter in fatigue life, particularly in the presence of extreme values or heavy tails. The GTBS distribution extends the classical two-parameter BS model by introducing a shape parameter and incorporating heavy-tailed behavior through the Student's t kernel via a degrees-of-freedom parameter, allowing for a possible modeling flexibility. As a result, the GTBS model offers a statistically rigorous and computationally tractable framework for representing the stochastic effects of material structure, particularly when direct physical measurements are unavailable. It is therefore especially well-suited for analyzing fatigue life data in advanced alloys, where conventional models such as the Weibull, gamma, lognormal, or BS distributions may be insufficient.

The remainder of this study is arranged as follows. Section 2 briefly overviews the proposed generalization for the BS distribution and its sub-models. Section 3 defines the previously named estimation approaches of the study. Section 4 reports the statistical analysis of two real datasets. Section 5 summarizes Monte Carlo simulation outcomes from which estimation efficiency is numerically examined and assessed. The study is concluded in Section 6 with a summary and future research directions.

2. A Generalized Student's *t*-Birnbaum–Saunders Distribution

A non-negative continuous random variable X is said to follow the generalized Student's t BS (GTBS) distribution with location parameter $\mu > 0$, scale parameter $\sigma > 0$, shape parameter $\theta > 0$, and degrees of freedom parameter $\nu > 0$ if the associated cumulative distribution function (CDF) is expressed as:

$$F(x;\boldsymbol{\theta}) = \Phi_\nu\left(\left[\frac{x-\mu}{\sigma}\right]^\theta - \left[\frac{\sigma}{x-\mu}\right]^\theta\right), \quad x > \mu, \tag{2}$$

such that $\Phi_\nu(\cdot)$ is the CDF of the standard Student's t distribution with degrees of freedom parameter $\nu > 0$, and $\boldsymbol{\theta} = (\mu,\sigma,\theta,\nu)$ is the vector of model parameters defined on a parameter space $\mathbb{R}^4_+$. The probability density function (PDF) of the GTBS distribution is obtained by differentiating (2) with respect to x, i.e.,

$$f(x;\boldsymbol{\theta}) = \frac{\theta}{(x-\mu)}\left(\left[\frac{x-\mu}{\sigma}\right]^\theta + \left[\frac{\sigma}{x-\mu}\right]^\theta\right)\phi_\nu\left(\left[\frac{x-\mu}{\sigma}\right]^\theta - \left[\frac{\sigma}{x-\mu}\right]^\theta\right), \quad x > \mu, \tag{3}$$

where $\phi_\nu(\cdot)$ is the PDF of the standard Student's t distribution with degrees of freedom parameter $\nu > 0$. For more details about Student's t distribution and its properties, see, for example, [44,45]. It is important to mention that most conducted studies about the BS distribution and its extensions and generalizations do not assume the existence of a location parameter; instead, it is typically assumed that the BS distribution has a shape parameter α and scale parameter β with CDF (1). Table 1 shows the sub-models of the GTBS distribution, while Figure 1 illustrates the shape of the PDF of the GTBS distribution for various values of θ and ν assuming the location parameter $\mu = 0$ and the scale parameter $\sigma = 1$ without loss of any generality. It is essential to note that some sub-models have counterintuitive (i.e., undefined) statistical aspects (e.g., the mean and variance). An example of a sub-model is the Cauchy BS (CBS) distribution.

Table 1. The sub-models of the GTBS lifetime distribution with model parameters vector $\boldsymbol{\theta} = (\mu,\sigma,\theta,\nu) \in \mathbb{R}^4_+$.

Model	$\boldsymbol{\theta}$
Location-scale Cauchy BS distribution (CBS)	$(\mu,\sigma,\frac{1}{2},1)$
Location-scale BS distribution (BS)	$(\mu,\sigma,\frac{1}{2},\infty)$
Location-scale Student's t BS distribution (TBS)	$(\mu,\sigma,\frac{1}{2},\nu)$
Generalized Cauchy BS distribution (GCBS)	$(\mu,\sigma,\theta,1)$
Generalized BS distribution (GBS)	$(\mu,\sigma,\theta,\infty)$

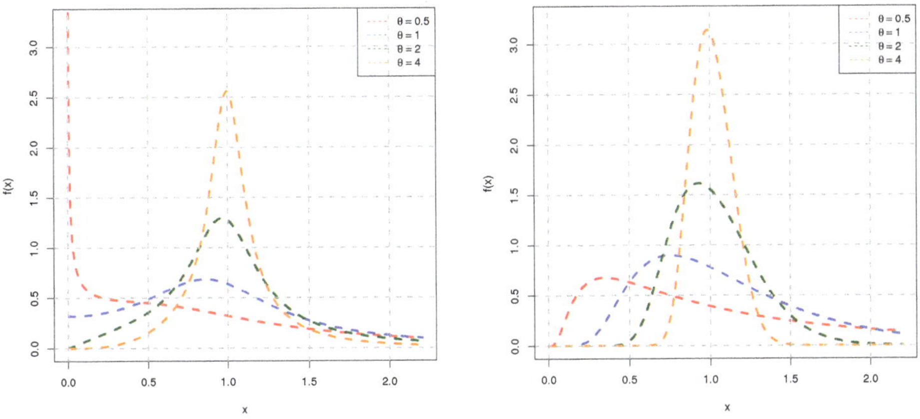

(a) (b)

Figure 1. *Cont.*

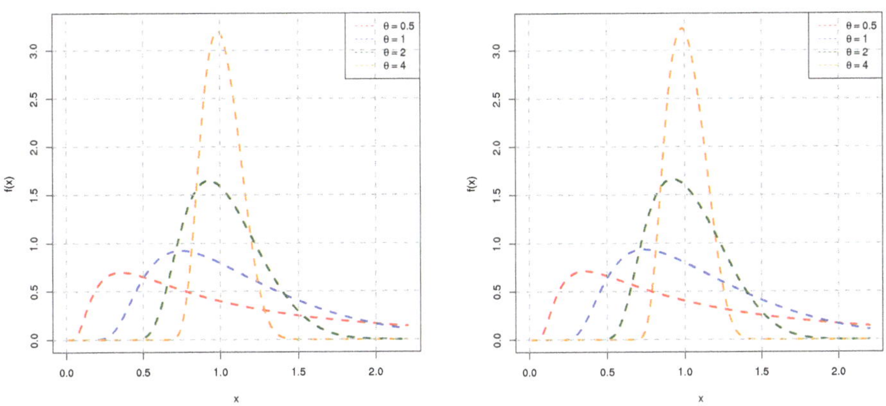

(c) (d)

Figure 1. The GTBS distribution PDF assuming the location parameter $\mu = 0$ and the scale parameter $\sigma = 1$, and different values of θ and ν: (**a**) $\nu = 1$. (**b**) $\nu = 10$. (**c**) $\nu = 30$. (**d**) $\nu \to \infty$.

3. Estimation Procedures

The estimation approaches of interest for the parameters of the GTBS distribution mentioned in the preceding introductory section are considered in this part of the study. It is essential to mention that this study considers the degrees of freedom parameter to be a natural number, i.e., $\nu = 1, 2, \ldots$ without loss of generality. Moreover, all estimation methods are obtained using the following profiling algorithm when estimating the model parameters:

1. For $\nu_1 = 1$ to ν_N by 1:
   (a) Determine the estimators of the model parameters μ, σ and θ by optimizing an objective function proposed in this section using suitable initial values;
   (b) Compete the value of the optimized objective function;

2. Using a convergence criterion, choose the value of ν that maximizes the objective function $g(\boldsymbol{\theta}|\mathbf{x})$ if the estimators are obtained by solving a maximization problem; otherwise, choose the value of ν that minimizes the objective function $g(\boldsymbol{\theta}|\mathbf{x})$ if the estimators are obtained by solving a minimization problem. For this study, the considered convergence criterion is given by:

$$\left|1 - \frac{g(\boldsymbol{\theta}_k|\mathbf{x})}{g(\boldsymbol{\theta}_{k+1}|\mathbf{x})}\right| < \epsilon,$$

where $\epsilon > 0$ is relatively small and $k = 1, \ldots, N$.

A similar approach is considered in the study by [24]. Furthermore, all mathematical derivatives with respect to the model parameters required by the upcoming optimization problems are provided in Appendix A for the sake of conciseness.

Much research is still required. For example, Bayesian estimation methods could be explored and compared with the frequentist approaches considered in this study. Furthermore, estimation efficiency and model fit could be assessed under censored data structures, such as progressive or hybrid censoring schemes.

3.1. Maximum Likelihood Estimation

The estimation theory of maximum likelihood is an extensively used statistical framework in which the parameters of lifetime models are estimated by maximizing a likelihood-based objective function; namely, the log-likelihood function. This type of estimation is versatile and asymptotically efficient; thus, the estimators obtained from this approach have become a fundamental tool in numerous scientific disciplines. Suppose $\mathbf{x} = [x_1, x_2, \ldots, x_n]$ is an observed random sample of size n from the GTBS distribution. The MLEs are obtained by solving the following maximization problem:

$$\begin{aligned} \text{maximize} \quad & \ell(\boldsymbol{\theta}|\mathbf{x}) = n\log\theta - \sigma\sum_{i=1}^{n}\log z_i + \sum_{i=1}^{n}\log\left(z_i^{\theta} + z_i^{-\theta}\right) + \sum_{i=1}^{n}\log\left[\phi_\nu\left(z_i^{\theta} - z_i^{-\theta}\right)\right] \\ \text{s.t.} \quad & \mu, \sigma, \theta, \nu \in \mathbb{R}_+^4, \end{aligned} \tag{4}$$

where $z_i = \sigma^{-1}(x_i - \mu)$.

3.2. Least-Squares Estimations

The least-squares estimators (LSEs) and the weighted least-squares estimators (WLSEs) are obtained using an approach similar to [46] in which the beta distribution parameters are estimated. Recall that $\mathbf{x} = [x_1, x_2, \ldots, x_n]$ is an observed random sample, and let $x_{1:n} < x_{2:n} < \cdots < x_{n:n}$ be the corresponding observed order statistics. Thus, $F(x_{1:n};\boldsymbol{\theta}), \ldots, F(x_{n:n};\boldsymbol{\theta})$ are the order statistics form a standard uniform distribution with expected values and variances given by:

$$\mathrm{E}[F(x_{i:n};\boldsymbol{\theta})] = \frac{i}{n+1} \quad \text{and} \quad \mathrm{V}[F(x_{i:n};\boldsymbol{\theta})] = \frac{i(n-i+1)}{(n+1)^2(n+1)},$$

respectively, such that $i = 1, \ldots, n$ [47]. The LSEs of $\boldsymbol{\theta}$ of the GTBS distribution are obtained by solving the following minimization problem:

$$\begin{aligned} \text{minimize} \quad & \sum_{i=1}^{n} w_i\left[\Phi_\nu\left(z_{i:n}^{\theta} - z_{i:n}^{-\theta}\right) - \frac{i}{n+1}\right]^2 \\ \text{s.t.} \quad & \mu, \sigma, \theta, \nu \in \mathbb{R}_+^4 \end{aligned} \tag{5}$$

such that $z_{i:n} = \sigma^{-1}(x_{i:n} - \mu)$ and $w_i = 1, \forall i$. However, if $w_i^{-1} = \mathrm{V}[F(x_{i:n};\boldsymbol{\theta})]$ in the above minimization problem, then the solution for the minimization problem is the WLSEs for $\boldsymbol{\theta}$.

3.3. Maximum Produce of Spacings Estimation

The maximum product of spacings estimation framework is another approach that requires solving a maximization problem to obtain maximum product of spacings estimators (MPSEs). The latter estimators rival those obtained under the maximum likelihood framework in terms of estimation effectiveness and asymptotic aspects [48–51]. An important note is that the calculation of the MPSEs becomes complicated if ties are present, as the standard method cannot be applied directly. Instead, a generalized maximum product of spacings approach proposed by [52] must be used. Recall that $x_{1:n} < x_{2:n} < \cdots < x_{n:n}$ represent the observed order statistics from a random sample of size n. The MPSEs for $\boldsymbol{\theta}$ are acquired numerically by solving the following maximization problem:

$$\begin{aligned} &\text{maximize} \quad \frac{1}{n+1}\sum_{i=1}^{n+1}\log(\Delta_i) \\ &\text{s.t.} \quad \mu,\sigma,\theta,\nu \in \mathbb{R}_+^4, \end{aligned} \tag{6}$$

where

$$\Delta_i = \begin{cases} \Phi_\nu\left(z_{1:n}^{\theta} - z_{1:n}^{-\theta}\right) & \text{if } i = 1 \\ \Phi_\nu\left(z_{i:n}^{\theta} - z_{i:n}^{-\theta}\right) - \Phi_\nu\left(z_{i-1:n}^{\theta} - z_{i-1:n}^{-\theta}\right) & \text{if } 1 < i \le n \\ 1 - \Phi_\nu\left(z_{n:n}^{\theta} - z_{n:n}^{-\theta}\right) & \text{if } i = n+1 \end{cases}.$$

3.4. Minimum Distance Estimations

The remaining three estimators, the CVMEs, ADEs, and RADEs, are obtained from the CVME, ADE, and RADE methods, respectively. These methods minimize goodness-of-fit statistics based on the theoretical CDF of the underlying model. The CVMEs of $\boldsymbol{\theta}$ are obtained by evaluating the following minimization problem recalling that $x_{1:n} < x_{2:n} < \cdots < x_{n:n}$ are the corresponding observed order statistics of a random sample of size n:

$$\begin{aligned} &\text{minimize} \quad \frac{1}{12n} + \sum_{i=1}^{n}\left[\Phi_\nu\left(z_{i:n}^{\theta} - z_{i:n}^{-\theta}\right) - \frac{2i-1}{2n}\right]^2 \\ &\text{s.t.} \quad \mu,\sigma,\theta,\nu \in \mathbb{R}_+^4 \end{aligned} \tag{7}$$

Alternatively , the ADEs and RADEs of $\boldsymbol{\theta}$ are obtained by solving:

$$\begin{aligned} &\text{minimize} \quad -n - \frac{1}{n}\sum_{i=1}^{n}(2i-1)\log\Phi_\nu\left(z_{i:n}^{\theta} - z_{i:n}^{-\theta}\right) \\ &\qquad -\frac{1}{n}\sum_{i=1}^{n}(2i-1)\log\left[1 - \Phi_\nu\left(z_{n+1-i:n}^{\theta} - z_{n+1-i:n}^{-\theta}\right)\right] \\ &\text{s.t.} \quad \mu,\sigma,\theta,\nu \in \mathbb{R}_+^4 \end{aligned} \tag{8}$$

and

$$\begin{aligned} \text{minimize } & \frac{n}{2} - 2\sum_{i=1}^{n} \Phi_\nu\left(z_{i:n}^{\theta} - z_{i:n}^{-\theta}\right) \\ & -\frac{1}{n}\sum_{i=1}^{n}(2i-1)\log\left[1 - \Phi_\nu\left(z_{n+1-i:n}^{\theta} - z_{n+1-i:n}^{-\theta}\right)\right] \\ \text{s.t. } & \mu, \sigma, \theta, \nu \in \mathbb{R}_{+}^{4}, \end{aligned} \quad (9)$$

respectively.

4. Applications

This section is bipartite. The first part of this section illustrates the applicability of the proposed model by analyzing real data and evaluating the goodness-of-fit compared to the sub-models under the maximum likelihood framework. In the second part of this section, the methods in the preceding section are used to estimate the main model parameters, and the acquired estimators are compared per some statistical aspects obtained using parametric bootstrapping. As a simulation technique, bootstrapping requires the ability to simulate random variates from the model of interest, which can be performed using the quantile function (QF). The latter function for the GTBS distribution is given by:

$$Q(u, \boldsymbol{\theta}) = \mu + \sigma\left\{\frac{\Phi_\nu^{-1}(u) + \sqrt{\left[\Phi_\nu^{-1}(u)\right]^2 + 4}}{2}\right\}^{\frac{1}{\theta}}, \quad 0 < u < 1, \quad (10)$$

such that $\Phi_\nu^{-1}(\cdot)$ is the inverse of the CDF of the standard Student's t distribution with degrees of freedom parameter $\nu > 0$.

The first dataset considered in this study was obtained from [53,54], containing stress-life (SN) data for welded specimens made of steel S690QL, assuming 144 megapascal (MPa). SN data are used to evaluate the relationship between stress and the number of cycles to failure. The considered data (in 100 cycles) are: 2826.80, 3095.70, 3551.00, 3983.19, 4560.20 4564.80, 4966.52, 5232.00, 5547.26, 6477.26, 6615.50, 6832.64, 7394.15, 8274.73, 8856.00, 8971.83, 9890.00, and 12,006.40.

The second dataset to be analyzed was acquired from [55,56], containing strain-life (eN) data for commercial pure titanium (CP-Ti), assuming a modulus of elasticity of 108.21 GPa and tensile strength (TS) of 418.23 MPa. The base material plates were made by rolling and the fatigue specimens were cut in the rolling direction. All specimens were ground and then polished using metallographic abrasive paper and a grinding rod to exclude the effect of outer and inner surface roughness. These treatments were applied before the fatigue tests. The considered data (in log cycles) are: 4.329682569, 4.230959556, 3.923658422, 3.798995734, 3.720076573, 3.77815125, 3.431685345, 3.343999069, 3.362859303, 3.418963831, 3.211921084, 3.121231455, 3.087071206, 3.077004327, 2.953276337, and 2.851258349.

All numerical analyses and results presented in this part of the study and the following one were performed using `R`, an environment for statistical computing [57]. To account for potential outliers in the data, robust summary statistics were computed for each dataset, including the minimum, first quartile (Q_1), median (Q_2), third quartile (Q_3), maximum, and the median absolute deviation (MAD). Moreover, the objective functions in the previous section were optimized using a limited-memory BFGS method [58]. Although it has a built-in implementation within `R` via the `optim` function, the `lbfgs` function from the `nloptr` is considered instead. The latter package solves optimization problems using an `R` interface to `NLopt`, a free/open-source library for nonlinear optimization. It has different free optimization routines available online as well as original implementations of various other algorithms [59]. The reason for choosing the limited-memory efficient optimization approach is that the optimization problems of interest have box constraints.

Before applying the limited-memory BFGS method to solve the optimization problems, it is essential to note that the location parameter μ is part of the domain of the random variable. This situation disrupts one of the standard regularity conditions required for maximum likelihood estimation. These conditions are crucial for ensuring the consistency, asymptotic normality, existence, and efficiency of MLEs; see [60,61] for further discussion. Consequently, a re-parameterization of the location parameter is given by $\mu = x_{1:n} - \delta$, where $0 < \delta < x_{1:n}$. This re-parameterization resolves the underlying issue.

4.1. Goodness-of-Fit Evaluation

The parameters of the GTBS distribution and its sub-models per Table 1 are estimated under the maximum likelihood framework as part of this study. The Kolmogorov–Smirnov (KS) [62,63], Anderson–Darling (AD) [64], and Cramér–von Mises (CVM) [65,66] statistics are used to check the goodness-of-fit of the fitted models. A parametric bootstrap approach with $B = 1000$ resamples is considered for calculating the p-values for these statistics. The latter statistical approach is implemented as follows. First, the GTBS model parameters μ, σ and θ are estimated from the observed data, producing $\hat{\mu}, \hat{\sigma}$ and $\hat{\theta}$, while the estimator of $\hat{\nu}$ is obtained via profiling. Afterward, for the sub-model, part of the model parameters is estimated since some of the parameters become known; for example, in the case of the CBS distribution, $\theta = 0.5$ and $\nu = 1$, while $\theta = 0.5$ and $\nu \to \infty$ in the case of the BS distribution. Consequently, the obtained estimators are then utilized to acquire B bootstrap samples from the underlying model. The goodness-of-fit statistics are computed for each bootstrap sample, and this procedure is repeated B times to approximate the distribution of test statistics. The bootstrap-based p-value is calculated as the proportion of bootstrap statistics exceeding the observed test statistic. For additional details and information about bootstrapping, readers may refer to [67–69].

Table 2 reports the SN and eN data summary statistics. These summary statistics reveal that the considered datasets seem to be from heavy-tailed populations since the gap between the median and Q_3 and the maximum is noticeably not small. Table 3 shows the data analysis for the SN dataset alongside the considered goodness-of-fit metrics, while Table 4 provides similar information but for the eN dataset. Both tables indicate that the sub-model of the GTBS distribution, i.e., the GBS distribution, fits the data better than its counterparts, followed by its generalization. The GBS and GTBS are very close in terms of goodness-of-fit. To strengthen the latter conclusion, Figures 2 and 3 visually support the claim that the GTBS model is suitable for both data sets.

Table 2. Summary statistics for SN and eN datasets. The distances between Q_2, Q_3, and the maximum value are noticeably large, which may suggest that the data follow a heavy-tail probabilistic model.

Data	Minimum	Q_1	Median	Q_3	Maximum	MAD
SN	2826.8	4561.35	6012.26	8054.585	12,006.4	2580.562
eN	2.8512583	3.1126914	3.3909116	3.7833624	4.3296826	0.4767095

Table 3. SN data analysis and goodness-of-fit. The GBS model achieved the smallest values for all goodness-of-fit statistics. Since the corresponding parametric bootstrap p-value exceeds the standard significance level, there is insufficient evidence to reject the hypothesis that the GTBS, GCBS, and GBS models describe the data better than their sub-models.

Model	$\hat{\mu}$	$\hat{\sigma}$	$\hat{\theta}$	$\hat{\nu}$	KS	p-Value	AD	p-Value	CVM	p-Value
GTBS	203.11	5595.92	1.2534	6	0.1084	0.9371	0.1687	0.9800	0.0251	0.9650
GCBS	2156.48	3927.38	1.2208	1 *	0.1206	0.8412	0.4357	0.6803	0.0605	0.6933
GBS	203.11	5595.92	1.1335	∞ *	0.1026	0.9600	0.1502	0.9960	0.0213	0.9930
TBS	2705.21	2802.40	0.5 *	3	0.1464	0.4805	0.5549	0.3477	0.0749	0.3776
CBS	2826.80	3138.20	0.5 *	1 *	0.2312	0.0260	1.5333	0.0949	0.1584	0.0519
BS	2395.51	2775.06	0.5 *	∞ *	0.1794	0.2537	0.5999	0.2637	0.1104	0.2118

* The values are fixed, not estimated.

Table 4. The analysis of the eN data and the corresponding goodness-of-fit. The GBS model achieved the smallest values for all goodness-of-fit statistics. Once again, the corresponding parametric bootstrap p-value was above the standard significance level, i.e., there is not enough evidence to reject the claim that the GTBS, GBS and BS distributions properly fit the data.

Model	$\hat{\mu}$	$\hat{\sigma}$	$\hat{\theta}$	$\hat{\nu}$	KS	p-Value	AD	p-Value	CVM	p-Value
GTBS	2.3208	1.0758	1.2843	20	0.1254	0.6194	0.2014	0.7882	0.0313	0.7353
GCBS	2.8513	0.5525	0.5051	1 *	0.2444	0.0310	1.2857	0.0170	0.1484	0.0420
GBS	2.2840	1.1144	1.2921	∞ *	0.1226	0.7143	0.1982	0.8392	0.0306	0.7982
TBS	2.7853	0.5089	0.5 *	16	0.1654	0.3996	0.5644	0.2478	0.0913	0.2398
CBS	2.8513	0.5521	0.5 *	1 *	0.2460	0.0300	1.2960	0.0999	0.1515	0.0549
BS	2.7745	0.5085	0.5 *	∞ *	0.1748	0.4006	0.5960	0.2857	0.1014	0.2657

* The values are fixed, not estimated.

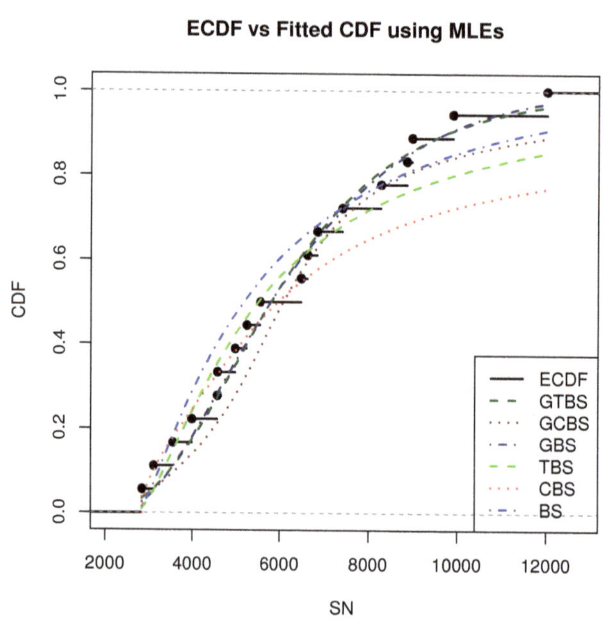

Figure 2. Comparing the empirical CDF (ECDF) to the fitting CDFs under the maximum likelihood framework based on the SN data.

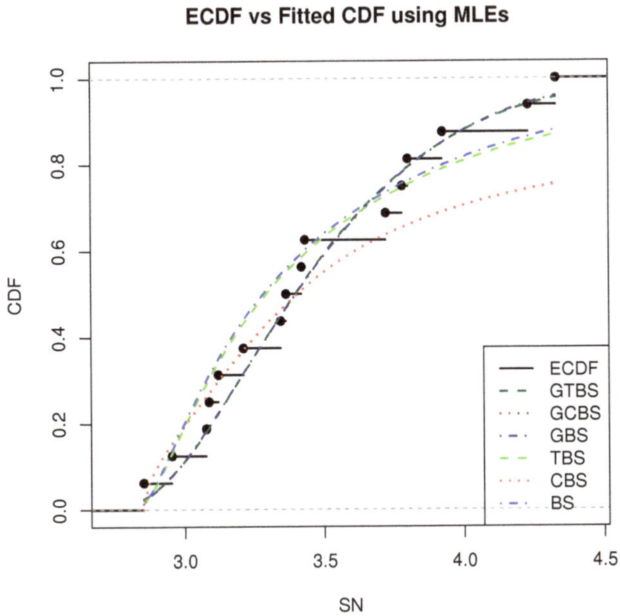

Figure 3. Comparing the ECDF to the fitting CDFs using the maximum likelihood theory based on the eN data.

4.2. Estimation Efficiency Assessment

In the previous part of the study, goodness-of-fit analysis has been conducted under the maximum likelihood framework; however, is it the best approach to estimate the considered model parameters? The joint relative efficiency (JRE) is calculated using bootstrap samples to answer this question. According to [70], one can define the joint relative efficiency (JRE) as:

$$\text{JRE} = \frac{\text{RMSE}^*(\tilde{\mu}) + \text{RMSE}^*(\tilde{\sigma}) + \text{RMSE}^*(\tilde{\theta})}{\text{RMSE}^*(\hat{\mu}) + \text{RMSE}^*(\hat{\sigma}) + \text{RMSE}^*(\hat{\theta})} \times 100\%,$$

such that $\text{RMSE}^*(\hat{\vartheta})$ and $\text{RMSE}^*(\tilde{\vartheta})$ are the bootstrap root mean square errors (RMSEs) of the MLEs and the other estimators under consideration for ϑ, where ϑ corresponds to μ, σ or θ. Note that the bootstrap estimator is acquired using an approach similar to the one mentioned previously. If the value of the JRE is larger than 100%, then the MLEs are jointly more efficient than the other estimators. Table 5 reveals that the MLEs did not outperform the considered estimation methods except for the RADEs in terms of JRE. In contrast, Table 6 shows that the MLEs are jointly more efficient than the LSEs, CVMEs, and RADEs. This might be due to the small sample sizes of the considered datasets.

Table 5. Estimation efficiency assessment of seven estimation methods in terms of JREs based on SN data. Note that $\text{RMSE}^*(\hat{\mu}) = 2185.23$, $\text{RMSE}^*(\hat{\sigma}) = 521.18$, and $\text{RMSE}^*(\hat{\theta}) = 0.6324$.

Model	$\text{RMSE}^*(\tilde{\mu})$	$\text{RMSE}^*(\tilde{\sigma})$	$\text{RMSE}^*(\tilde{\theta})$	JRE (%)
LSE	2268.99	308.50	0.6619	95.2388
WLSE	2079.67	529.08	0.5303	96.3884
MPSE	1634.82	641.45	0.5106	84.1059
CVME	2418.33	281.60	0.7406	99.7649
ADE	2488.87	174.21	0.7881	98.4053
RADE	2550.32	412.58	0.7846	109.4804

Table 6. Estimation efficiency evaluation of seven estimation methods in terms of JREs based on eN data. Note that RMSE*$(\hat{\mu}) = 1.0309$, RMSE*$(\hat{\sigma}) = 1.0758$, and RMSE*$(\hat{\theta}) = 1.6558$.

Model	RMSE*$(\tilde{\mu})$	RMSE*$(\tilde{\sigma})$	RMSE*$(\tilde{\theta})$	JRE (%)
LSE	1.1746	1.2281	1.3998	101.06
WLSE	1.1207	1.1789	1.3209	96.22
MPSE	1.0806	1.1184	1.4709	97.54
CVME	1.2483	1.3001	1.8118	115.88
ADE	1.0598	1.1130	1.4767	97.00
RADE	1.1976	1.3149	1.7335	112.85

5. Monte Carlo Simulation Outcomes

Simulation studies have a vital role in evaluating the performance of estimation procedures across different settings, offering a reliable framework for assessing their effectiveness and precision. Researchers frequently employ Monte Carlo simulation studies to analyze and compare estimation efficiency when multiple estimators are considered for the parameters of a given model in practical applications. Thus, numerical insights from both statistical and sometimes computational viewpoints are provided.

This part of the study reports the Monte Carlo simulation outcomes. These results reflect the performance of the considered estimation methods, assuming different sample sizes and parameters with different values. That is, the simulation study results help identify the most effective and consistent estimation method for the GTBS model parameters. All numerical outcomes are based on $N = 1000$ random samples generated from the GTBS model. The options for the sample size and the model parameters are $n = 10, 25, 50, 100, 250$, $\mu = 0, \sigma = 1, \theta = 0.25, 1, 2, 4$, and $v = 5, 10, 15, 30$ without loss of generality. The comparison and evaluation metric of choice for the estimation efficiency is the RMSE, and it is defined similarly to what was mentioned in the preceding application section. For instance, the RMSE for the estimator of μ is given by:

$$\mathrm{RMSE}(\hat{\mu}) = \sqrt{\frac{1}{N}\sum_{i=1}^{N}(\hat{\mu}_i - \mu)^2},$$

where $N = 1000$ and $\hat{\mu}_i$ is an estimator of the model parameter μ based on simulation run i. As shown in the previous section, goodness-of-fit analysis for the fitted model parameters alongside estimation efficiency provides additional insights on performance. The mean absolute difference between the actual and estimated CDFs (D_{abs}) and the maximum absolute difference between the actual and estimated CDFs ($D_{\max}$) are two goodness-of-fit metrics that are given by:

$$D_{\text{abs}} = \frac{1}{n \times N}\sum_{i=1}^{N}\sum_{j=1}^{n}\left|F(x_j;\boldsymbol{\theta}) - F(x_j;\hat{\boldsymbol{\theta}}_i)\right|,$$

and

$$D_{\max} = \frac{1}{N}\sum_{i=1}^{N}\max_{j=1,\ldots,n}\left|F(x_j;\boldsymbol{\theta}) - F(x_j;\hat{\boldsymbol{\theta}}_i)\right|,$$

respectively, where $F(x;\boldsymbol{\theta})$ is given by (2), and $\hat{\boldsymbol{\theta}}_i$ is an estimator of the vector of model parameters $\boldsymbol{\theta}$ based on simulation run i. For computational convenience, min-max normalization is considered for all metrics to help in data visualization and interpretation and to reduce the impact of outliers (if any); see, for example, ref. [71] for further details.

When the normalized RMSEs approach zero as the sample size increases, the estimators are computationally efficient, regardless of the underlying probability distribution.

Moreover, estimators are deemed effective if the goodness-of-fit metrics D_{abs} and D_{max} are minimized and tend to zero as the sample size increases and remains robust to the other factors, such as the heaviness of the distribution's tail or the existence of outliers, for example. Figures 4–8 reveal the following observations:

- The estimators of μ behave similarly for small values of θ. As the value of θ starts to increase, both MLE and MPSE of μ perform better than the remaining estimation procedures; nevertheless, when the value of θ surges, some of the other estimation methods provide better estimates for μ.
- The MLEs of σ and θ are less efficient than the other estimators for small values of θ; however, once the value of θ increases, the methods behave similarly, especially when the sample size is large enough. MLEs, MPSEs, and ADEs of σ and θ surpass the other estimates for larger values of θ in terms of RMSEs.
- For the goodness-of-fit measurements D_{abs} and D_{max}, MLEs performed well compared to the remaining estimators for small values of θ, while MLEs, MPSEs, and ADEs slightly did not perform well for large values of θ unless the sample size is large enough.
- Overall, all methods behave similarly when the sample size is large enough as expected in terms of estimation efficiency and goodness-of-fit.

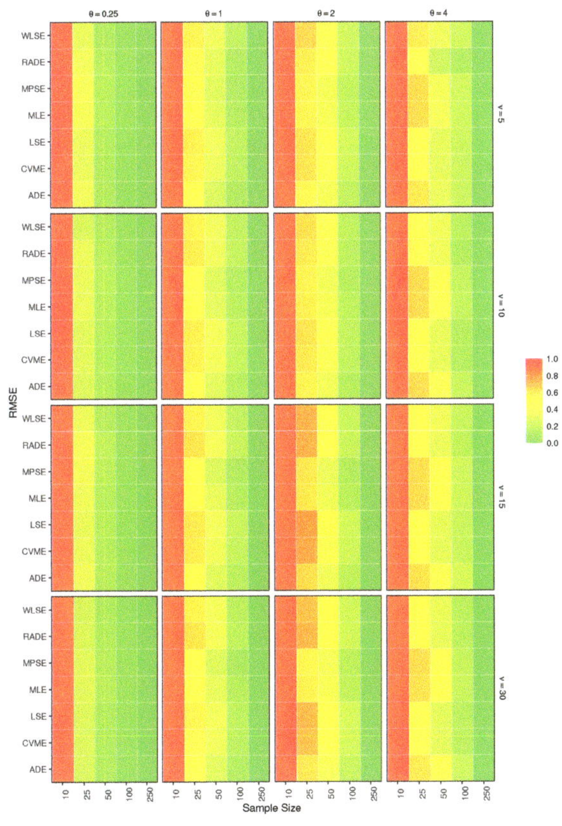

Figure 4. RMSEs for the estimators of μ obtained using simulation. All methods behave similarly for small values of θ. When the value of θ starts to increase, both MLE and MPSE outperform the remaining estimation techniques; however, when the value of θ dramatically increases, some of the other estimation methods provide better estimates than the MLE and MPSE.

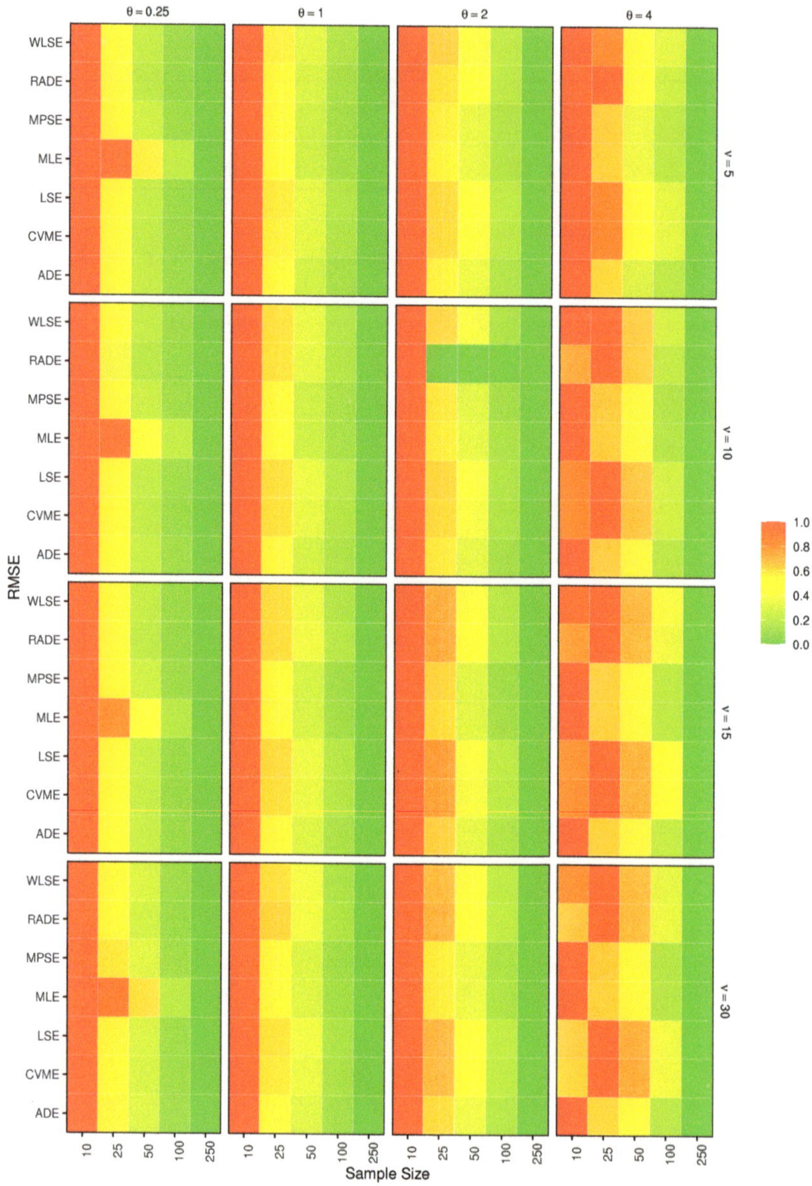

Figure 5. RMSEs for the estimators of σ obtained using simulation. For small values of θ, the MLE is less efficient than the other estimators. When the value of θ increases, the methods act similarly, especially when the sample size is large enough. Nevertheless, MLE, MPSE, and ADE outperform the other estimates for larger values of θ.

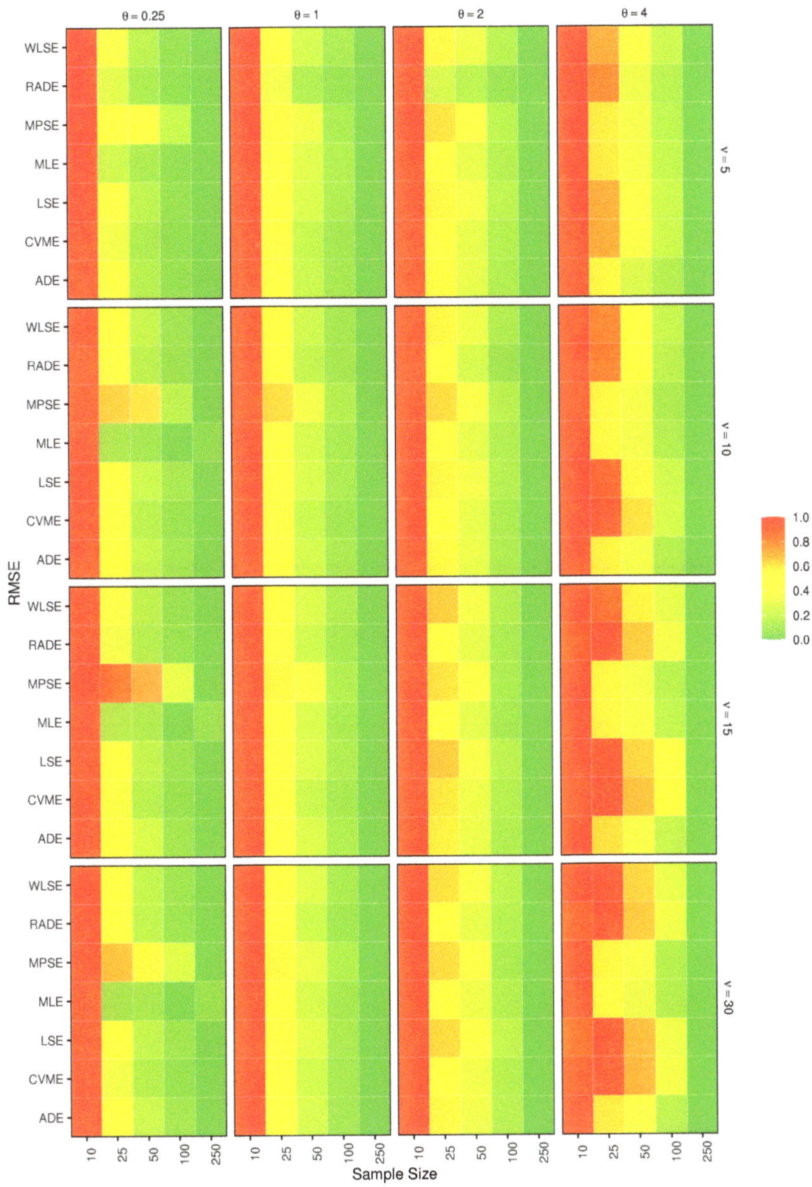

Figure 6. RMSEs for the estimators of θ obtained using simulation. MLE is less efficient than the other estimators for small values of θ. As the value of θ increases, the methods behave similarly, especially when the sample size is large enough. However, MLE, MPSE, and ADE outperform the other estimates for larger values of θ.

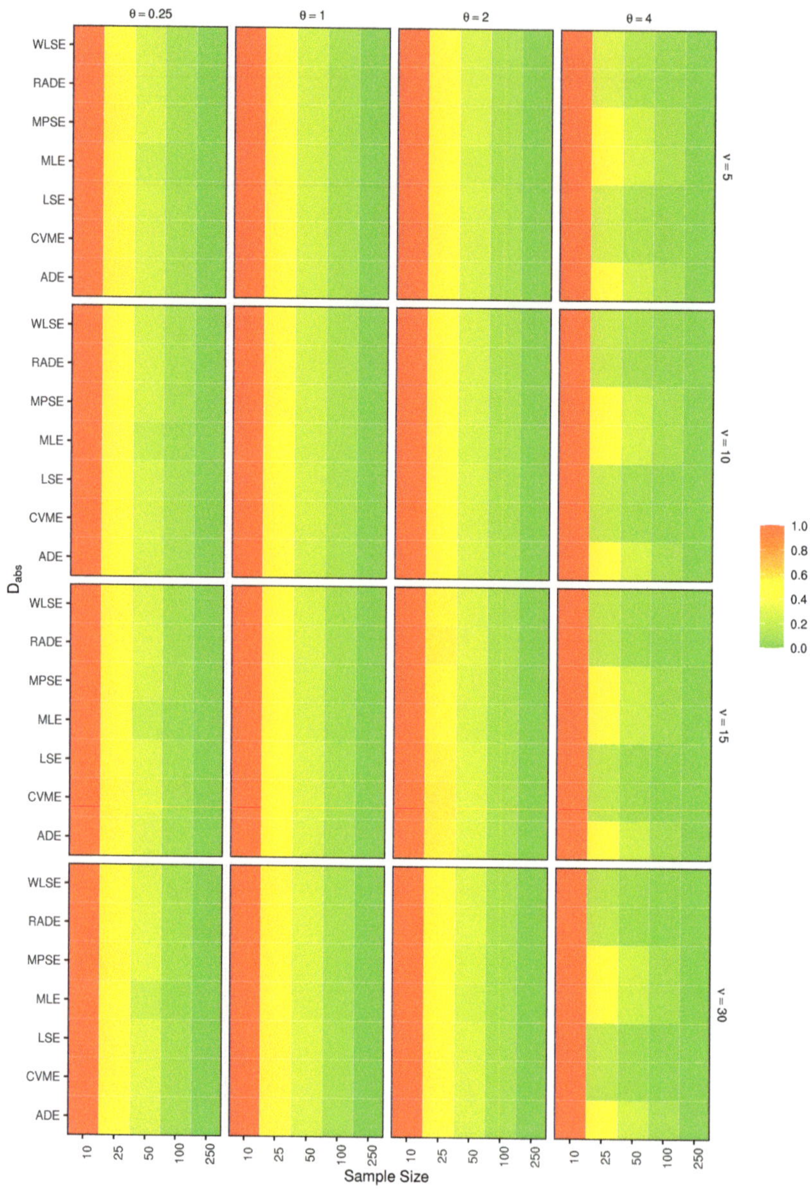

Figure 7. Simulated mean absolute difference between the actual and fitted CDFs. MLEs outperform the remaining estimators for small values of θ, while MLEs, MPSEs, and ADEs slightly did not perform well for large values. For large sample sizes, all methods perform similarly.

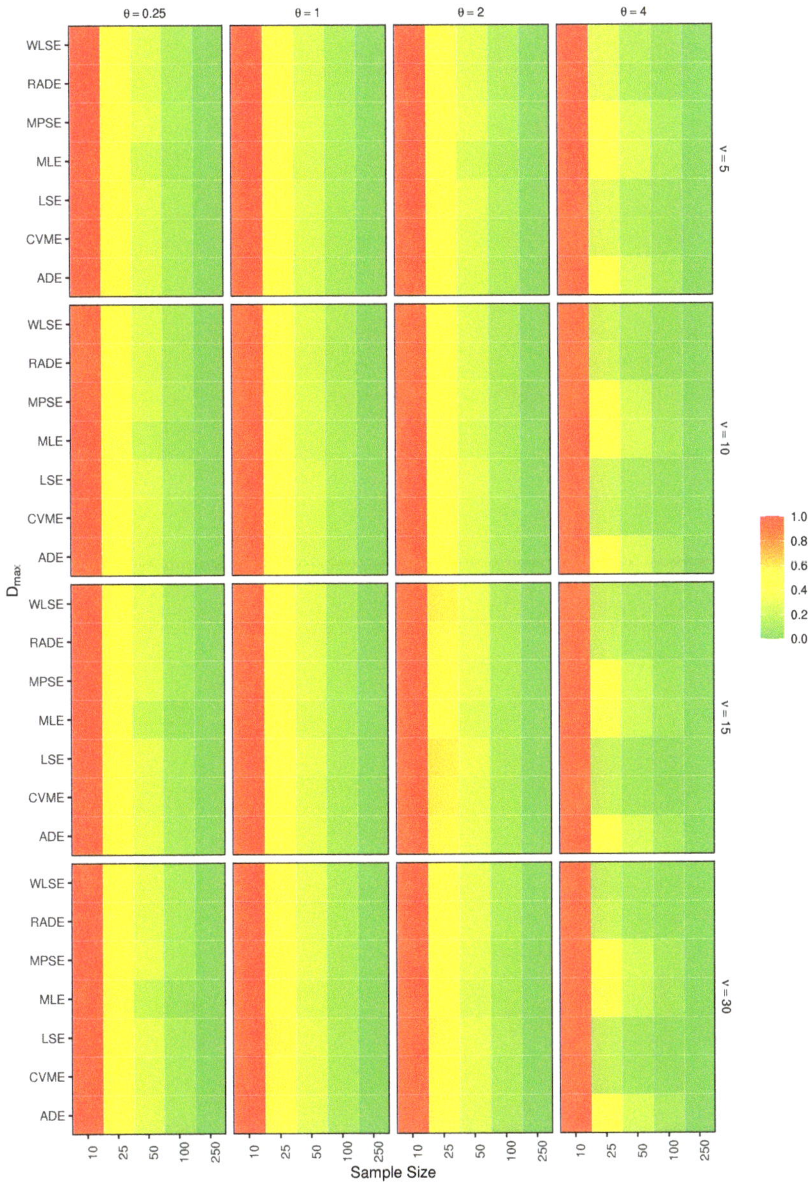

Figure 8. Simulated maximum absolute difference between the actual and fitted CDFs. Again, for small values of θ, MLEs surpass the remaining estimates, while for large values, MLEs, MPSEs, and ADEs slightly did not perform well. For large sample sizes, all methods perform similarly.

6. Conclusions

This paper addressed one potential approach to modeling fatigue data using a statistical model called the GTBS family of lifetime distributions. The main contributions and findings of this study can be summarized as follows:

- A generalization of the conventional two-parameter BS distribution was considered. The model of interest, known as the GTBS distribution, extends the classical BS family by incorporating a shape parameter and heavy-tailed behavior via a Student's t kernel.
- The model parameters were estimated using seven frequentist methods, including maximum likelihood, least squares, and spacing- and distance-based approaches.
- The validity of the model and its sub-models was verified using real fatigue data. Specifically, two real fatigue datasets were analyzed, and the GTBS model provided superior goodness-of-fit compared to its sub-models, as evidenced by bootstrapped KS, AD, and CVM statistics. Nevertheless, the GTBS model is expected to perform even better when the sample size is sufficiently large.
- In addition to the practical applications, estimation efficiency was demonstrated through Monte Carlo simulations. The results showed that the GTBS model, particularly under the MLE and MPSE methods, performs well in terms of RMSE and goodness-of-fit metrics across a range of parameter settings and sample sizes.

Author Contributions: Conceptualization, F.M.A.A.; methodology, A.M.D.; software, F.M.A.A.; validation, F.K.; formal analysis, F.M.A.A.; investigation, F.K.; writing—original draft preparation, F.M.A.A. and A.M.D.; writing—review and editing, F.K.; visualization, F.M.A.A.; supervision, F.M.A.A. All authors have read and agreed to the published version of the manuscript.

Funding: This research received no external funding.

Institutional Review Board Statement: Not applicable.

Informed Consent Statement: Not applicable.

Data Availability Statement: No new data were created or analyzed in this study. Data sharing is not applicable to this article.

Acknowledgments: The authors sincerely appreciate the editors and the two anonymous reviewers for their valuable and insightful comments and recommendations, which have significantly enhanced the quality of this article. The authors also extend their sincere thanks to their colleague, Mazen Nassar, for his assistance and constructive feedback. Additionally, the authors certify that OpenAI's ChatGPT 4.0 was used during the preparation of this manuscript for tasks such as sentence rephrasing, grammar checking, proofreading, and copyediting. All generated content has been thoroughly reviewed and edited by the authors, who take full responsibility for the final version of this publication.

Conflicts of Interest: The authors declare no conflicts of interest.

Appendix A. Mathematical Derivatives

This appendix contains the first-order derivatives for the CDF (2) and the log-likelihood function $\ell(\boldsymbol{\theta}|\mathbf{x})$ with respect to the model parameters, which are part of the gradients of the objective functions required by the limited-memory BFGS algorithm to perform the optimization process. Note that $z_i = \sigma^{-1}(x_i - \mu)$, and it is easy to show that $\frac{\partial \ell}{\partial \delta} = -\frac{\partial \ell}{\partial \mu}$ since $\mu = x_{1:n} - \delta$.

Appendix A.1. First-Order Derivatives of the CDF

$$\frac{\partial}{\partial \mu} F(x;\boldsymbol{\theta}) = -f(x;\boldsymbol{\theta}),$$

$$\frac{\partial}{\partial\sigma}F(x;\boldsymbol{\theta}) = -\sigma^{-1}(x-\mu)f(x;\boldsymbol{\theta}),$$

$$\frac{\partial}{\partial\theta}F(x;\boldsymbol{\theta}) = \log\left(\frac{x-\mu}{\sigma}\right)\left(\left[\frac{x-\mu}{\sigma}\right]^{\theta} + \left[\frac{\sigma}{x-\mu}\right]^{\theta}\right)\phi_{\nu}\left(\left[\frac{x-\mu}{\sigma}\right]^{\theta} - \left[\frac{\sigma}{x-\mu}\right]^{\theta}\right).$$

Appendix A.2. First-Order Derivatives of the Log-Likelihood Function

$$\frac{\partial}{\partial\mu}\ell(\boldsymbol{\theta}|\mathbf{x}) = \sum_{i=1}^{n}\frac{1}{\sigma z_i} - \frac{\theta}{\sigma}\sum_{i=1}^{n}\frac{\left(z_i^{\theta}-z_i^{-\theta}\right)}{z_i\left(z_i^{\theta}+z_i^{-\theta}\right)} + \frac{(\nu+1)\theta}{\sigma}\sum_{i=1}^{n}\frac{\left(z_i^{\theta}-z_i^{-\theta}\right)\left(z_i^{\theta}+z_i^{-\theta}\right)}{z_i\left[\nu+\left(z_i^{\theta}-z_i^{-\theta}\right)^2\right]},$$

$$\frac{\partial}{\partial\sigma}\ell(\boldsymbol{\theta}|\mathbf{x}) = -\frac{\theta}{\sigma}\sum_{i=1}^{n}\frac{\left(z_i^{\theta}-z_i^{-\theta}\right)}{\left(z_i^{\theta}+z_i^{-\theta}\right)} + \frac{(\nu+1)\theta}{\sigma}\sum_{i=1}^{n}\frac{\left(z_i^{\theta}-z_i^{-\theta}\right)\left(z_i^{\theta}+z_i^{-\theta}\right)}{\left[\nu+\left(z_i^{\theta}-z_i^{-\theta}\right)^2\right]},$$

$$\frac{\partial}{\partial\theta}\ell(\boldsymbol{\theta}|\mathbf{x}) = \frac{n}{\theta} + \sum_{i=1}^{n}\log(z_i)\frac{\left(z_i^{\theta}-z_i^{-\theta}\right)}{\left(z_i^{\theta}+z_i^{-\theta}\right)} - (\nu+1)\sum_{i=1}^{n}\frac{\log(z_i)\left(z_i^{\theta}-z_i^{-\theta}\right)\left(z_i^{\theta}+z_i^{-\theta}\right)}{\left[\nu+\left(z_i^{\theta}-z_i^{-\theta}\right)^2\right]}.$$

References

1. Ghofrani, F.; Pathak, A.; Mohammadi, R.; Aref, A.; He, Q. Predicting rail defect frequency: An integrated approach using fatigue modeling and data analytics. *Comput. Aided Civ. Infrastruct. Eng.* **2019**, *35*, 101–115. [CrossRef]
2. Gary Harlow, D. Fatigue Life Distribution Estimation. In *Reliability and Statistical Computing*; Springer International Publishing: Berlin/Heidelberg, Germany, 2020; pp. 1–16. [CrossRef]
3. Wu, Y.L.; Zhu, S.P.; Liao, D.; Correia, J.A.F.O.; Wang, Q. Probabilistic fatigue modeling of notched components under size effect using modified energy field intensity approach. *Mech. Adv. Mater. Struct.* **2021**, *29*, 6379–6389. [CrossRef]
4. Zhang, Q.; Liu, J.; Jiang, W. Tensile fatigue behaviour and life distribution model of the pultruded fibre reinforced composites. *Polym. Polym. Compos.* **2022**, *30*, 096739112210837. [CrossRef]
5. Chen, J.; Liu, Y. Fatigue modeling using neural networks: A comprehensive review. *Fatigue Fract. Eng. Mater. Struct.* **2022**, *45*, 945–979. [CrossRef]
6. Gu, T.; Stopka, K.S.; Xu, C.; McDowell, D.L. Modeling the statistical distribution of fatigue crack formation lifetime in large volumes of polycrystalline microstructures. *Acta Mater.* **2023**, *247*, 118715. [CrossRef]
7. Cameron, Z.A.; Krantz, T.L. Statistical distribution of gear surface fatigue lives at high reliability. *Int. J. Fatigue* **2023**, *167*, 107350. [CrossRef]
8. Fernández-Canteli, A.; Castillo, E.; Díaz-Salamanca, D.; Muñiz-Calvente, M.; Seitl, S. Advances in low cycle fatigue probabilistic modeling. *Theor. Appl. Fract. Mech.* **2024**, *133*, 104611. [CrossRef]
9. Schijve, J. A normal distribution or a weibull distribution for fatigue lives. *Fatigue Fract. Eng. Mater. Struct.* **1993**, *16*, 851–859. [CrossRef]
10. Pascual, F.; Meeker, W. Analysis of Fatigue Data with Runouts Based on a Model with Nonconstant Standard Deviation and a Fatigue Limit Parameter. *J. Test. Eval.* **1997**, *25*, 292–301. [CrossRef]
11. Pascual, F.G.; Meeker, W.Q. Estimating Fatigue Curves With the Random Fatigue-Limit Model. *Technometrics* **1999**, *41*, 277–289. [CrossRef]
12. Ryan, K.J. Estimating Expected Information Gains for Experimental Designs with Application to the Random Fatigue-Limit Model. *J. Comput. Graph. Stat.* **2003**, *12*, 585–603. [CrossRef]
13. Babuška, I.; Sawlan, Z.; Scavino, M.; Szabó, B.; Tempone, R. Bayesian inference and model comparison for metallic fatigue data. *Comput. Methods Appl. Mech. Eng.* **2016**, *304*, 171–196. [CrossRef]

14. Zou, Q.; Wen, J. Bayesian model averaging for probabilistic S-N curves with probability distribution model form uncertainty. *Int. J. Fatigue* **2023**, *177*, 107955. [CrossRef]
15. Birnbaum, Z.; Saunders, S. A new family of life distributions. *J. Appl. Probab.* **1969**, *6*, 319–327. [CrossRef]
16. Birnbaum, Z.; Saunders, S. Estimation for a family of life distributions with applications to fatigue. *J. Appl. Probab.* **1969**, *6*, 328–347. [CrossRef]
17. Desmond, A. Stochastic models of failure in random environments. *Can. J. Stat.* **1985**, *13*, 171–183. [CrossRef]
18. Balakrishnan, N.; Kundu, D. Birnbaum-Saunders distribution: A review of models, analysis, and applications. *Appl. Stoch. Model. Bus. Ind.* **2018**, *35*, 4–49. [CrossRef]
19. Bdair, O.M. Inference for Two-Parameter Birnbaum–Saunders Distribution Based on Type-II Censored Data with Application to the Fatigue Life of Aluminum Coupon Cuts. *Mathematics* **2025**, *13*, 590. [CrossRef]
20. Park, C.; Wang, M. A Goodness-of-Fit Test for the Birnbaum-Saunders Distribution Based on the Probability Plot. In *Reliability Analysis and Maintenance Optimization of Complex Systems*; Springer Nature: Cham, Switzerland, 2025; pp. 345–359. [CrossRef]
21. Díaz-García, J.A.; Leiva-Sánchez, V. A new family of life distributions based on the elliptically contoured distributions. *J. Stat. Plan. Inference* **2005**, *128*, 445–457. [CrossRef]
22. Díaz-García, J.A.; Leiva-Sánchez, V. Erratum to "A new family of life distributions based on the elliptically contoured distributions". *J. Stat. Plan. Inference* **2007**, *137*, 1512–1513. [CrossRef]
23. Barros, M.; Paula, G.A.; Leiva, V. A new class of survival regression models with heavy-tailed errors: Robustness and diagnostics. *Lifetime Data Anal.* **2008**, *14*, 316–332. [CrossRef] [PubMed]
24. Balakrishnan, N.; Leiva, V.; Sanhueza, A.; Vilca, F. Estimation in the Birnbaum-Saunders distribution based on scale-mixture of normals and the EM-algorithm. *SORT-Stat. Oper. Res. Trans.* **2010**, *33*, 171–192.
25. Alosaimi, B.S.; Alam, F.M.; Baaqeel, H.M. Modeling Air Pollution Data Using a Generalized Birnbaum-Saunders Distribution with Different Estimation Procedures. In *Mathematical Modeling in Physical Sciences*; Springer Nature: Cham, Switzerland, 2024; pp. 587–618. [CrossRef]
26. ArruVé, J.; Arellano-Valle, R.B.; Gómez, H.W.; Leiva, V. On a new type of Birnbaum-Saunders models and its inference and application to fatigue data. *J. Appl. Stat.* **2019**, *47*, 2690–2710. [CrossRef]
27. Saulo, H.; Leao, J.; Leiva, V.; Vila, R.; Tomazella, V. A bivariate fatigue-life regression model and its application to fracture of metallic tools. *Braz. J. Probab. Stat.* **2021**, *35*, 119–137. [CrossRef]
28. Patil, S.M.; Kulkarni, H.V. Analysis of medians under two-way model with and without interaction for Birnbaum–Saunders distributed response. *J. Appl. Stat.* **2022**, *50*, 2717–2738. [CrossRef]
29. Basu, S.; Kundu, D. Model misspecification of Log-Normal and Birnbaum-Saunders distributions. *Commun. Stat. Simul. Comput.* **2023**, *54*, 1125–1145. [CrossRef]
30. Sha, N. Classical and Bayesian inference for an extended generalized Birnbaum-Saunders distribution. *Commun. Stat. Simul. Comput.* **2023**, *52*, 5615–5632. [CrossRef]
31. Jayalath, K.P. Improved Bayesian Inferences for Right-Censored Birnbaum–Saunders Data. *Mathematics* **2024**, *12*, 874. [CrossRef]
32. Sawlan, Z.; Scavino, M.; Tempone, R. Modeling Metallic Fatigue Data Using the Birnbaum–Saunders Distribution. *Metals* **2024**, *14*, 508. [CrossRef]
33. Busababodhin, P.; Phoophiwfa, T.; Volodin, A.; Suraphee, S. On the Asymptotic Normality of the Method of Moments Estimators for the Birnbaum–Saunders Distribution with a New Parametrization. *Mathematics* **2025**, *13*, 636. [CrossRef]
34. Gupta, R.D.; Kundu, D. Generalized exponential distribution: Different method of estimations. *J. Stat. Comput. Simul.* **2001**, *69*, 315–337. [CrossRef]
35. Alkasasbeh, M.R.; Raqab, M.Z. Estimation of the generalized logistic distribution parameters: Comparative study. *Stat. Methodol.* **2009**, *6*, 262–279. [CrossRef]
36. Mazucheli, J.; Louzada, F.; Ghitany, M. Comparison of estimation methods for the parameters of the weighted Lindley distribution. *Appl. Math. Comput.* **2013**, *220*, 463–471. [CrossRef]
37. do Espirito Santo, A.; Mazucheli, J. Comparison of estimation methods for the Marshall–Olkin extended Lindley distribution. *J. Stat. Comput. Simul.* **2014**, *85*, 3437–3450. [CrossRef]
38. Balakrishnan, N.; Alam, F.M.A. Maximum likelihood estimation of the parameters of Student's *t* Birnbaum-Saunders distribution: A comparative study. *Commun. Stat. Simul. Comput.* **2019**, *51*, 793–822. [CrossRef]
39. Nassar, M.; Okasha, H.; Albassam, M. E-Bayesian estimation and associated properties of simple step–stress model for exponential distribution based on type-II censoring. *Qual. Reliab. Eng. Int.* **2020**, *37*, 997–1016. [CrossRef]
40. Alam, F.M.A.; Nassar, M. On Modeling Concrete Compressive Strength Data Using Laplace Birnbaum-Saunders Distribution Assuming Contaminated Information. *Crystals* **2021**, *11*, 830. [CrossRef]
41. Alam, F.M.A.; Almalki, A.M. The hazard rate function of the logistic Birnbaum-Saunders distribution: Behavior, associated inference, and application. *J. King Saud Univ. Sci.* **2021**, *33*, 101580. [CrossRef]

42. Moloy, D.J.; Ali, M.A.; Alam, F.M.A. Modeling Climate data using the Quartic Transmuted Weibull Distribution and Different Estimation Methods. *Pak. J. Stat. Oper. Res.* **2023**, *19*, 649–669. [CrossRef]
43. Alam, F.M.A. On Modeling X-Ray Diffraction Intensity Using Heavy-Tailed Probability Distributions: A Comparative Study. *Crystals* **2025**, *15*, 188. [CrossRef]
44. Johnson, N.L.; Kotz, S.; Balakrishnan, N. *Continuous Univariate Distributions, Volume 1, 2nd Edition*; Wiley: Hoboken, NJ, USA, 1994. Available online: https://www.wiley.com/en-us/Continuous+Univariate+Distributions%2C+Volume+1%2C+2nd+Edition-p-9780471584957 (accessed on 13 June 2025).
45. Johnson, N.L.; Kotz, S.; Balakrishnan, N. *Continuous Univariate Distributions, Volume 2, 2nd Edition*; Wiley: Hoboken, NJ, USA, 1995. Available online: https://www.wiley.com/en-us/Continuous+Univariate+Distributions%2C+Volume+2%2C+2nd+Edition-p-9780471584940 (accessed on 13 June 2025).
46. Swain, J.; Venkatraman, S.; Wilson, J. Least squares estimation of distribution function in Johnson's translation system. *J. Stat. Comput. Simul.* **1988**, *29*, 271–297. [CrossRef]
47. Arnold, B.C.; Balakrishnan, N.; Nagaraja, H.N. *A First Course in Order Statistics*; Society for Industrial and Applied Mathematics: Philadelphia, PA, USA, 2008. [CrossRef]
48. Cheng, R.C.H.; Amin, N.A.K. *Maximum Product-of-Spacings Estimation with Applications to the Lognormal Distribution*; Technical Report; Department of Mathematics, University of Wales IST: Cardiff, UK, 1979.
49. Cheng, R.C.H.; Amin, N.A.K. Estimating Parameters in Continuous Univariate Distributions with a Shifted Origin. *J. R. Stat. Soc. Ser. B Stat. Methodol.* **1983**, *45*, 394–403. [CrossRef]
50. Ranneby, B. The Maximum Spacing Method. An Estimation Method Related to the Maximum Likelihood Method. *Scand. J. Stat.* **1984**, *11*, 93–112.
51. Ghosh, K.; Jammalamadaka, S. A general estimation method using spacings. *J. Stat. Plan. Inference* **2001**, *93*, 71–82. [CrossRef]
52. Murage, P.; Mung'atu, J.; Odero, E. Optimal Threshold Determination for the Maximum Product of Spacing Methodology with Ties for Extreme Events. *Open J. Model. Simul.* **2019**, *7*, 149–168. [CrossRef]
53. Toasa Caiza, P.D.; Ummenhofer, T. Consideration of the runouts and their subsequent retests into S-N curves modelling based on a three-parameter Weibull distribution. *Int. J. Fatigue* **2018**, *106*, 70–80. [CrossRef]
54. Fakoor, H.; Alizadeh Kaklar, J. A modification in Weibull parameters to achieve a more accurate probability distribution function in fatigue applications. *Sci. Rep.* **2023**, *13*, 17537. [CrossRef]
55. Ma, T.H.; Chang, L.; Guo, S.; Kong, L.R.; He, X.H.; Zhou, C.Y. Comparison of multiaxial low cycle fatigue behavior of CP-Ti under strain-controlled mode at different multiaxial strain ratios. *Int. J. Fatigue* **2020**, *140*, 105818. [CrossRef]
56. Chen, S.; Bai, Y.; Zhou, X.; Yang, A. A deep learning dataset for metal multiaxial fatigue life prediction. *Sci. Data* **2024**, *11*, 1027. [CrossRef]
57. R Core Team. *R: A Language and Environment for Statistical Computing*; R Foundation for Statistical Computing: Vienna, Austria, 2024.
58. Liu, D.C.; Nocedal, J. On the limited memory BFGS method for large scale optimization. *Math. Program.* **1989**, *45*, 503–528. [CrossRef]
59. Johnson, S.G. The NLopt Nonlinear-Optimization Package. 2008. Available online: https://nlopt.readthedocs.io/en/latest/Citing_NLopt/ (accessed on 13 June 2025).
60. Self, S.G.; Liang, K.Y. Asymptotic Properties of Maximum Likelihood Estimators and Likelihood Ratio Tests under Nonstandard Conditions. *J. Am. Stat. Assoc.* **1987**, *82*, 605–610. [CrossRef]
61. Casella, G.; Berger, R.L. *Statistical Inference*, 2nd ed.; Duxbury Press: Boston, MA, USA, 2002.
62. Kolmogorov, A. Sulla determinazione empirica di una legge di distribuzione. *G. dell'Istituto Ital. Attuari* **1933**, *4*, 83–91.
63. Smirnov, N. Table for Estimating the Goodness of Fit of Empirical Distributions. *Ann. Math. Stat.* **1948**, *19*, 279–281. [CrossRef]
64. Anderson, T.W.; Darling, D.A. Asymptotic Theory of Certain "Goodness of Fit" Criteria Based on Stochastic Processes. *Ann. Math. Stat.* **1952**, *23*, 193–212. [CrossRef]
65. Cramér, H. On the Composition of Elementary Errors. *Scand. Actuar. J.* **1928**, *1928*, 13–74. [CrossRef]
66. von Mises, R. *Wahrscheinlichkeit, Statistik und Wahrheit*; Julius Springer: Vienna, Austria, 1928.
67. Efron, B.; Tibshirani, R.J. *An Introduction to the Bootstrap*; Chapman and Hall/CRC: Boca Raton, FL, USA, 1994.
68. Davison, A.C.; Hinkley, D.V. *Bootstrap Methods and Their Applications*; Cambridge University Press: Cambridge, UK, 1997.
69. Chernick, M.R. *Bootstrap Methods: A Guide for Practitioners and Researchers*, 2nd ed.; Wiley-Interscience: Hoboken, NJ, USA, 2008.
70. Balakrishnan, N.; Cutler, C.D. Maximum Likelihood Estimation of Laplace Parameters Based on Type-II Censored Samples. In *Statistical Theory and Applications*; Springer: New York, NY, USA, 1996; pp. 145–151. [CrossRef]
71. Han, J.; Kamber, M.; Pei, J. *Data Mining: Concepts and Techniques*; Elsevier: Amsterdam, The Netherlands, 2012. [CrossRef]

crystals

MDPI

Article

Optimization of Critical Parameters in Friction Stir Spot Welding of AA5052 Aluminum Alloy Using Response Surface Methodology

Mohamed M. El-Sayed Seleman [1,2], Sabbah Ataya [3,*], Nashmi H. Alrasheedi [3], Mohamed M. Z. Ahmed [4], Hagar A. Reyad [2], Ashraf Bakkar [5] and Ramy A. Fouad [6]

1 Department of Metallurgical and Materials Engineering, Faculty of Petroleum and Mining Engineering, Suez University, Suez 43221, Egypt; mohamed.elnagar@suezuniv.edu.eg
2 Suez and Sinai Metallurgical and Materials Research Center of Scientific Excellence (SSMMR-CSE), Faculty of Petroleum and Mining Engineering, Suez University, Suez 43221, Egypt; hagar.reeb@pme.suezuni.edu.eg
3 Department of Mechanical Engineering, College of Engineering, Imam Mohammad Ibn Saud Islamic University (IMSIU), Riyadh 11432, Saudi Arabia; nhrasheedi@imamu.edu.sa
4 Mechanical Engineering Department, College of Engineering at Al Kharj, Prince Sattam Bin Abdulaziz University, Al Kharj 11942, Saudi Arabia; moh.ahmed@psau.edu.sa
5 Department of Civil and Environmental Engineering, College of Engineering and Computers, Umm Al-Qura University, Al-Lith 28425, Saudi Arabia; atbakkar@uqu.edu.sa
6 Mechanical Department, Faculty of Technology and Education, Suez University, Suez 43221, Egypt; ramy_fouad12@suezuniv.edu.eg
* Correspondence: smataya@imamu.edu.sa

Abstract

Understanding and optimizing the relationship between critical processing parameters (rotational speed and dwell time) and the resulting weld performance is crucial for the effective application of friction stir spot welding (FSSW) in joining aluminum alloys. FSSW is an increasingly important solid-state, clean technology alternative for joining lightweight alloys such as AA5052-H32 in various industries. To optimize this technique for lap joint configurations, the current study examines the influence of rotational speeds (500, 1000, and 1500 rpm) and dwell times (1, 2, and 3 s) on the heat input energy, hardness across weld zones, and tensile/shear load, using a full factorial Design-Expert (DOE) analysis. The FSSW responses of the numerical model were validated using the experimental results for the spot-welded joints. The findings indicate that the dwell time significantly affected the mechanical properties, while the tool rotational speed had a substantial effect on the heat input energy and mechanical properties. Fracture surfaces predominantly exhibited ductile failure with diverse dimple morphologies, consistent with the enhanced tensile properties under optimal parameters. The presence of finer dimples suggests a mixed-mode fracture involving shear.

Keywords: AA 5052-H32; rotational speed; dwell time; response surface methodology (RSM); mechanical behavior; statistical modeling; heat input energy

1. Introduction

Aluminum alloys are highly regarded as lightweight materials due to their low density, high specific strength, corrosion resistance, impact resistance, and recyclability [1]. Their growing use in the transportation industry, replacing traditional steel, has led to significant weight reduction in vehicles [2]. This improves performance and fuel efficiency and reduces emissions, supporting global sustainability efforts [3]. Therefore, the need to join aluminum alloys in numerous applications, such as automotive, aerospace, and marine industries, has driven the development of advanced welding methods and the optimization of existing welding process parameters [4]. Traditional methods for spot-joining aluminum alloys, such as mechanical fastening (riveting) [5] and electrical resistance welding [5,6], present several limitations. Riveting is associated with low production efficiency and the use of harsh environmental conditions [4], while resistance spot welding suffers from high energy consumption, excessive current requirements, and significant joint distortion [7,8].

While friction stir welding (FSW) and friction stir spot welding (FSSW) both employ a rotating tool to generate frictional heat and plasticize material, their mechanics and applications differ significantly. FSW produces continuous linear welds by traversing the tool along a joint line, making it suitable for long-seam applications. In contrast, FSSW creates discrete nugget-style joints by plunging the tool at a single spot without linear movement, making it ideal for lap-sheet assemblies that require localized joining with minimal thermal distortion [9,10]. In fact, friction stir spot welding (FSSW) [11–14] derived from friction stir welding (FSW) [10,15], where it is used to join overlapping similar or dissimilar material sheets. Moreover, it is effective for joining materials with lower energy consumption that are difficult to weld using traditional joining methods. The produced welds have high strength, fatigue resistance, and lower defects such as porosity, cracks, and sheet distortion [16–18]. FSSW is a promising technology for engineering industries seeking lightweight, high-strength, and defect-free joints, especially in applications involving non-ferrous materials [19–22]. These applications include the automotive, aerospace, electronics, and shipbuilding industries [23,24].

Zhang et al. [1] investigated applying two variants of FSSW: conventional FSSW and walking FSSW. They used AA5052-H112 alloy sheets with a 1 mm thickness and examined the effect of FSSW parameters in terms of rotational speed and dwell time on the microstructure and mechanical properties of joints. The welding parameters included rotational speeds of 1541 and 2256 rpm and dwell times of 5, 10, and 15 s. They reported that the highest tensile/shear load of 2847.7 N was achieved in joints welded at 1541 rpm and 5 s. Additionally, the FSSW joint strength decreased with increasing rotational speed at a constant dwell time, which can be attributed to the higher heat input energy and the resulting softening of the joint material.

Mitlin et al. [25] studied the effect of the plunge depth on the microstructure and mechanical properties of FSSW joints of AA 6111-T4 alloy. The AA 6111-T4 alloy dimensions were 1 mm in thickness, 25 mm in width, and 100 mm in length. Their findings indicated that the tool plunge depth had a minor effect on the FSSW joint strength. Additionally, they observed that increasing the plunge depth reduced the thickness of the upper sheet.

Tiwan et al. [26] studied the influence of FSSW variables, including the pin geometry (cylindrical and step pin) and tool rotational speed (900, 1400, and 1800 rpm), on the mechanical behavior of AA5052-H112 welded joints. They found that the highest tensile strength was obtained using the cylindrical pin at rotational speeds of 900 and 1400 rpm. Furthermore, the shear and cross-tension loads of welded joints initially improved as the rotational speed increased from 900 to 1400 rpm but decreased when the speed reached 1800 rpm. Conversely, joints processed with a step pin at higher rotational speeds showed reduced shear loads but increased cross-tension loads.

Response Surface Methodology (RSM) utilizing designs such as Face-Centered Composite Design (FCCD) offers a statistical approach to maximizing the yield of a specific material through the optimization of its processing parameters. The analysis of variance (ANOVA) response surface method investigates the relationships between many process variables and one or more response variables [27,28]. FCCD based on RSM offer a statistical approach to maximizing the yield of a specific material through the optimization of its processing parameters [29]. RSM has recently become popular for formulation optimization, particularly when paired with a suitable design of experiments [30]. Unlike traditional approaches, the RSM methodology can identify how process factors interact with one another [31]. Tutar et al. [32] employed a mathematical model based on Taguchi orthogonal array for optimizing the spot welding parameters of 3 mm thick AA3003-H12 in terms of plunge depths (3.2, 4, and 4.8 mm), tool rotational speeds (1500, 2000, and 2500 rpm), and dwell times (1.5, 2, and 2.5 s) with a threaded cylindrical pin and concave shoulder. Their results showed that plunge depth had the strongest influence on joint strength (69.26% contribution), followed by rotational speed and dwell time. The optimal welding parameters were a dwell time of 2 s, a plunge depth of 4.8 mm, and a rotational speed of 1500 rpm. However, their study had two key limitations: (1) the use of a short pin (3 mm) required an excessive plunge depth (~1.8 mm insertion beyond the pin length), potentially causing keyhole defects in the 6 mm thick lap joint; and (2) their model was specifically developed for AA3003-H12 and cannot be directly applied to other aluminum series such as 5XXX alloys, due to differences in the chemical composition and thermal properties, particularly when joining thinner sheets.

Despite previous studies [1,4,8] on the friction stir spot welding (FSSW) of AA5052 alloys, a critical gap remains in the systematic optimization of two key parameters, the tool rotational speed and welding time (dwell time), particularly in lower parameter ranges (rotational speed $\leq$ 1500 rpm; dwell time < 5 s) for AA5052-H32 alloy. Prior studies have predominantly focused on high-energy regimes (rotational speed > 1500 rpm, dwell time > 5 s [1]), leaving the thermo-mechanical effects and microstructure–property correlations at moderate parameters underexplored. Ahmed et al., in two separate studies [4,8], explored more practical 2 mm AA5052-H32 sheets but employed restrictive one-factor-at-a-time approaches. In the first work [4], they varied only the dwell time (1–3 s) at a fixed rotational speed of 500 rpm to produce spot joints in AA5052 aluminum alloy, while in the second work [8], they examined only the effect of the rotational speed (500–1500 rpm) at a fixed dwell time of 2 s on joint performance. These studies represent isolated investigations, examining only one parameter at a time, and both ignored statistical optimization and the microstructure–property correlations. Furthermore, the synergistic interaction between the rotational speed and dwell time, which critically governs the heat input, grain refinement, and joint strength, lacks rigorous statistical modeling via RSM. Addressing this gap is essential to expand the industrial applicability of FSSW, especially for thin-walled AA5052 applications that demand precise thermal management combined with high productivity, where a reduced welding time and minimized heat input are desirable.

While RSM has demonstrated effectiveness in optimizing FSW [33,34] and FSSW [35] processes for various materials, its specific application to the FSSW of AA5052-H32, specifically for investigating the interaction effects between the tool rotational speed and dwell time in low-to-moderate ranges ($\leq$1500 rpm, <5 s) remains underexplored in the literature.

Thus, the novelty of this work lies in developing the first integrated statistical response methodology based on FCCD for AA5052-H32 FSSW that systematically correlates the welding parameters in terms of tool rotational speed (500–1500 rpm) and dwell time (1–3 s) with multi-objective performance metrics critical for lightweight automotive and aerospace applications. Unlike previous fragmented studies, our RSM approach simultaneously models three key industrial response variables: (1) heat input energy (governing process efficiency); (2) stir zone (SZ) hardness (linked to microstructure evolution); and (3) maximum shear load (determining joint reliability), capturing their complex interdependencies through ANOVA-validated mathematical models. This represents a paradigm shift from conventional one-factor-at-a-time studies by providing (1) a physics-informed statistical framework for parameter optimization, (2) quantitative prediction of the mechanical behavior under varying energy inputs, and (3) a scientific basis for industrial process control in friction stir spot welding applications.

2. Materials and Methods

2.1. Starting Materials and FSSW Procedure

In the present work, spot welds were produced in a lap joint configuration using as-received AA5052-H32 aluminum alloy sheets with dimensions of 150 mm (length) × 40 mm (width) × 2 mm (thickness). Tables 1 and 2 illustrate the chemical composition and mechanical performance, respectively, of the AA5052-H32 alloy.

Table 1. Chemical composition of the AA5052-H32 alloy.

Chemical Composition								
Element	**Mn**	**Si**	**Mg**	**Zn**	**Fe**	**Cu**	**Cr**	**Al**
Wt.%	0.042	0.080	2.21	0.029	0.218	0.0037	0.152	Bal.

Table 2. Mechanical characteristics of the AA5052-H32 alloy.

Mechanical Properties					
Property	**UTS (MPa)**	**Yield Strength (MPa)**	**Elongation (%)**	**Shear Strength (MPa)**	**Hardness (Hv)**
AA5052-H32	228 ± 2	193	12	138	68 ± 2

An FSSW machine (Model-EG-FSW-M1, Suez, Egypt) [36] was employed to weld AA5052 lap joints (Figure 1a). The designed tool dimensions were a shoulder diameter of 20 mm, a pin diameter of 5 mm, and a pin length of 3 mm, as shown in Figure 1b. This tool was fabricated from H13 steel. Based on numerous preliminary experiments and our previous findings [4,8], as well as additional testing and investigation, this study focused on optimizing key FSSW parameters for AA5052-H32 (Table 3) while maintaining the following constant conditions: 3 mm plunge depth, 0.1 mm/s plunge rate, and a 0-degree tilt angle [4,8].

Table 3. The processing variables for the experimental design.

FSSW Parameters	**Units**	**Codes**		
		−1	**0**	**1**
Rotational speed	rpm	500	1000	1500
Dwell time

| S | 1 | 2 | 3 |

2.2. Characterization of FSSW Joints

Weld cross-sections were prepared according to ASTM E384-73 for microstructural and hardness analysis. The hardness distribution in the weld zones (SZ: stir zone, TMAZ: thermomechanical affected zone, and HAZ: heat affected zone) was evaluated using a Vickers hardness tester (HWDV-75, TTS Unlimited, Osaka, Japan) with a 5 N load and 15 s indentation time.

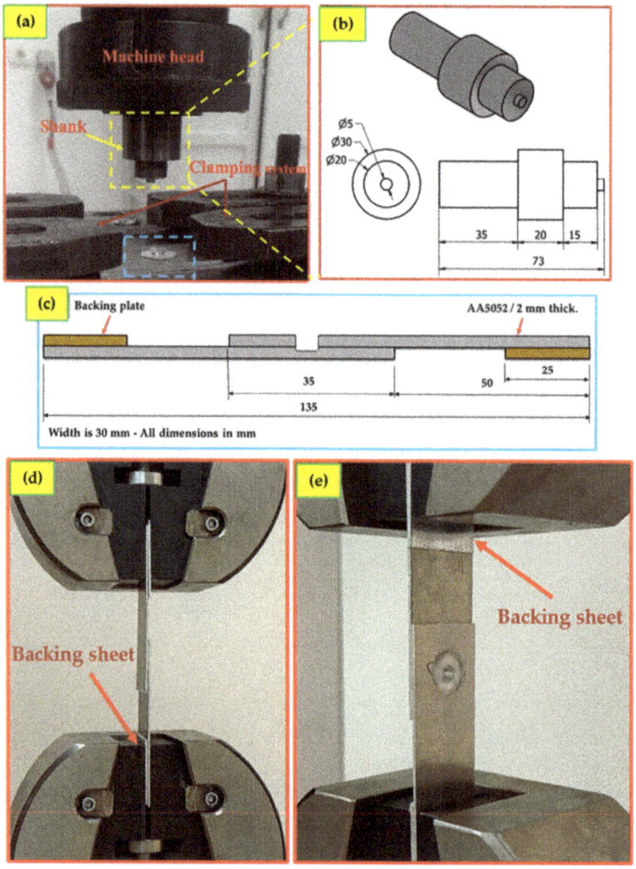

Figure 1. (**a**) FSSW process and clamping system, (**b**) FSSW tool setup used, (**c**) FSSW specimens (all dimensions in mm) with two backing sheets, (**d**) side and (**e**) front view of AA5052-H32 FSSW specimen under tensile–shear testing.

In addition, tensile/shear testing was conducted on the spot-welded joints fabricated at various rotational speeds and dwelling times, using the specimen lap joint configuration (Figure 1c). The tensile/shear test was performed at the usual quasi-static strain rate of 0.05 mm/s using a universal testing machine (Type-WDW-300D, Jinan, China), as shown in Figure 1d,e. The welding factors are shown in Table 3. For microstructural evaluation for both the base metal (BM) and the spot-welded specimens, the cross-sectioned specimens were investigated via an electron backscatter diffraction (EBSD) technique using a Quanta FEG 250 SEM (FEI Company, Hillsboro, OR, USA). The specimens were mechanically polished and subsequently electropolished for 60 s at −15 °C and 14 V. In addition, the fractured surfaces were investigated using the same Quanta FEG 250 SEM (Manufacturer is Field Electron and Ion Company, Hillsboro, OR, USA).

2.3. Regression Model

The RSM used in this part utilizes an FCCD and has three autonomous parameters that are block- and repeat-free.

Based on Equation (1), the design expert software (Version 10) prediction for the experimental result is given below [27,34]:

$$Y = \beta_o + \sum_{i=1}^{k} \beta_i X_i + \sum_{i=1}^{k} \beta_{ii} X^2{}_i + \sum_i \sum_j \beta_{ij} X_i X_j + \varepsilon \quad (1)$$

where Y, β_o, $\beta_i X_i$, $\beta_{ii} X^2{}_i$, $\beta_{ij} X_i X_j$, k, and ε are the expected response, steady coefficient, linear major impact, non-linear impact, and the interaction impact of variables, reflecting the number of process variables and error, respectively. There are three basic steps in this method. In the initial stage, numerous FCCD method runs are performed, and different FSSW factor values are experimented. Regression modeling is the second phase, which involves creating and developing a compatible mathematical model of the interaction between the FSSW response and experimental input components. The final improvement is to characterize how each FSSW factor contributes to the correct output FSSW response. The mathematical factors were estimated using ANOVA and FCCD. The important aspects of the FSSW technique are evaluated using the RSM data analysis based on the two-factor, three-level FCCD method (L27 orthogonal array). The analysis revealed significant correlations between the process parameters (rotational speed and dwell time) and the response variables, confirming that the friction stir spot welding (FSSW) conditions could be effectively modeled using second-order polynomial equations within the current statistical framework. The complete design matrix with corresponding response characteristics is presented in Table 4.

Table 4. Input FSSW parameters against the response of mechanical properties for AA5052-H32.

	Input FSSW Variables		Response of Mechanical Properties				
Exp. Run	**A: Rotational Speed, Rpm**	**B: Dwell Time, S**	**Heat Input Energy, J**	**SZ Hardness, Hv**	**TMAZ Hardness, Hv**	**HAZ Hardness, Hv**	**Maximum Shear Load, N**
1	1	1	6137	62.8	55.8	59.1	1986
2	1	0	4500	78.6	74.1	62.4	2461
3	−1	−1	771	88.2	77.8	66.1	3696
4	1	0	4500	81	76	63.8	2569
5	1	−1	2366	85	82.4	61.4	3301
6	0	1	3892	84.2	80.9	64.1	2493
7	1	−1	2366	83	81.6	64.2	3190
8	0	−1	1316	81.5	77.5	65.7	2897
9	−1	−1	771	86.6	76	67	3660
10	−1	1	2092	97.6	84.1	63.7	3957
11	0	0	2609	86.9	82.1	64.7	2935
12	−1	0	1525	107.1	91.2	64.6	4421
13	0	0	2609	89	84	63.1	3000
14	1	−1	2366	81.7	79.6	63.9	3215
15	−1	1	2092	94.3	80	65.4	3820
16	1	0	4500	79.4	75.6	60.5	2511
17	0	0	2609	92.4	86.3	62	3215
18	−1	−1	771	84.7	74.3	69.3	3594
19	1	1	6137	66	59.7	57.3	2113
20	−1	0	1525	102.3	87	66.3	4330
21	0	1	3892	86.7	82.1	62.4	2668
22	0	1	3892	87.8	83.7	60.7	2781
23	0	−1	1316	84.2	79.2	64.8	3156
24	1	1	6137	64.8	57.3	58.4	2074
25	0	−1	1316	85.1	81.7	62.6	3234
26	−1	0	1525	100.5	85.9	67.9	4269
27	−1	1	2092	92.4	78.6	66.8	3712

The R^2 coefficient of determination was used to quantify the reasonable degree of the experimental response to the polynomial mathematical modeling equation [37]. Therefore, the F-test was subsequently applied to assess the significance of each term in the polynomial equation, demonstrating 99.08% confidence in the model's validity. The analysis revealed significant correlations between the process parameters (rotational speed and dwell time) and the response variables, confirming that the friction stir spot welding (FSSW) conditions could be effectively modeled using second-order polynomial equations within the current statistical framework. The complete design matrix with corresponding response characteristics is presented in Table 4.

3. Results

3.1. Mechanical Properties of AA 5052-H32 Spot-Welded Specimens

To assess the joint's strength, the efficiency of the joining process, and its effect on the material's internal structure, the hardness and tensile/shear load of the fabricated joints were analyzed. The measured SZ hardness values are presented in Figure 2. At all constant dwell times (1 s, 2 s, and 3 s), the average hardness values of the SZ of the FSSW AA5052-H32 joint decrease with the increase in rotational speed from 500 to 1500 rpm. This consistent trend can be attributed to the higher welding heat input energy and the consequent increased width of the softened area as the rotational speed increases [38]. Otherwise, the highest hardness values of 102.3 Hv, 89 Hv, and 85 Hv are found at rotational speed of 500, 1000, and 1500 rpm, respectively, at a fixed dwell time of 2 s. It can be concluded that the optimum spot welding parameters to have the highest hardness are a 2 s holding time and a 500 rpm rotational speed. Regarding the effect of different dwell times, the FSSW joint spot welded at the lowest dwell time of 1 s generates insufficient heat input energy to create good mixing and wide SZ size. Conversely, the highest dwell time generates high heat input energy, which causes softening of the welded zone area and a decrease in the area beneath the shoulder. From these results, it can be noted that increasing the dwell time causes a decrease in the hardness, which is in good agreement with the previous study [12]. Figure 3 illustrates the relationship between the maximum tensile/shear strength and the applied rotational speeds and dwell times. At a constant dwell time of 2 s, the highest tensile/shear load is noted at a rotational speed of 500 rpm. Furthermore, the maximum tensile/shear load values decrease with an increase in the rotational speed from 500 rpm to 1500 rpm at a constant dwell time. This reduction in maximum tensile/shear load can be attributed to higher heat input energy, which results in the softening of the welded areas of the FSSW joints. Moreover, this result is in agreement with the measured hardness values. Hence, we can conclude that the optimum processing parameters for spot welding to achieve the highest mechanical properties are a 500 rpm rotational speed and a dwell time of 2 s. From the discussed results, it is clear that both a high rotational speed and an extended dwell time negatively impact both the hardness and joint strength [26,39]. After tensile–shear testing, the fracture surfaces of both the 5052-H32 base alloy and the produced FSSW joints were investigated using an ETD-SEM detector, as shown in Figure 4. It can be noted that the primary failure mechanism was ductile fracture, evidenced by the presence of dimples with various sizes and forms, indicative of severe plastic deformation prior to rupture [40]. Furthermore, the observation of finer dimples suggests a superimposed shear component to this ductile fracture, potentially influenced by the refined microstructure achieved through the optimized FSSW parameters.

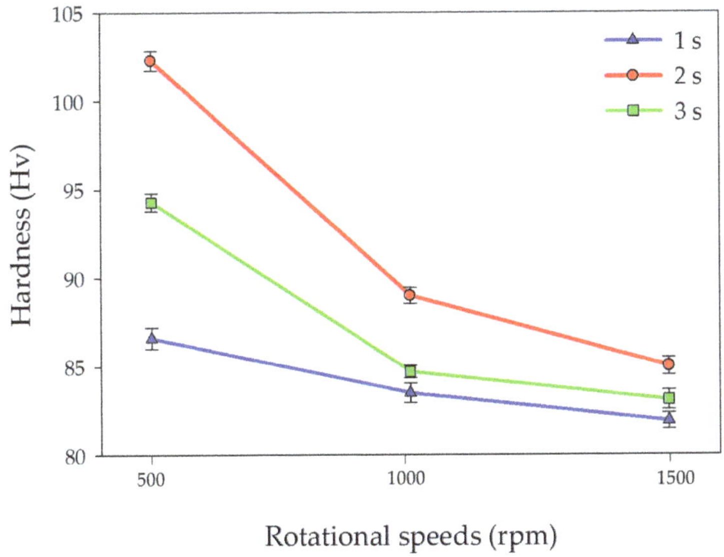

Figure 2. The average hardness values measured in SZ of the produced spot welds processed at various rotational speed (500, 1000, 1500 rpm) and dwell times.

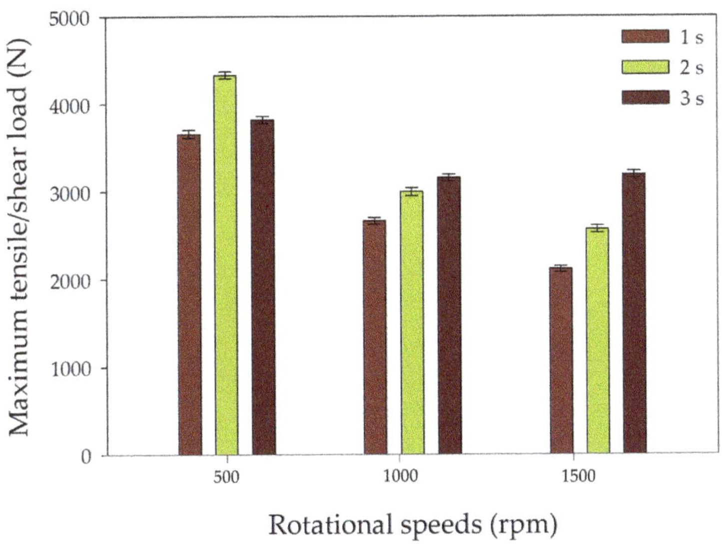

Figure 3. Provides a comparison of the maximum tensile/shear strength of the produced spot welds across a range of rotational speeds (500–1500 rpm) and dwell times (1, 2, and 3 s).

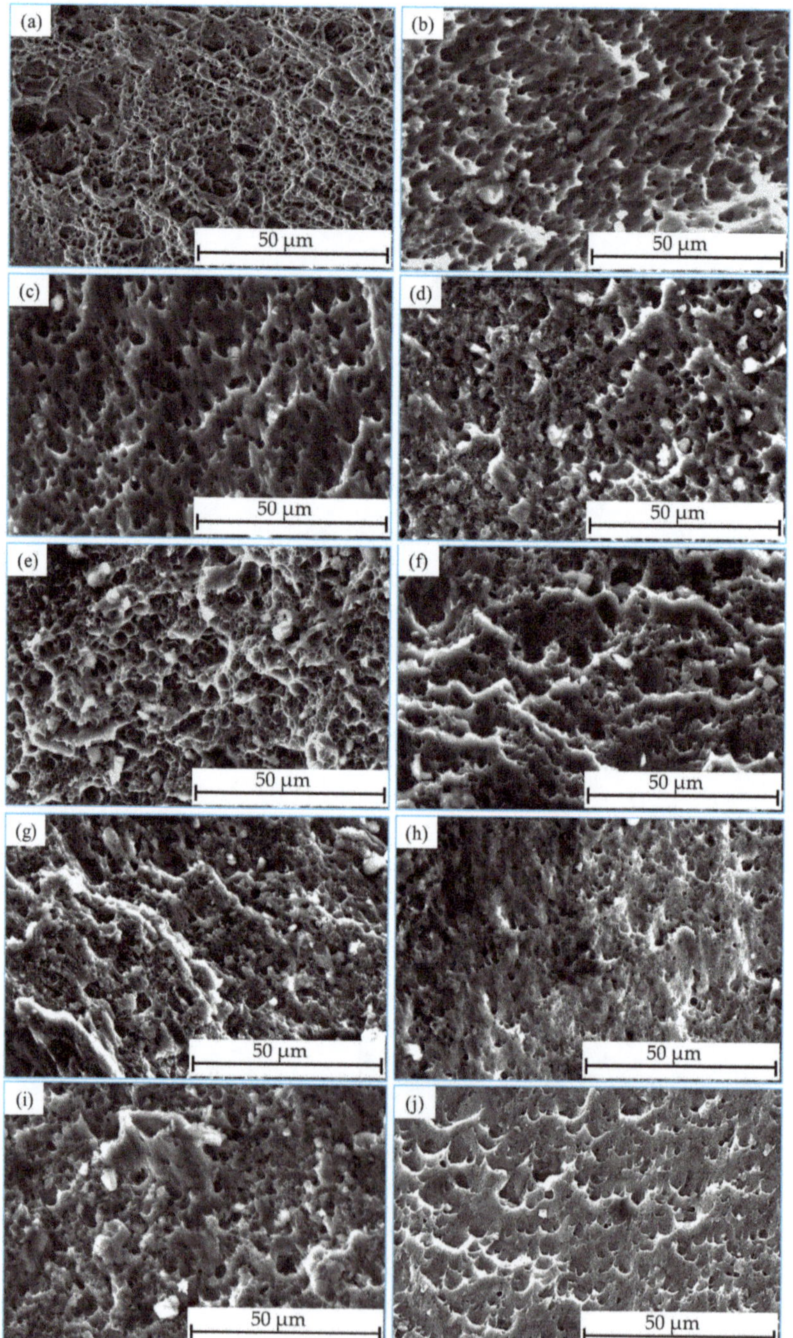

Figure 4. The fracture surface for the BM (**a**) and spot-welded joints processed at various dwell times and rotational speeds: (**b**) 500 rpm and 1 s, (**c**) 500 rpm and 2 s, (**d**) 500 rpm and 3 s, (**e**) 1000 rpm and 1 s, (**f**) 1000 rpm and 2 s, (**g**) 1000 rpm and 3 s, (**h**) 1500 rpm and 1 s, (**i**) 1500 rpm and 2 s, and (**j**) 1500 rpm and 3 s.

3.2. Microstructure of AA 5052-H32 Spot-Welded Specimen

Based on the mechanical behavior of the produced FSSW joints, microstructural characterization was conducted via EBSD on optimally selected samples, including the AA5052 base alloy and SZ of spot-welded specimens. EBSD analysis was performed for welds produced at a constant dwell time of 2 s with varying rotational speeds (500, 1000, and 1500 rpm). The results are presented as (1) inverse pole figure (IPF)-colored maps with corresponding grain size distribution analyses (Figure 5) and (2) grain boundary maps with associated misorientation angle distributions (Figure 6).

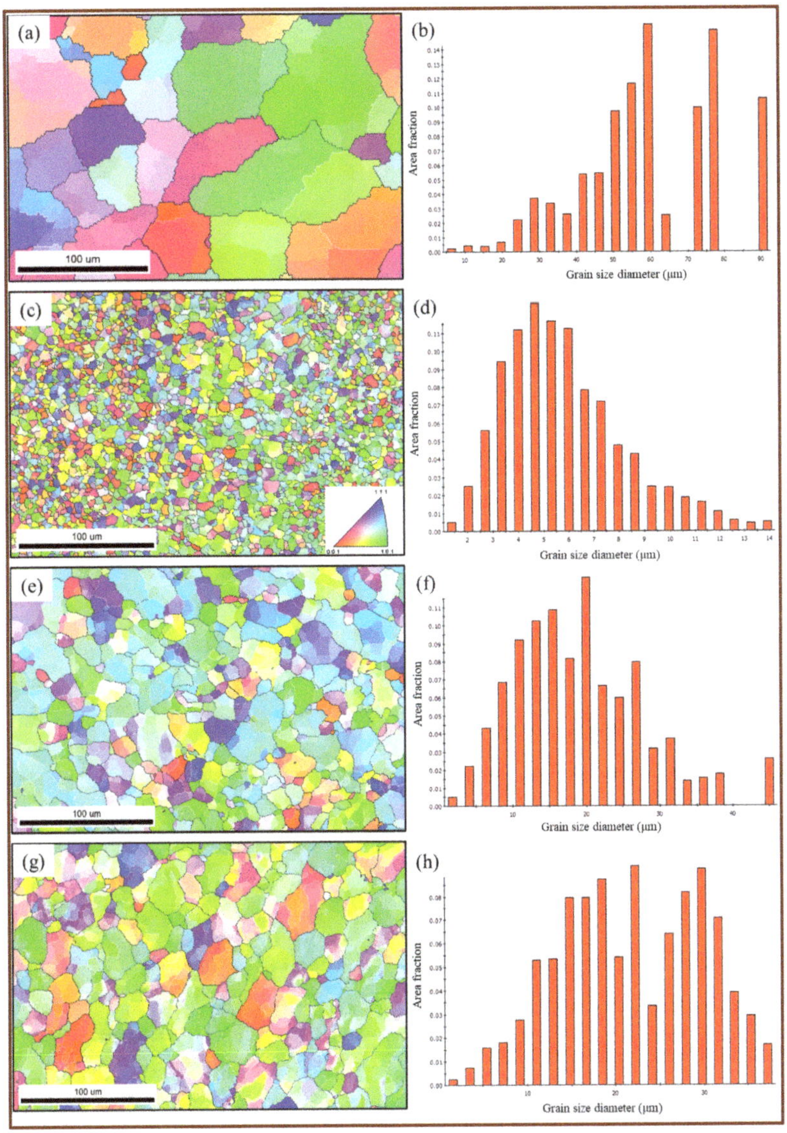

Figure 5. IPF-colored maps and grain size analysis of the BM and SZ microstructures of friction stir welded samples at different tool rotational speeds and a fixed dwell time of 2 s: (**a**,**b**) BM, (**c**,**d**) 500 rpm, (**e**,**f**) 1000 rpm, and (**g**,**h**) 1500 rpm.

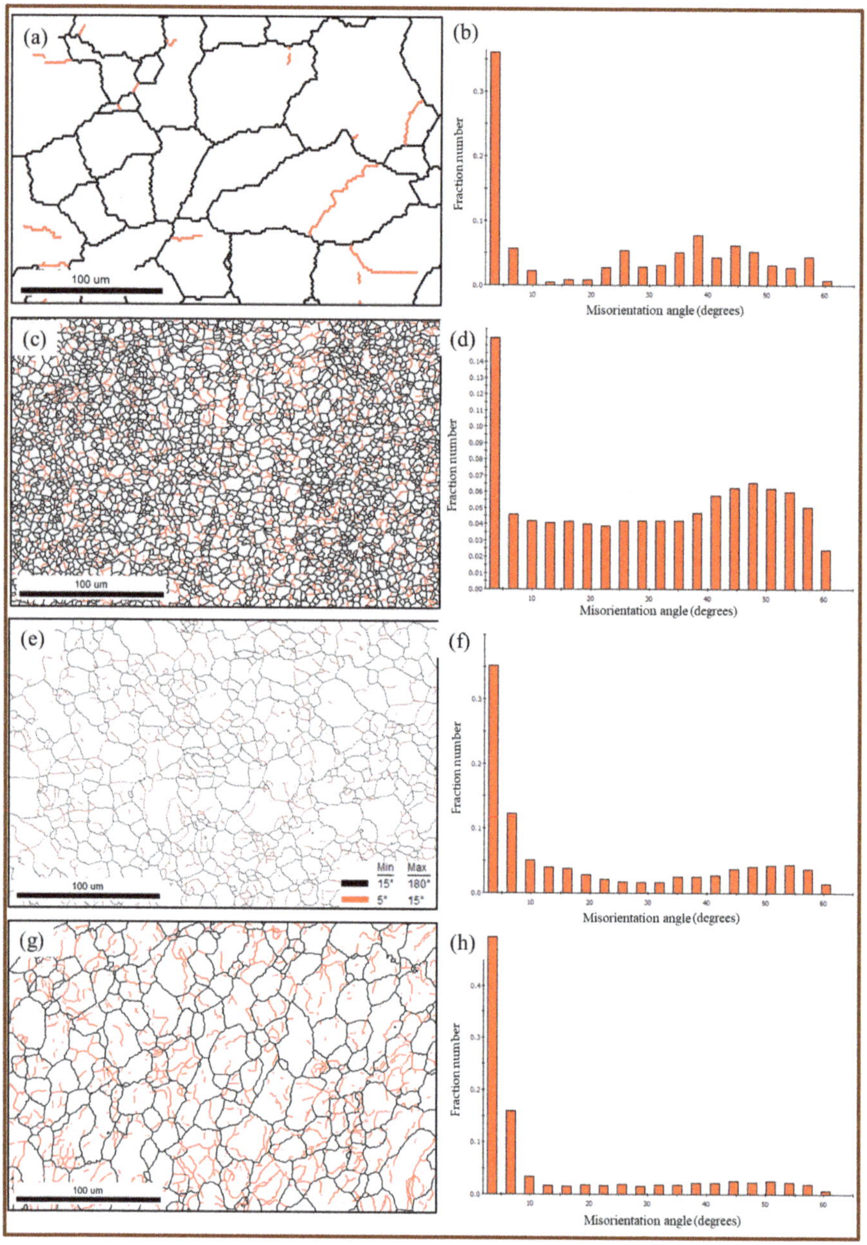

Figure 6. Represented grain boundary maps and misorientation angle analysis of the BM and SZ microstructures of friction stir welded samples at different rotational speeds and a fixed dwell time of 2 s: (**a**,**b**) BM, (**c**,**d**) 500 rpm, (**e**,**f**) 1000 rpm, and (**g**,**h**) 1500 rpm.

From Figure 5, it can be noted that the coarse equiaxed grains of the AA5052-H32 BM (Figure 5a,b) have an approximately average grain size diameter of 38.5 μm. As a result of the lap spot welding of AA5052-H32 sheets, the grains are refined and the microstructure is dominated by dynamic recrystallization due to high-temperature plastic deformation. Moreover, the finest grain size diameter of 4.1 μm is found in the spot-welded joint processed at a dwell time of 2 s and a rotational speed of 500 rpm (Figure 5c,d). However, the plasticized, deformed grains become coarser with an increase in rotational speed from 500 rpm to 1500 rpm. This can be attributed to the higher heat input energy and increased plastic deformation energy. The spot-welded joint at the highest rotational speed has an average grain size diameter of 11.2 μm (Figure 5g,h).

Figure 6 reveals that the grain boundary map of the AA5052-H32 BM predominantly features high-angle grain boundaries (>15°), as shown in Figure 6a,b, which are represented by black lines. As a result of the spot welding of AA5052-H32, the density of low-angle grain boundaries (<15°) increases with increasing rotational speed from 500 rpm to 1500 rpm. This increase in low-angle grain boundaries (<15°) can be ascribed to the high heat input energy generated and severe dynamic plastic deformation (Figure 6c–h).

3.3. Development of RSM for AA 5052-H32 Specimens

The experimental results have undergone analysis of variance (ANOVA) to create a regression model and identify significant input parameters and interactions. The R^2 value is used to analyze the fit between model-predicted values and experimentally acquired values and to determine the model's suitability. The correlation between them is better when the R^2 value is higher. These situations have high enough R^2 values, which show a strong fit between the expected and measured data. As a result, appropriate response models are created using ANOVA analysis. The results of the ANOVA analysis are used to create the final models for heat input energy, SZ hardness, TMAZ hardness, HAZ hardness, and maximum shear load. Additionally, using the backward elimination procedure while considering the *p*-values, the insignificant terms from the ANOVA analysis are eliminated. When the *p*-values against each term of variables (linear, square, and interaction impact) decrease below 0.05 (i.e., in this example, the 99.08% confidence interval), larger *p*-values that signify the least significance are gradually eliminated. Tables 5–9 display the findings of the final ANOVA for heat input energy, SZ hardness, and maximum shear load, respectively. Thus, the R^2 and R^2-predicted values for heat input energy in Table 5 are 99.08% and 98.89%, respectively, indicating that the predictions are accurate, and the design space has been thoroughly examined. Additionally, *p*-values for both the rotational speed and dwell time model factors are <0.0001. They indicate that both factors have a significant effect on heat input energy. Furthermore, the R^2 value indicates a good fit of the model with the data, and the low *p*-values confirm the statistical significance.

Table 5. ANOVA for heat input energy.

Source	Sum of Squares	Df	Mean Squares	F-Value	*p*-Value	
Model	7.101×10^7	3	2.367×10^7	829.33	<0.0001	Significant
A-Rotational Speed	3.711×10^7	1	3.711×10^7	1300.20	<0.0001	
B-Dwell Time	2.940×10^7	1	2.940×10^7	1030.06	<0.0001	
AB	4.502×10^6	1	4.502×10^6	157.73	<0.0001	
Residual	6.564×10^5	23	28,541.18			
Lack of Fit	6.564×10^5	5	1.313×10^5			
Pure Error	0.0000	18	0.0000			
Cor Total	7.167×10^7	26				
Std. Dev.	168.94		R^2		0.9908	
Mean	2800.89		Adjusted R^2		0.9896	
CV%	6.03		Predicted R^2		0.9889	
			Adeq Precision		83.4697	

Table 6. ANOVA for SZ hardness.

Source	Sum of Squares	Df	Mean Squares	F-Value	p-Value	
Model	2675.69	7	382.24	62.72	<0.0001	Significant
A-Rotational Speed	837.80	1	837.80	137.47	<0.0001	
B-Dwell Time	10.40	1	10.40	1.71	0.2070	
AB	545.40	1	545.40	89.49	<0.0001	
A^2	7.11	1	7.11	1.17	0.2935	
B^2	351.65	1	351.65	57.70	<0.0001	
A^2B	61.62	1	61.62	10.11	0.0049	
AB^2	47.38	1	47.38	7.77	0.0117	
Residual	115.80	19	6.09			
Lack of Fit	29.56	1	29.56	6.17	0.0231	Significant
Pure Error	86.24	18	4.79			
Cor Total	2791.49	26				
Std. Dev.	2.47		R^2		0.9585	
Mean	85.70		Adjusted R^2		0.9432	
CV%	2.88		Predicted R^2		0.9194	
			Adeq Precision		27.6805	

Table 7. ANOVA for TMAZ hardness.

Source	Sum of Squares	Df	Mean Squares	F-Value	p-Value	
Model	1785.38	7	255.05	48.58	<0.0001	Significant
A-Rotational Speed	245.76	1	245.76	46.81	<0.0001	
B-Dwell Time	11.48	1	11.48	2.19	0.1556	
AB	607.76	1	607.76	115.77	<0.0001	
A^2	177.85	1	177.85	33.88	<0.0001	
B^2	232.71	1	232.71	44.33	<0.0001	
A^2B	147.22	1	147.22	28.04	<0.0001	
AB^2	13.94	1	13.94	2.65	0.1197	
Residual	99.75	19	5.25			
Lack of Fit	26.01	1	26.01	6.35	0.0214	Significant
Pure Error	73.74	18	4.10			
Cor Total	1885.13	26				
Std. Dev.	2.29		R^2		0.9471	
Mean	78.31		Adjusted R^2		0.9276	
CV%	2.93		Predicted R^2		0.8954	
			Adeq Precision		23.2206	

The R^2 and R^2-predicted values for the SZ hardness in Table 6 are 95.85% and 91.94%, respectively. These values, being close to 1, indicate the model's adequacy. From Table 6, the p-values for both the rotational speed and dwell time factors are <0.0001 and 0.2070, respectively. These indicate that the rotational speed, as a model factor, has a significant effect on SZ hardness, but the dwell time does not have a significant effect on SZ hardness.

Table 8. ANOVA for HAZ hardness.

Source	Sum of Squares	Df	Mean Squares	F-Value	p-Value	
Model	164.47	3	54.82	25.02	<0.0001	Significant
A-Rotational Speed	118.07	1	118.07	53.89	<0.0001	
B-Dwell Time	40.80	1	40.80	18.62	0.0003	
AB	5.60	1	5.60	2.56	0.1234	
Residual	50.39	23	2.19			
Lack of Fit	8.27	5	1.65	0.7063	0.6262	Not significant
Pure Error	42.13	18	2.34			
Cor Total	214.86	26				
Std. Dev.	1.48		R^2		0.7655	
Mean	63.64		Adjusted R^2		0.7349	
CV%	2.33		Predicted R^2		0.6767	
			Adeq Precision		14.2759	

Table 9. ANOVA for a maximum shear load.

Source	Sum of Squares	Df	Mean Squares	F-Value	p-Value	
Model	1.145×10^7	6	1.909×10^6	50.44	<0.0001	Significant
A-Rotational Speed	8.052×10^6	1	8.052×10^6	212.75	<0.0001	
B-Dwell Time	3.015×10^5	1	3.015×10^5	7.97	0.0105	
AB	1.382×10^6	1	1.382×10^6	36.51	<0.0001	
A^2	6.938×10^5	1	6.938×10^5	18.33	0.0004	
B^2	2.781×10^5	1	2.781×10^5	7.35	0.0135	
A^2B	2567.11	1	2567.11	0.0678	0.7972	
Residual	7.569×10^5	20	37,847.44			
Lack of Fit	5.413×10^5	2	2.707×10^5	22.60	<0.0001	Significant
Pure Error	2.156×10^5	18	11,977.89			
Cor Total	1.221×10^7	26				
Std. Dev.	194.54		R^2		0.9380	
Mean	3157.70		Adjusted R^2		0.9194	
CV%	6.16		Predicted R^2		0.8925	
			Adeq Precision		21.6217	

Table 7 shows that the R^2 and R^2-predicted values for TMAZ hardness are 94.71% and 89.54%, respectively. From this table, it can be noted that the p-values for both the rotational speed and dwell time factors are <0.0001 and 0.1556, respectively. These indicate that the rotational speed, as a model factor, has a significant effect on TMAZ hardness, but the dwell time does not have a significant effect on TMAZ hardness.

Moreover, the R^2 and R^2-predicted values for the HAZ hardness in Table 8 are 76.55% and 67.67%. Table 9 shows that the R^2 and R^2-predicted values for the maximum shear load are 93.8% and 89.25%, respectively. Furthermore, the p-values for both the rotational speed and dwell time factors are <0.0001 and 0.0105, respectively. It can be noted that these factors have a significant effect on the FSSW joint strength. These results are in agreement with other published data [32].

A mathematical second-order equation was formulated using regression analysis to describe the relationship between the input parameters of the FSSW process and the resulting output responses. Therefore, the regression model coefficients for the heat input energy, SZ hardness, TMAZ hardness, HAZ hardness, and maximum shear load were calculated with Design-Expert (version 10) software at a 99.08% confidence level. According to the summary of model statistics, the quadratic model is the most suitable for prediction and was therefore employed to predict the responses. Equations (2) to (6) represent the final regression models for the heat input energy, SZ hardness, TMAZ hardness, HAZ hardness, and tensile/shear load based on ANOVA analysis, respectively:

$$\text{Heat input energy (KJ)} = -176.77778 + (0.421667 \times \text{A}) + (53 \times \text{B}) + (1.225 \times \text{A} \times \text{B}) \quad (2)$$

$$\begin{aligned}\text{SZ Hardness (Hv)} = {} & +54.44815 - (0.023222 \times \text{A}) + (57.25556 \times \text{B}) - (0.009617 \times \text{A} \times \text{B}) + \\ & \left(0.000027 \times \text{A}^2\right) - \left(14.53889 \times \text{B}^2\right) - \left(0.000016 \times \text{A}^2 \times \text{B}\right) + \left(0.006883 \times \text{A} \times \text{B}^2\right)\end{aligned} \quad (3)$$

$$\begin{aligned}\text{TMAZ Hardness (Hv)} = {} & +54.57407 - (0.022911 \times \text{A}) + (31.19444 \times \text{B}) + (0.019367 \times \text{A} \times \text{B}) + \\ & \left(0.000027 \times \text{A}^2\right) - \left(9.96111 \times \text{B}^2\right) - \left(0.000024 \times \text{A}^2 \times \text{B}\right) + \left(0.003733 \times \text{A} \times \text{B}^2\right)\end{aligned} \quad (4)$$

$$\text{HAZ Hardness (Hv)} = +69.03704 - (0.002389 \times \text{A}) - (0.138889 \times \text{B}) - (0.001367 \times \text{A} \times \text{B}) \quad (5)$$

$$\begin{aligned}\text{Maximum shear load (N)} = {} & +4204.96296 - (3.10611 \times \text{A}) + (1214.27778 \times \text{B}) - (0.476 \times \text{A} \times \text{B}) + \\ & \left(0.001563 \times \text{A}^2\right) - \left(215.27778 \times \text{B}^2\right) - \left(0.000101 \times \text{A} \times \text{B}^2\right)\end{aligned} \quad (6)$$

Figure 7a–e compares the measured and predicted values for the heat input energy (Figure 7a), SZ hardness (Figure 7b), TMAZ hardness (Figure 7c), HAZ hardness (Figure 7d), and maximum shear load for AA5052-H32 joints (Figure 7e). All measured and predicted points lie close to the line, indicating that the error is normally distributed. As a result, the points are close to the measured values, indicating that the predicted values agree well with the actual values. Figure 7a depicts the heat input energy relationship between predicted and measured values, revealing a good agreement between the measured and predicted values. This result is in good agreement with the R^2 value of 0.9908, which indicates the model's adequacy.

Additionally, Figure 7b,c depicts comparison graphs between the predicted and measured SZ hardness and TMAZ hardness values. The predicted points and the actual points are close, indicating that the predicted values agree well with the actual values for both the SZ hardness and TMAZ hardness, and this confirms the model's adequacy. These results are also in good agreement with the R^2 values of 0.9585 for SZ hardness and 0.9471 for TMAZ hardness, which indicates the model's good fitting and adequacy. While some discrepancies between the measured and predicted HAZ hardness values are observed in Figure 7b, these are primarily attributed to the complex thermal gradients and varying cooling rates inherent to the HAZ's transitional position. Despite these localized variations, which make the HAZ response particularly sensitive, the hardness values remain within expected ranges for the alloy system and do not compromise the overall validity of our optimization approach.

The relationship between the predicted and actual maximum shear load values is given in Figure 7e. Furthermore, the predicted values from the maximum shear load model appear to be in good agreement with the actual values in this figure. In addition, this result is confirmed by an R^2 value of 0.9380, which demonstrates the model's good fit and adequacy.

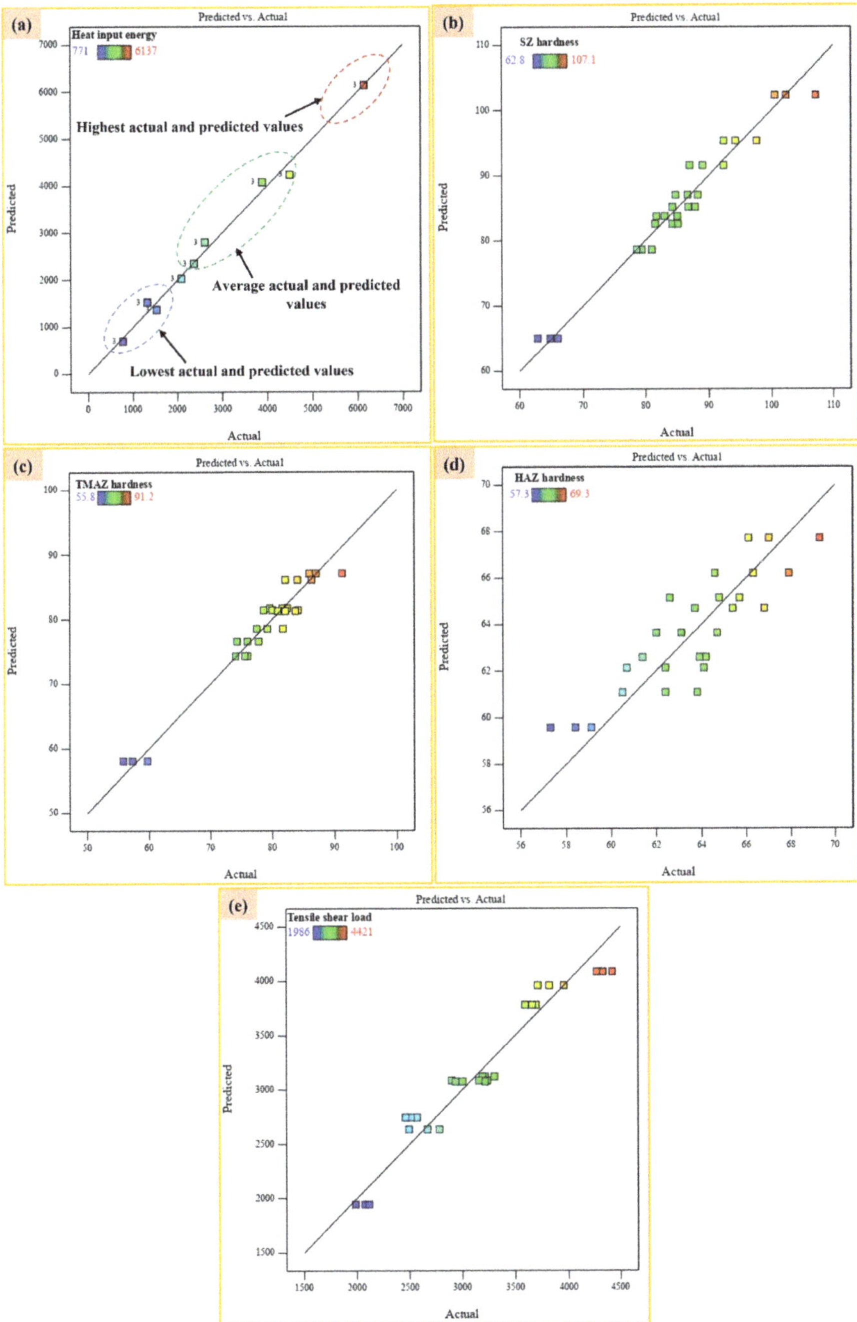

Figure 7. The scatter graph of the (**a**) heat input energy, (**b**) SZ hardness, (**c**) TMAZ hardness, (**d**) HAZ hardness, and (**e**) tensile/shear load.

3.4. *Effects of Variables on the Output Responses*

The major effects of the heat input energy, SZ hardness, and maximum shear load are depicted in Figure 8, as are the individual effects of the rotational speed and dwell time. In this figure, the actual measured data points are represented by dots, while their average values are shown by square symbols. Dashed lines illustrate the estimated trend behavior between the measured data points, and continuous lines represent the predicted results derived from the model. A red cross specifically marks the mean of the measured data at the selected factors (rotational speed of 1000 rpm and dwell time of 2 s), which were chosen based on their representation of optimum measured data. Figure 8a shows that increasing the rotational speed and dwell time increases the heat input energy. As a result, the highest heat input energy was obtained at a 1500 rpm rotational speed and a 3 s dwell time. The hardness of the SZ, TMAZ, and HAZ decreases as the heat input energy increases, but increasing the dwell time increases the hardness, as shown in Figure 8b–d. The best results for SZ and TMAZ hardness are obtained at a rotational speed of 500 rpm and a dwell time of 2 s, while the lowest results are attained at a tool rotational speed of 1500 rpm and a spot-welding time of 1 s. On the other hand, the highest results for the HAZ hardness are found at 500 rpm and 1 s, while the lowest results are obtained at a tool rotational speed of 1500 rpm and a dwell time of 3 s. In Figure 8e, the maximum shear load decreases with an increase in the rotational speed from 500 to 1500 rpm and the dwell time from 1 s to 3 s.

3.5. Influences of Two-Factor Interaction on Heat Input Energy, SZ Hardness, and Maximum Tensile–Shear Load

The dual interaction of the rotational speed and spot-welding time governs the heat input energy, the SZ hardness and that of adjacent zones (TMAZ, HAZ), and the maximum tensile shear load of the FSSW joint. Figure 9 depicts a 3D response surface plot of the effect of the dwell time and rotational speed on the heat input energy, SZ hardness, TMAZ hardness, HAZ hardness, and maximum tensile shear load for the FSSW 5052-H32 aluminum alloy joints. The impact of the rotational speed and dwell time on the heat input energy is shown in Figure 9a. The heat input energy is minimized (771 J) at a low rotational speed of 500 rpm and a dwell time of 1 s. Conversely, it reaches its maximum value (up to 6137 J) at the highest rotational speed of 1500 rpm and a dwell time of 3 s. As a result, at all dwell times, the heat input energy increases as the rotational speed increases. Additionally, this figure demonstrates that the rotational speed and dwell time have a greater impact on the heat input. Figure 9b illustrates how the rotational speed and dwell duration influence the hardness within the SZ of the produced FSSW samples. It is worth noting that the SZ hardness reaches its lowest value of 81 Hv when the rotational speed is 1000 rpm and the dwell time is 1 s. With a rotational speed of 500 rpm and a dwell time of 2 s, the highest SZ hardness value of 110 Hv is obtained. As a result, as the rotational speed increases, the hardness of the SZ decreases. Hence, the rotational speed has a greater effect on the SZ hardness than the dwell time. The main focus of the maximum tensile shear load is to assess the strength and plasticity of FSSW materials and to investigate the effect of processing parameters on the performance of the produced specimens. As shown in Figure 9e, the higher shear load is obtained at the mean rotational speed of 500 rpm and a 2 s dwell time, whereas the lowest shear load is obtained at the rotational speed of 1500 rpm and dwell time of 3 s. The findings demonstrate that the duration of the dwell time significantly influences the maximum tensile shear load.

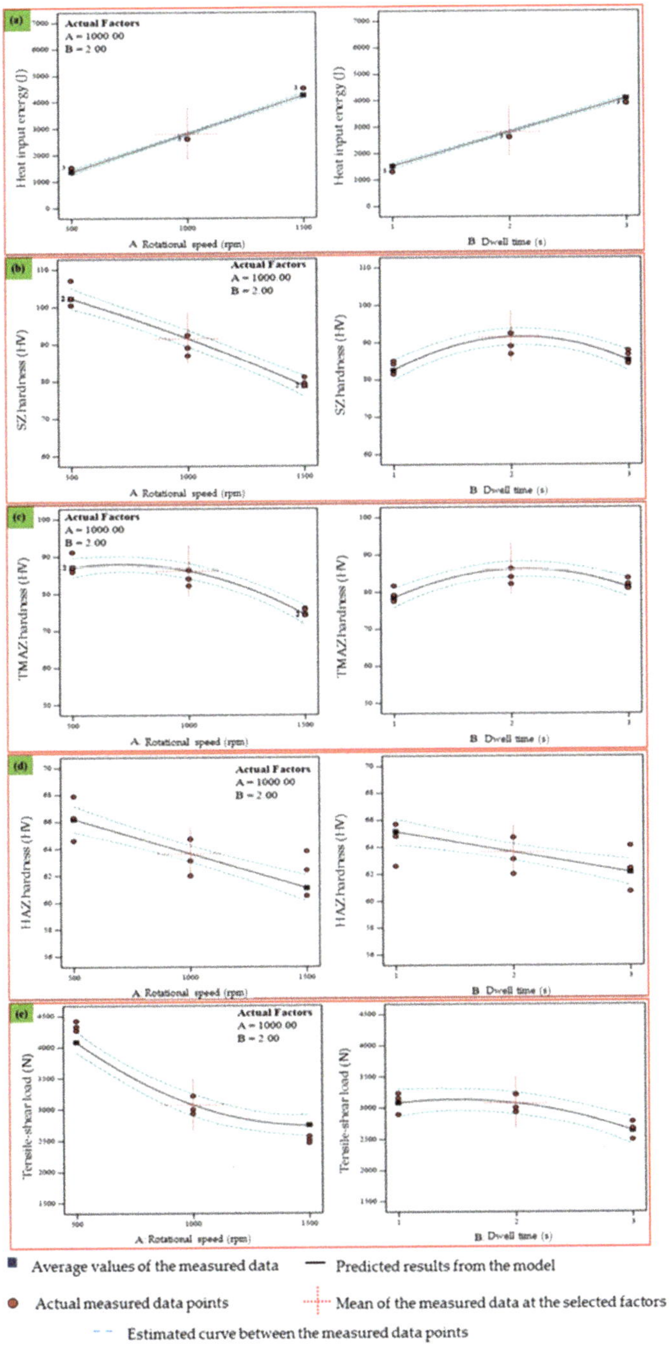

Figure 8. Effect of FSSW factors on (**a**) heat input energy, (**b**) SZ hardness, (**c**) TMAZ hardness, (**d**) HAZ hardness, and (**e**) tensile/shear load.

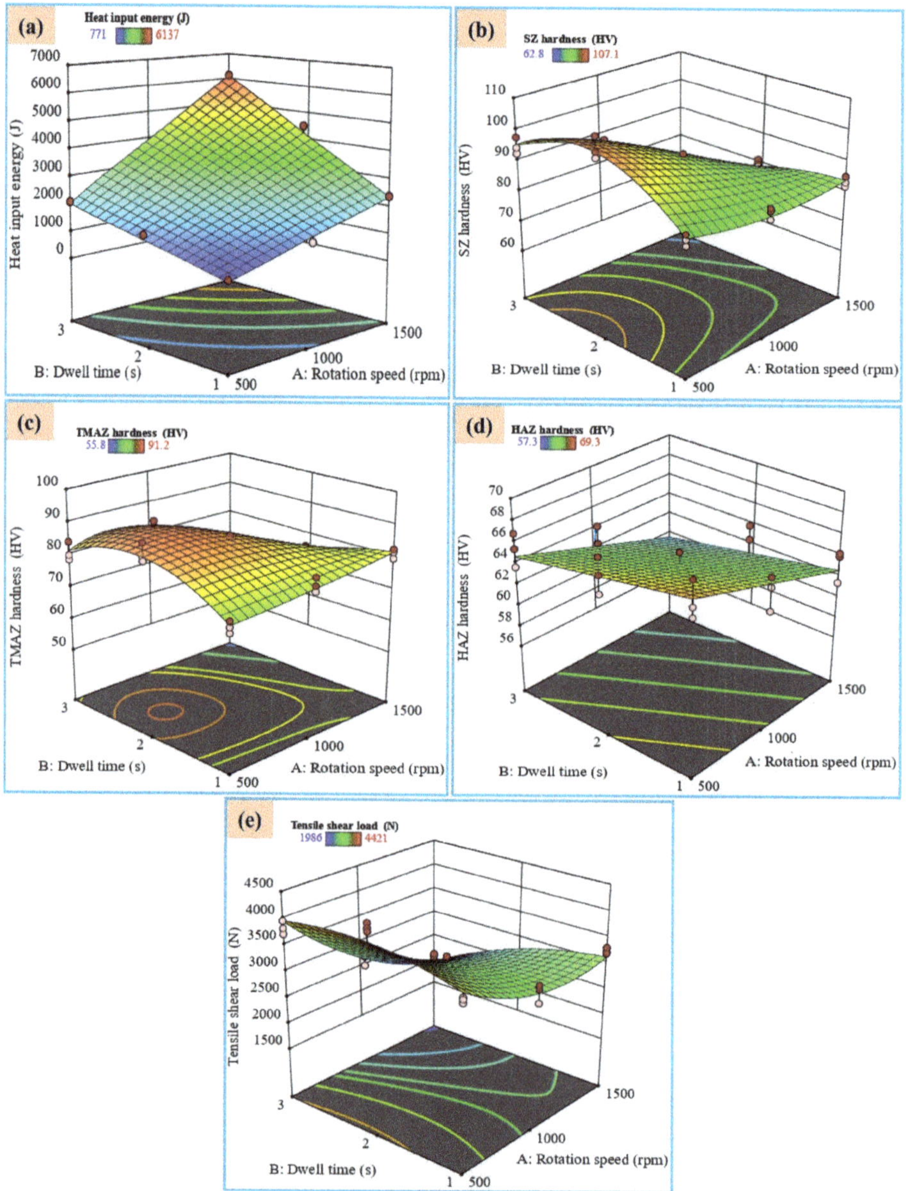

Figure 9. 3D surface plots for the (**a**) heat input energy, (**b**) SZ hardness, (**c**) TMAZ hardness, (**d**) HAZ hardness, and (**e**) tensile/shear load.

3.6. Optimization of FSSW Variables via the Desirability Approach

Many statistical strategies can be used to solve optimization problems, but due to their generality and software simplicity, the desirability method is often preferred. This method determines the best operating conditions to achieve the desired response values. The optimization of a desirability function involves three key stages: Initially, the desired outcomes for each measured variable, including the welding parameters, should be defined. Next, the acceptable range (minimum and maximum values) for each of these variables should be specified. Finally, the desired operating conditions should be inputted to see the model's predicted outcomes. Table 10 depicts the optimization criteria utilized in the current study. All responses are converted to an individual degree of freedom on a scale of 0 to 1. The combination of FSSW variables with the greatest desirability is considered optimal. A maximum desirability of approximately 1.00 was accomplished using a rotational speed of 627.32 rpm and a dwell time of 2.49 s. The predicted heat input energy, SZ hardness, TMAZ hardness, HAZ hardness, and maximum shear load for the best combination are 2132.48 J, 99.3187 Hv, 87.6418 Hv, 64.8096 Hv, and 3717.63 N, respectively. The results are obtained by calculating the tensile–shear load of the weld lap joint using the ramp chart, as shown in Figure 10.

Table 10. FSSW optimization criteria.

Constraint Name	Units	Goal	Limits		Weights		Importance
			Lower	Upper	Lower	Upper	
Tool rotational speed	rpm	in range	500	1500	1	1	3
Dwell time	s	in range	1	3	1	1	3
Heat input energy	KJ	maximize	771	6137	1	1	3
SZ hardness	Hv	maximize	62.8	107.1	1	1	3
TMAZ	Hv	maximize	55.8	91.2	1	1	3
HAZ	Hv	maximize	57.3	69.3	1	1	3
Maximum shear load	N	maximize	1986	4421	1	1	3

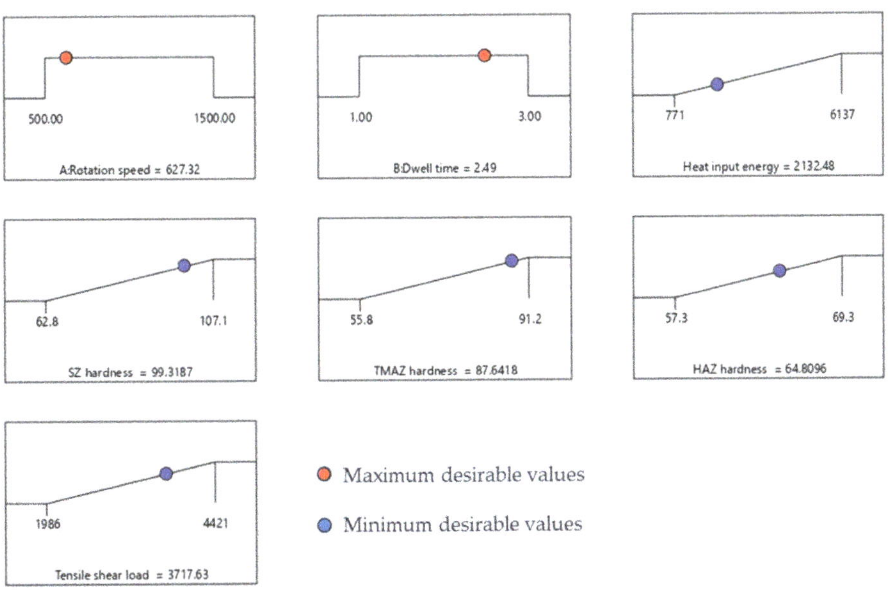

Figure 10. Ramp chart with optimized FSSW variables for maximizing heat input energy, SZ hardness, TMAZ hardness, HAZ hardness, and maximum shear load.

3.7. Confirmation Test of the Developed Mathematical Models

To validate the regression models, two confirmation tests were performed. The values used in the validation tests were all within the specified range. However, these validation points were distinct from the combinations used to build the initial model (the FCCD matrix). Hence, the percentage error was calculated by Equation (7). Table 11 presents the actual and model-predicted results for these validation tests. The observed difference between the experimental and predicted values was within the 95% confidence interval, indicating the model's reliability and good agreement between predictions and experimental results. As a result, it can be concluded that the developed models are suitable and applicable to all values within the designed space. Additionally, the developed models yield nearly accurate results.

$$Percentage\ error\ (E\%) = \left| \frac{Actual\ value - Predicted\ value}{Predicted\ value} \right| * 100 \tag{7}$$

Table 11. Confirmation of test results for mathematical models.

Exp. No.	Rotational Speed, Rpm	Dwell Time, S		Heat Input Energy, J	SZ Hardness, Hv	TMAZ Hardness, Hv	HAZ Hardness, Hv	Maximum Shear Load, N
1.	750	1.5	Actual	1653.32	94.09	81.76	66.66	4037.22
			Predicted	1597.1	92.7	78.5	64.0	3960
			∣Error∣, %	3.52	1.51	3.67	4.15	1.95
2.		2.5	Actual	2660.51	86.68	81.73	65,73	4260.39
			Predicted	2568.8	84.6	78.4	62.3	4150
			∣Error∣, %	3.57	2.46	3.77	5.50	2.66
1.	1250	1.5	Actual	2815.32	120.64	85.55	62.92	2446.17
			Predicted	2726.7	118.4	83.1	60.5	2390
			∣Error∣, %	3.25	1.89	2.95	4.00	2.35
2.		2.5	Actual	4460.06	83.17	82.68	61.43	2342.55
			Predicted	4310.9	81.1	79.8	59.1	2300
			∣Error∣, %	3.46	2.55	3.61	3.95	1.85

4. Conclusions

1. A higher generated frictional heat input decreases the mechanical properties and increases the material softening of the produced FSSW joints.
2. RSM was proven to be a precise technique for optimizing the FSSW process to achieve optimal mechanical properties of the AA5052-H32 weldments. Based on FCCD, RSM can effectively model and predict the heat input energy, SZ hardness, and maximum shear load for friction stir spot welding technology.
3. The ANOVA values showed that the significant terms (F-values) used to build and develop the mathematical models were 829.33, 62.72, 48.58, 25.02, and 50.44 for the heat input energy, SZ hardness, TMAZ hardness, HAZ hardness, and maximum shear load, respectively, indicating that all models were statistically significant.
4. As a result of the mathematical model, the rotational speed has a significant influence on the SZ hardness values of AA5052-H32 welded joints. It was noticed that the dwell time of 1.5 s at rotational speeds of 750 rpm and 1250 rpm possesses the highest SZ hardness value of 92.7 Hv and 118.4 Hv, respectively.
5. The predominant fracture mode observed was ductile fracture, characterized by dimples of varying sizes and shapes. Additionally, the presence of fine dimples suggests a mixed fracture mechanism involving both shear and ductile modes.

Author Contributions: Conceptualization, M.M.E.-S.S., S.A., R.A.F., M.M.Z.A. and H.A.R.; methodology, S.A., R.A.F., M.M.E.-S.S. and H.A.R.; software, M.M.Z.A., S.A. and N.H.A.; investigation, N.H.A., S.A. and A.B.; resources, M.M.E.-S.S. and A.B.; data curation, H.A.R., A.B. and M.M.Z.A.; writing—original draft preparation, R.A.F., H.A.R., M.M.Z.A. and S.A.; writing—review and editing, R.A.F., N.H.A., M.M.E.-S.S., S.A. and M.M.Z.A.; visualization, H.A.R., A.B. and S.A.; supervision, M.M.E.-S.S., S.A., A.B., H.A.R. and M.M.Z.A.; project administration, N.H.A., S.A. and. M.M.E.-S.S.;

funding acquisition, S.A. and N.H.A. All authors have read and agreed to the published version of the manuscript.

Funding: This work was supported and funded by the Deanship of Scientific Research at Imam Mohammad Ibn Saud Islamic University (IMSIU) (grant number IMSIU-DDRSP2503).

Data Availability Statement: The original contributions presented in this study are included in the article. Further inquiries can be directed to the corresponding author.

Conflicts of Interest: The authors declare no conflicts of interest.

References

1. Zhang, Z.; Yang, X.; Zhang, J.; Zhou, G.; Xu, X.; Zou, B. Effect of Welding Parameters on Microstructure and Mechanical Properties of Friction Stir Spot Welded 5052 Aluminum Alloy. *Mater. Des.* **2011**, *32*, 4461–4470. [CrossRef]
2. Mothilal, M.; Kumar, A. Optimization of Friction Stir Welding Process Parameter in the Joining of AA7075-T6/AA5083-O Dissimilar Aluminum Alloy Using Response Surface Methodology. *Int. J. Press. Vessel. Pip.* **2024**, *211*, 105282. [CrossRef]
3. Piccini, J.M.; Svoboda, H.G. Effect of the Tool Penetration Depth in Friction Stir Spot Welding (FSSW) of Dissimilar Aluminum Alloys. *Procedia Mater. Sci.* **2015**, *8*, 868–877. [CrossRef]
4. Ahmed, M.M.Z.; El-Sayed Seleman, M.M.; Sobih, A.M.E.S.; Bakkar, A.; Albaijan, I.; Touileb, K.; Abd El-Aty, A. Friction Stir-Spot Welding of AA5052-H32 Alloy Sheets: Effects of Dwell Time on Mechanical Properties and Microstructural Evolution. *Materials* **2023**, *16*, 2818. [CrossRef]
5. Wang, B.; Xu, L.; Guo, B.; Zhang, H. Process and Performance Characteristics of an Improved Friction-Stir Riveting Process. *J. Manuf. Process* **2021**, *62*, 234–246. [CrossRef]
6. Ragab, A.E.; Alsaty, A.; Alsamhan, A.; Al-Tamimi, A.A.; Dabwan, A.; Sayed, A.; Alghilan, W. Open-Source Real-Time Monitoring System of Temperature and Force during Friction Stir Spot Welding. *J. Eng. Res.* **2024**, *13*, 84–96. [CrossRef]
7. Jambhale, S.; Kumar, S.; Kumar, S. Characterization and Optimization of Flat Friction Stir Spot Welding of Triple Sheet Dissimilar Aluminium Alloy Joints. *Silicon* **2022**, *14*, 815–830. [CrossRef]
8. Ahmed, M.M.Z.; El-Sayed Seleman, M.M.; Albaijan, I.; Abd El-Aty, A. Microstructure, Texture, and Mechanical Properties of Friction Stir Spot-Welded AA5052-H32: Influence of Tool Rotation Rate. *Materials* **2023**, *16*, 3423. [CrossRef]
9. Mubiayi, M.P. Current Developments in Friction Stir Welding (FSW) and Friction Stir Spot Welding (FSSW) of Aluminium and Titanium Alloys. *Eng. Proc.* **2023**, *56*, 184. [CrossRef]
10. Kilic, S.; Ozturk, F.; Demirdogen, M.F. A Comprehensive Literature Review on Friction Stir Welding: Process Parameters, Joint Integrity, and Mechanical Properties. *J. Eng. Res.* **2024**, *13*, 122–130. [CrossRef]
11. Anton Savio Lewise, K.; Edwin Raja Dhas, J.; Pandiyarajan, R.; Sabarish, S. Metallurgical and Mechanical Investigation on FSSWed Dissimilar Aluminum Alloy. *J. Alloys Metall. Syst.* **2023**, *2*, 100010. [CrossRef]
12. Jo, D.S.; Kahhal, P.; Kim, J.H. Optimization of Friction Stir Spot Welding Process Using Bonding Criterion and Artificial Neural Network. *Materials* **2023**, *16*, 3757. [CrossRef] [PubMed]
13. Fan, W.; Yang, X.; Chu, Q.; Li, W. An Efficient Synergistic Double-Sided Friction Stir Spot Welding Method: A Case Study on Process Optimization, Interfacial Characteristics and Mechanical Properties of 2198-T8 Aluminum-lithium Alloy Joints. *J. Manuf. Process* **2024**, *131*, 213–232. [CrossRef]
14. Meng, T.; Yang, X.; Su, Y.; Ma, S.; Xu, H.; Guo, Z.; Ma, T.; Li, W. Formation Mechanism and Failure Behavior in Synergistic Double-Sided Probeless Friction Stir Spot Welded Joints of 6061 Aluminum Alloy. *Mater. Charact.* **2025**, *219*, 114631. [CrossRef]
15. El-Sayed Seleman, M.M.; Alateyah, A.I.; Mahmoud, E.A.E.; Elsoeudy, R.I.; Ahmed, M.M.Z.; Hafez, K.M.; El-Garaihy, W.H.; A.S, A. Friction Stir Welding of 2507 Super-Duplex Stainless Steel: Feasibility of Butt Joint Groove Filling at Different Process Parameters. *Adv. Mater. Process. Technol.* **2024**, 1–23. [CrossRef]
16. Yang, X.W.; Fu, T.; Li, W.Y. Friction Stir Spot Welding: A Review on Joint Macro- and Microstructure, Property, and Process Modelling. *Adv. Mater. Sci. Eng.* **2014**, *2014*, 697170. [CrossRef]
17. Shen, Z.; Ding, Y.; Gerlich, A.P. Advances in Friction Stir Spot Welding. *Crit. Rev. Solid. State Mater. Sci.* **2020**, *45*, 457–534. [CrossRef]
18. Pan, T.Y. *Friction Stir Spot Welding (FSSW)—A Literature Review*; SAE Technical Paper 2007-01-1702; SAE International: Warrendale, PA, USA, 2007. [CrossRef]
19. Badwelan, A.; Al-Samhan, A.M.; Anwar, S.; Hidri, L. Novel Technique for Enhancing the Strength of Friction Stir Spot Welds through Dynamic Welding Parameters. *Metals* **2021**, *11*, 280. [CrossRef]
20. Suryanarayanan, R.; Sridhar, V.G. Influence of Welding Parameters on the Weld Properties in Friction Stir Spot Welding of Aluminium Alloys of Varying Thicknesses. *Mater. Today Proc.* **2021**, *46*, 8525–8531. [CrossRef]

21. Guo, Y.; Zhao, H.; Ai, C.; Zhao, J.; Su, H.; Chen, J.; Zhao, G. Parameter Optimization of Friction Stir Spot Welded Al 6061 and CFRTP PA6 with Surface Treatment and Interfacial Adhesion. *Thin-Walled Struct.* **2024**, *197*, 111585. [CrossRef]
22. Shekhawat, R.S.; Nadakuduru, V.N. Friction Stir Spot Welding and Characterization of Bulk Aluminum Alloy 6061 Synthesized Using Powder Metallurgical Route. *Powder Metall. Met. Ceram.* **2024**, *63*, 240–249. [CrossRef]
23. Suryanarayanan, R.; Sridhar, V.G. Materials Today: Proceedings Studies on the Influence of Process Parameters in Friction Stir Spot Welded Joints—A Review. *Mater. Today Proc.* **2021**, *37*, 2695–2702. [CrossRef]
24. Yöntemi, T.; Sürtünme, K.; Aisi, B. Optimization of Welding Parameters of AISI 431 and AISI 1020 Joints Joined by Friction Welding Using Taguchi Method. *Bilecik Şeyh Edebali Üniv. Fen Bilim. Derg.* **2022**, *9*, 453–470.
25. Mitlin, D.; Radmilovic, V.; Pan, T.; Chen, J.; Feng, Z.; Santella, M.L. Structure-Properties Relations in Spot Friction Welded (Also Known as Friction Stir Spot Welded) 6111 Aluminum. *Mater. Sci. Eng. A* **2006**, *441*, 79–96. [CrossRef]
26. Tiwan; Ilman, M.N.; Kusmono. Microstructure and Mechanical Properties of Friction Stir Spot Welded AA5052-H112 Aluminum Alloy. *Heliyon* **2021**, *7*, e06009. [CrossRef]
27. Reji, M.; Kumar, R. Response Surface Methodology (RSM): An Overview to Analyze Multivariate Data. *Indian J. Microbiol. Res.* **2022**, *9*, 241–248.
28. Chen, W.-H.; Carrera Uribe, M.; Kwon, E.E.; Lin, K.-Y.A.; Park, Y.-K.; Ding, L.; Saw, L.H. A Comprehensive Review of Thermoelectric Generation Optimization by Statistical Approach: Taguchi Method, Analysis of Variance (ANOVA), and Response Surface Methodology (RSM). *Renew. Sustain. Energy Rev.* **2022**, *169*, 112917. [CrossRef]
29. Kumar, D.; Mandal, A. Response Surface Method-Based Optimisation of Advanced Mechanochemical Approach for Bead Minimisation in Bamboo Fiber Extraction, and Improving Hydrophobicity via Diisopropanolamine Treatment. *Biomass Convers. Biorefinery* **2024**, *14*, 23743–23759. [CrossRef]
30. Ren, G.; Heo, S.; Kim, T.H.; Cheong, C. Response Surface Method-Based Optimization of the Shroud of an Axial Cooling Fan for High Performance and Low Noise. *J. Mech. Sci. Technol.* **2013**, *27*, 33–42. [CrossRef]
31. Albaijan, I.; Ahmed, M.M.Z.; El-Sayed Seleman, M.M.; Touileb, K.; Habba, M.I.A.; Fouad, R.A. Optimization of Bobbin Tool Friction Stir Processing Parameters of AA1050 Using Response Surface Methodology. *Materials* **2022**, *15*, 6886. [CrossRef]
32. Tutar, M.; Aydin, H.; Yuce, C.; Yavuz, N.; Bayram, A. The Optimisation of Process Parameters for Friction Stir Spot-Welded AA3003-H12 Aluminium Alloy Using a Taguchi Orthogonal Array. *Mater. Des.* **2014**, *63*, 789–797. [CrossRef]
33. Al-Sabur, R. Tensile Strength Prediction of Aluminium Alloys Welded by FSW Using Response Surface Methodology–Comparative Review. *Mater. Today Proc.* **2021**, *45*, 4504–4510. [CrossRef]
34. Kadaganchi, R.; Gankidi, M.R.; Gokhale, H. Optimization of Process Parameters of Aluminum Alloy AA 2014-T6 Friction Stir Welds by Response Surface Methodology. *Def. Technol.* **2015**, *11*, 209–219. [CrossRef]
35. Rajendran, C.; Sonar, T.; Ivanov, M.; Kumar, P.S.; Amarnath, V.; Lokanadham, R. Optimization of Friction Stir Spot Welding Parameters for Joining Dissimilar AZ31B Magnesium Alloy and AA6061 Aluminium Alloy Using Response Surface Methodology. *Int. J. Interact. Des. Manuf. (IJIDeM)* **2025**, *19*, 115–126. [CrossRef]
36. El-Sayed Seleman, M.M.; Elnaml, A.M.; Ataya, S.; Ahmed, M.M.Z.; Elnekhaily, S.A.; Hafez, K.M. A Hybrid Analytical-Numerical Model of Heat Generation and Distribution in Friction Stir Welded AA2024 Butt Joints. *Int. J. Integr. Eng.* **2024**, *16*, 140–161. [CrossRef]
37. Chatterjee, S.; Mahapatra, S.S.; Bharadwaj, V.; Upadhyay, B.N.; Bindra, K.S.; Thomas, J. Parametric Appraisal of Mechanical and Metallurgical Behavior of Butt Welded Joints Using Pulsed Nd:YAG Laser on Thin Sheets of AISI 316. *Opt. Laser Technol.* **2019**, *117*, 186–199. [CrossRef]
38. Zhao, Y.Q.; Liu, H.J.; Chen, S.X.; Lin, Z.; Hou, J.C. Effects of Sleeve Plunge Depth on Microstructures and Mechanical Properties of Friction Spot Welded Alclad 7B04-T74 Aluminum Alloy. *Mater. Des.* **2014**, *62*, 40–46. [CrossRef]
39. Tozaki, Y.; Uematsu, Y.; Tokaji, K. Effect of Processing Parameters on Static Strength of Dissimilar Friction Stir Spot Welds between Different Aluminium Alloys. *Fatigue Fract. Eng. Mater. Struct.* **2007**, *30*, 143–148. [CrossRef]
40. Ibrahim, I.J.; Yapici, G.G. Application of a Novel Friction Stir Spot Welding Process on Dissimilar Aluminum Joints. *J. Manuf. Process* **2018**, *35*, 282–288. [CrossRef]

crystals

MDPI

Article

Phase-Field Simulation of the Creep Mechanism in the AZ31 Magnesium Alloy Under Discontinuous Dynamic Recrystallization Dominance

Alireza Rezvani [1], Ramin Ebrahimi [1,*] and Ebad Bagherpour [2,*]

[1] Department of Materials Science and Engineering, School of Engineering, Shiraz University, Shiraz 7134851154, Iran; a.rezvani@shirazu.ac.ir

[2] Brunel Centre for Advanced Solidification Technology (BCAST), Brunel University of London, Uxbridge UB8 3PH, UK

* Correspondence: ebrahimy@shirazu.ac.ir (R.E.); ebad.bagherpour@brunel.ac.uk (E.B.)

Abstract

Discontinuous dynamic recrystallization is a critical microstructural evolution mechanism during high-temperature deformation, influencing material properties significantly. This study develops a two-dimensional phase-field model to predict steady-state creep rates in the AZ31 magnesium alloy, focusing on DRX during creep. To enhance simulation accuracy, initial microstructures are generated from optical microscopy data, enabling simulations at larger scales with higher representativeness. A novel nucleation methodology is implemented, eliminating the need for nuclei order parameter adaptation, improving computational efficiency. Finite element analysis (FEA) is integrated to capture initial instantaneous deformation. The Kocks–Mecking model is employed to describe the evolution of average dislocation density, accounting for work hardening and dynamic recovery within the initial polycrystalline microstructure. Instead of conventional creep testing, impression creep, a cost-effective alternative, is used for validation. This method provides constant stress and steady penetration velocity, simulating creep conditions effectively. The model accurately predicts recrystallization kinetics and microstructural evolution, exhibiting a strong correlation with experimental results, with an error of approximately 5%. This research provides a robust and efficient approach for predicting creep behavior in high-temperature applications, vital for optimizing material selection and predicting component lifespan in industries. The methodology offers a significant advancement in understanding and predicting DRX-driven creep behavior.

Keywords: impression creep; phase-field; dynamic recrystallization; magnesium alloy AZ31; microstructural evolution

1. Introduction

Dynamic recrystallization (DRX) [1,2], meta-dynamic recrystallization (MDRX) [3–5], and static recrystallization (SRX) [6–8] are the primary mechanisms of microstructural evolution in metals and alloys. Among them, DRX occurs during plastic deformation at elevated temperatures, where the formation of new, strain-free grains replaces the deformed microstructure. However, it is important to note that DRX does not occur in all materials; it is particularly prevalent in materials with low-to-medium stacking fault energies [9]. This process is driven by the accumulation of dislocations and the resulting stored energy, which promotes the nucleation and growth of new grains. DRX is particularly significant in hot working processes, such as rolling, forging, and extrusion, as it helps to refine the grain structure, enhance mechanical properties, and reduce flow stress [10]. Consequently, the elucidation of grain structure evolution during dynamic recrystallization (DRX) is paramount for the rational design of thermomechanical processing regimes, enabling the attainment of targeted microstructural morphologies and the subsequent modulation of alloy material properties. To this end, predictive models have been formulated to describe microstructural transformations and DRX kinetics. These include, but are not limited to, the Johnson–Mehl–Avrami–Kolmogrov (JMAK) phenomenological model [9] and physically based models that employ internal state variables such as dislocation density and subgrain size to characterize DRX evolution [11,12]. The advancement of computational resources has enhanced the adoption of microstructure-based simulation methodologies. Peczak et al. [13] and Rollett et al. [14] utilized the Monte Carlo (MC) method to simulate various aspects of dynamic recrystallization (DRX) during thermomechanical deformation. Goetz et al. [15] employed a cellular automaton (CA) approach to model DRX, noting its capacity for calibration to spatial and temporal scales, a feature distinct from the MC method. Ding et al. [16] developed a cellular automaton (CA) model, integrating microstructural evolution with plastic flow behavior and establishing a correlation between microstructural properties and the Zener–Hollomon parameter. This model has demonstrated broad applicability across diverse materials and phenomena, including magnesium alloys [17–19]. Phase-field (PF) models have emerged as a powerful tool for investigating microstructural evolution during recrystallization across diverse materials and deformation conditions. Notably, PF models offer an advantage over cellular automaton (CA) and Monte Carlo (MC) methods by inherently incorporating interfacial energy and curvature, enabling a more accurate representation of complex microstructural morphologies. These PF models, encompassing both static and dynamic recrystallization (SRX and DRX), have integrated various physical mechanisms and demonstrated broad applicability across a wide range of materials [20–26]. Takaki et al. [27,28] developed a PF model of DRX and extended the model by replacing the flow-stress model with an elastic–plastic finite element model in polycrystalline copper [28]. Similar models were developed by Zhao et al. [29,30] and Chen et al. [31].

Dynamic recrystallization (DRX) also plays a critical role in influencing the mechanical behavior of materials during high-temperature deformation. It induces significant softening of the crystalline structure, which is manifested by a pronounced stress drop in stress–strain curves and a corresponding increase in creep rate observed in creep curves [32].

The extent of this phenomenon is governed by the kinetics of DRX and the resulting microstructure, which determine the rate at which the recrystallized state is achieved. The increase in creep rate is particularly dependent on the interplay between the initial microstructure and the evolution of the recrystallized grains, as well as the rate at which new, strain-free grains form and grow under applied stress. Thus, the creep behavior is intrinsically linked to the dynamics of DRX, underscoring the importance of recrystallization kinetics in controlling the mechanical response of materials under high-temperature deformation conditions [33,34].

Creep and hot deformation are two high-temperature deformation processes that differ primarily in their loading conditions and experimental approaches. In creep, a constant external stress is applied to the material, and the resulting strain is measured as a function of time, reflecting the material's time-dependent deformation behavior. In contrast, hot deformation involves imposing a constant strain rate on the material, and the corresponding stress response is observed. This distinction in loading conditions—constant stress in creep versus constant strain rate in hot deformation—leads to different macroscopic mechanical responses and testing methodologies. However, at the microscopic scale, both processes share fundamental similarities in terms of dislocation activity and plastic deformation mechanisms. Thus, while the macroscopic loading conditions differ, the underlying microscopic mechanisms of plastic deformation are essentially the same [35].

Given the similarities in the underlying microscopic mechanisms of plastic deformation between creep and hot deformation, the models developed to predict discontinuous dynamic recrystallization (DDRX) in hot deformation can also be applied to the DDRX process occurring during creep. However, certain modifications are necessary to account for the specific loading conditions and time-dependent nature of creep. For instance, the strain rate in creep is not constant but evolves over time to a steady-state condition. Therefore, while the core principles of DDRX models remain valid, adjustments must be made to incorporate the unique aspects of creep. By adapting these models, it becomes possible to accurately predict DDRX behavior in creep, leveraging the established framework of hot deformation while addressing the distinct characteristics of creep deformation.

The primary objective of this investigation is to develop a two-dimensional phase-field model for the prediction of the steady-state creep rate in the AZ31 magnesium alloy, achieved through the simulation of microstructural evolution during discontinuous dynamic recrystallization (DDRX). To enhance the representativeness of the simulations, initial microstructures are generated from optical microscopy data, rather than relying on synthetic or limited-scale electron backscatter diffraction (EBSD) datasets. This approach facilitates larger-scale simulations, more closely approximating industrial processing conditions, while maintaining cost-effectiveness and accessibility. Furthermore, a novel and efficient nucleation methodology is implemented, eliminating the holding time required for nuclei order parameter adaptation within the system. Additionally, finite element analysis (FEA) is integrated to determine the initial instantaneous deformation. The phase-field model of discontinuous dynamic recrystallization (PF-DDRX) incorporates the Kocks–Mecking (KM) model [36,37] to describe the evolution of average dislocation density, accounting for work hardening and dynamic recovery (DRV) within the initial polycrystalline microstructure. Creep deformation is simulated by adjusting the finite difference grid size, maintaining a constant simulation area.

However, instead of the expensive conventional creep testing, the impression creep technique is used. Impression creep is a small specimen testing technique that has been employed for many years as an alternative to the costly and time-consuming conventional uniaxial tensile creep test. It is a modified indentation test wherein the conical or ball indenter is replaced by a cylindrical flat-bottomed one. This replacement provides constant stress and steady penetration velocity at a constant load [38,39].

Magnesium alloys, including AZ31, exhibit considerable potential for lightweighting applications in the automotive and aerospace sectors due to their superior specific strength and stiffness [40]. However, a critical engineering bottleneck hindering their wider adoption, particularly in high-temperature environments such as automotive powertrain components and certain aerospace structures, is their inherent limitation in creep resistance [41,42]. Overcoming this limitation necessitates accurate prediction of the steady-state creep rate, which is paramount for the design and long-term performance assessment of high-temperature structural components in numerous industrial applications, including power generation, aerospace, and chemical processing. In these environments, materials are subjected to sustained loads and elevated temperatures, leading to time-dependent deformation that can ultimately result in component failure. Precise creep rate prediction allows engineers to optimize material selection, establish safe operating limits, and predict component lifespan, thereby ensuring structural integrity, preventing catastrophic failures, and minimizing costly downtime. Reliable creep rate estimations are therefore essential for the development of robust and efficient high-temperature systems [43].

To validate the predictive capabilities of the developed model, impression creep tests were conducted at varying impression depths, accompanied by microscopic characterization of the resulting microstructures. Comparative analysis of experimental and simulated results demonstrated a strong correlation, thereby substantiating the model's robustness in accurately predicting recrystallization kinetics and microstructural evolution.

2. Experimental Procedure

2.1. Microstructural Characterization

The material employed in this study is a commercial AZ31 magnesium alloy with a chemical composition determined by quantometric analysis and presented in Table 1.

Table 1. Chemical composition of the AZ31 magnesium alloy used in this study (wt. (%)).

Al	Zn	Mn	Si	Cu	Fe	Be	Mg
3.11	1.02	0.18	0.07	0.012	0.019	0.0008	Bal.

To establish a stable initial microstructure for subsequent creep testing, the as-received material was homogenized at 450 °C for 5 h and then furnace-cooled. The homogenized microstructure of the AZ31 alloy is depicted in Figure 1a. Microstructural analysis was performed using optical microscopy after mechanical polishing and chemical etching with an acetic picral solution (5 g picric acid, 5 mL acetic acid, 10 mL distilled water, and 100 mL ethyl alcohol). A fully homogenized microstructure, comprising 91 grains with a mean size of 32.4 μm, was observed. This characterization was achieved through a custom MATLAB R2023a script, which performed binary image conversion and subsequent analysis, including grain identification, indexing, and size determination. The resulting indexed and color-coded grain map and grain size distribution are illustrated in Figures 1b and 1c, respectively.

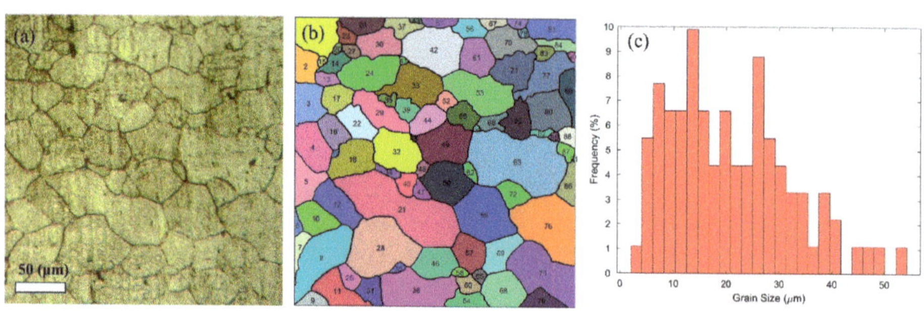

Figure 1. (**a**) Microstructure of the as-homogenized AZ31 Mg alloy. (**b**) Digitized and indexed matrix. (**c**) Grain size distribution.

2.2. Impression Creep Testing

The impression creep testing apparatus utilized in this study is comprehensively detailed in prior research [44]. The system is an in-house designed testing machine, functionally analogous to those reported by Zhang et al. [45] and Kim [46]. It is equipped with a flat-bottomed cylindrical tungsten carbide indenter, measuring 2 mm in diameter. Specimens with dimensions of 15 mm × 15 mm × 10 mm were prepared from the previously homogenized material using wire-cut electrical discharge machining (EDM).

The experimental procedure for the impression creep tests is depicted in Figure 2. During the tests, the specimens were initially heated to a temperature of 250 °C at a rate of

0.6 K/s. Subsequently, an isothermal dwell period of 300 s was implemented to ensure thermal homogeneity and minimize temperature gradients within the specimens. Following the stabilization period, a constant stress of 200 MPa was applied. Upon completion of the creep test, specimens were rapidly quenched in a water bath. Throughout the experimental procedure, the temperature was maintained within a tolerance of ±1 °C of the target value, ensuring precise thermal control.

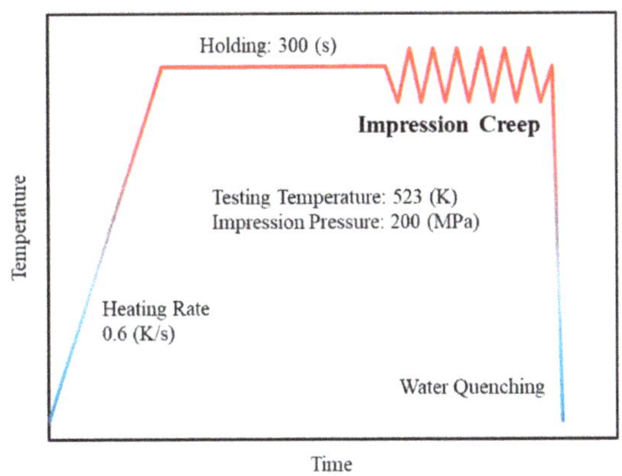

Figure 2. The experimental scheme of the impression creep test.

To elucidate the discontinuous dynamic recrystallization (DDRX) mechanisms operative during impression creep, microstructural analysis was focused on the region directly beneath the center of the indenter. This zone, characterized by unidirectional downward material flow [44,47,48], represents the preferential nucleation site for DDRX. For microstructural characterization, deformed specimens were sectioned axially, bisecting the impression cavity (as depicted in Figure 3). The sectioned surfaces were then prepared using standard metallographic techniques, including sequential grinding and polishing, suitable for optical microscopy.

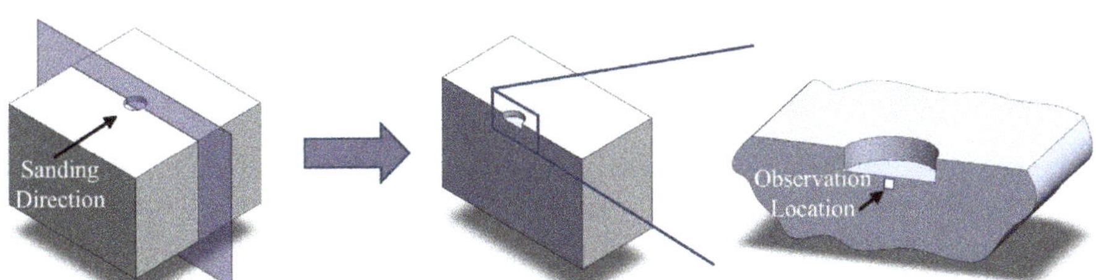

Figure 3. Schematic representation of the specimen preparation process, highlighting the axial cross-section of the impression hole for microstructural analysis.

2.3. Finite Element Analysis

The impression creep test has been subjected to extensive scrutiny via finite element analysis (FEA) in prior research [47,49,50]. While a comprehensive FEA approach has been established within the existing literature, this current investigation will utilize FEA solely for the determination of the instantaneous impression depth resulting from the initial application of a constant load to the indenter. The focus is limited to the elastic–plastic response immediately following loading, rather than the time-dependent creep deformation evolution.

The axisymmetric geometric model of the specimen, along with the rigid indenter, was imported into the Deform-3D ver. 11.0 simulation environment. The indenter was designed with a 0.1 mm (5% of the indenter diameter) fillet to avoid stress concentration at the edges [50]. Material properties, including temperature-dependent flow stress data obtained from Liu et al. [51], were incorporated into the model using a custom key to ensure consistency in all material properties used in this study. Boundary conditions were applied to replicate the experimental setup, including fixed supports, applied force on the indenter, and thermal boundary conditions. Friction at the contact interfaces was modeled using a shear friction model with a coefficient of 0.35 [44]. The simulation parameters, such as time step size and remeshing criteria, were optimized to ensure numerical stability.

3. Multi-Phase-Field DDRX Model

To investigate grain structure evolution during discontinuous dynamic recrystallization (DDRX) under deformation, a phase-field model with multi-order parameters was utilized. The model's total free energy, representing the inhomogeneous system, is a functional of the orientation field and its gradients. It includes contributions from bulk free energy (F_{bulk}), interfacial free energy (F_{int}) [52], and stored energy field (F_s) due to plastic deformation from dislocation generation [21,23]. This total free energy is defined as follows:

$$\mathcal{F} = F_{bulk} + F_{int} + F_s = \int_V [f_{bulk}(\{\eta_i\}) + f_{int}(\{\eta_i\}, \{\nabla\eta_i\}) + f_s(\{\eta_i\})]\, dV \tag{1}$$

where V is the volume of the system and η_i denotes the orientation field variables, also known as order parameters [53], used to differentiate individual grains. Two distinct sets of order parameters are employed: $\eta_{i_{DEF}}$ $(i = 1, \ldots, g_d)$ to describe the initial grains that deform during the simulation and $\eta_{i_{DRX}}$ $(i = 1, \ldots, g_n)$ to represent the DRX nuclei. g_d denotes the total number of grains before discontinuous dynamic recrystallization, while g_n represents the number of DDRX nuclei, which increases as the nucleation process proceeds.

The bulk energy term establishes uniform potential wells within grain interiors, driving grain growth and the subsequent filling of the simulation domain. Conversely, the gradient energy density is localized to grain boundary regions [54]. These energy contributions are mathematically represented as follows:

$$f_{bulk} + f_{int} = m_0\left[\sum_{i=1}^{g}\left(\frac{\eta_i^4}{4} - \frac{\eta_i^2}{2}\right) + \gamma\sum_{i=1}^{g}\sum_{j>i}^{g}\eta_i^2\eta_j^2 + \frac{1}{4}\right] + \frac{\kappa}{2}\sum_{i=1}^{g}|\nabla\eta_i|^2 \tag{2}$$

where g is the total number of order parameters, m_0 is the energy density coefficient, and γ is a positive constant set to 1.5 for symmetrical order parameter values across the grain boundaries. The gradient energy penalty, κ, affects both the grain boundary energy and thickness [55].

The driving force for DRX nuclei growth, arising from the high dislocation density in deformed grains, is provided by the stored plastic deformation energy density term. This term is approximated by the stress field of dislocations generated during plastic deformation, as follows [56]:

$$f_S = \frac{1}{2}\rho G b^2 \tag{3}$$

where b is the magnitude of the burgers vector and G is the shear modulus. The dislocation density field, ρ, is determined at each point from the order parameters and their respective dislocation densities [22] through the application of Moelans' interpolation function [57,58]:

$$\rho(\eta_1, \eta_2, \ldots, \eta_g) = \frac{\sum_{i=1}^{g}\eta_i^2\rho_g}{\sum_{j=1}^{g}\eta_j^2} \tag{4}$$

The dislocation density at each grid point, ρ_g, is obtained from the dislocation evolution inside the microstructure during the plastic deformation process.

The microstructural evolution is governed by the time-dependent Ginzburg–Landau equation [53], also referred to as the Allen–Cahn equation [59], which is formulated using a non-conserved order parameter:

$$\frac{\partial \eta_i(r,t)}{\partial t} = -L\frac{\delta \mathcal{F}}{\delta \eta_i(r,t)} \tag{5}$$

where L is the relaxation coefficient related to the interfacial mobility, t is time, and r is the spatial position of order parameters in the system. Utilizing the total free energy function, as defined in Equation (1), the evolution equation governing the system's behavior is expressed as follows [22]:

$$\frac{\partial \eta_i}{\partial t} = -L\left[m_0\left(-\eta_i + \eta_i^3 + 2\gamma\eta_i \sum_{j\neq i}^{g} \eta_j^2\right) - \kappa\nabla^2\eta_i + Gb^2\frac{\eta_i}{\sum_{j=1}^{g}\eta_j^2}(\rho_g - \rho)\right] \quad i = 1,2,\ldots,g \tag{6}$$

The spatiotemporal evolution of the dynamic microstructure is determined by numerically solving Equation (6), coupled with constitutive equations describing dislocation evolution and discontinuous dynamic recrystallization nucleation. A finite difference method is employed to obtain numerical solutions to this coupled system of kinetic equations. This method discretizes the continuous system onto a discrete lattice [53]. The phase-field equations are similarly discretized, yielding a system of algebraic equations. The solution to these algebraic equations provides the phase-field variables at each lattice node. A common and computationally efficient approach utilizes a finite difference discretization with uniform lattice spacing, employing a central second-order spatial discretization and a forward Euler temporal discretization.

To reduce computational cost and resource consumption, active parameter tracking [60,61] is employed. The computational complexity arising from the numerous DRX nuclei within the system necessitates an efficient approach. To address this, a sparse data structure is implemented, exploiting the localized activity of phase-field variables. By restricting computations to the non-zero order parameters at each grid point, a significant reduction in computational effort is achieved. The determination of the active order parameters at each location is defined as follows:

$$\mathcal{P}(r,t) = \{(i,\eta_i) : \eta_i(r,t) > \delta\} \tag{7}$$

where δ is chosen to be a small positive threshold value, wherein only order parameters bigger than that are stored.

The following relations were derived for relating the simulation coefficients to the grain boundary energy, γ_{gb}, and grain boundary mobility, M_{gb}, as follows [54]:

$$\kappa = \frac{3}{4}\gamma_{gb} l_{gb} \tag{8}$$

$$\mathrm{L} = \frac{4M_{gb}}{3l_{gb}} \tag{9}$$

$$m_0 = \frac{6\gamma_{gb}}{l_{gb}} \tag{10}$$

where l_{gb} is the grain boundary width, which is a model parameter.

The boundary mobility for high angle grain boundaries (HAGB), M_{gb} $\left(m^4J^{-1}s^{-1}\right)$, is defined as follows [62]:

$$M_{gb} = M_0 \exp\left(\frac{-Q_b}{RT}\right) \tag{11}$$

where Q_b is the activation energy for the migration of a HAGB, R is the universal gas constant, and M_0 is the pre-exponential factor. In view of the solute drag effect, the strain rate also has an effect on grain boundary mobility. Therefore, M can be improved as follows [63].

$$M_{gb} = \dot{\varepsilon} M_0 \exp\left(\frac{-Q_b}{RT}\right) \tag{12}$$

The parameters used for the simulation are listed in Table 2.

Table 2. Material properties of the AZ31 magnesium alloy employed in the phase-field model.

Parameter	Value	Parameter	Value
Q_{act} $\left[kJ\ mol^{-1}\right]$	134 [51]	ρ_0 $\left[m^{-2}\right]$	10^{10} [51]
Q_b $\left[kJ\ mol^{-1}\right]$	121.791 [51]	ρ_c $\left[m^{-2}\right]$	2.14×10^{14}
R $\left[J\ mol^{-1}K^{-1}\right]$	8.314	k_1 $\left[m^{-1}\right]$	3.88×10^9 [51]
M_{gb} $\left(m^4J^{-1}s^{-1}\right)$	1.7×10^{-13} [17]	k_2	153.37 [51]
γ_{gb} $\left[J\ m^{-2}\right]$	0.332 [64]	$\Delta x = \Delta y$ $[\mu m]$	0.26
G [GPa]	14.7 [41]	Δt [s]	5×10^{-2}
b $\left[\dot{A}\right]$	3.21 [65]	l_{gb} $[\mu m]$	$4\Delta x$
l $[\mu m]$	0.363 [64]	δ	10^{-6}
τ [N]	6.82×10^{-10} [64]	r_{seed}	$14.5\Delta x$
C	8.5×10^{18} [51]		

The phase-field method's inherent diffuse interface necessitates the creation of a finite width at grain boundaries for initial microstructural representation in discontinuous dynamic recrystallization simulations. To prevent artificial changes in grain area during this process, which is especially critical for small grains, a fine grid resolution was implemented. This ensured that the diffuse interface's characteristic length scale remained substantially smaller than the minimum grain size, thus avoiding numerical dissolution of small grains, which results from the interface width becoming comparable to or larger than grain dimensions. Consequently, to achieve this resolution, the grid size was set to 940×940, and the grid spacing, Δx, considering the scale of the original microstructure and the phase-field grid, was set to 0.266 μm. To ensure sufficient numerical resolution across the diffuse interface, the grain boundary width, l_{gb}, was set to $4\Delta x$. The simulation time step, Δt, was set to 5×10^{-2} (s) to maintain numerical stability during simulation. The 2D microstructure evolution was computed at each grid point using finite difference, with an eighth-order scheme for spatial derivatives and the Euler forward method for time integration.

A unique order parameter field was generated for each grain, wherein regions occupied by grains were assigned a value of 1, and the remaining regions were assigned a value of 0. This binary field represented the initial sharp interface condition. To initiate diffuse interface formation, a brief temporal evolution of the order parameter field was performed according to Equation (6). This evolution was terminated upon the attainment of a symmetric diffuse interface profile, ensuring a consistent representation of grain boundaries throughout the microstructure. Figure 4 illustrates the processed microstructure, visualized using order parameters and Equation (13) [66]. As shown, the diffuse interfaces are accurately formed throughout the simulation domain.

$$\varphi(r,t) = \sum_{i=1}^{g} \eta_i^2(r,t) \tag{13}$$

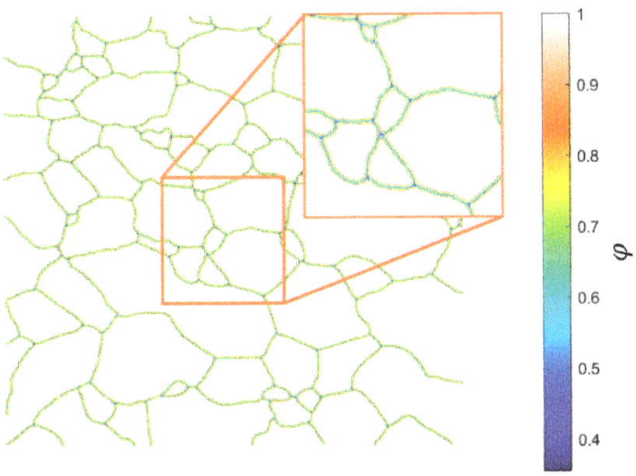

Figure 4. Grain topology of the microstructure showing diffuse interface development after holding time in phase-field model.

3.1. Dislocation Density Evolution

During thermomechanical processing, the dislocation density evolves as a result of the dynamic interaction between work hardening (WH), dynamic recovery (DRV), and dynamic recrystallization (DRX) [67,68]. Initially, work hardening, driven by dislocation generation and entanglement, causes an increase in flow stress with strain. Subsequently, dynamic recovery, involving dislocation slip and climb [69], counteracts the stress increase from work hardening. Finally, dynamic recrystallization commences when the dislocation density (ρ) reaches a critical value (ρ_c), determined by the following [70]:

$$\rho_c = \left(\frac{20\gamma_{gb}\dot{\varepsilon}}{3blM_{gb}\tau^2} \right)^{1/3} \tag{14}$$

where $\dot{\varepsilon}$ is the strain rate, l is the mean free path of the dislocation, and τ is the linear energy of the dislocation. Dynamic recrystallization induces a substantial decrease in dislocation density within newly formed grains through dislocation annihilation, leading to flow stress softening [71,72]. Therefore, the observed flow stress behavior during hot deformation is a result of the competing mechanisms of work hardening (WH), DRV, and DRX. The Kocks–Mecking (KM) model provides a framework for describing the relationship between true strain (ε) and local dislocation density (ρ_g) within a grain, incorporating dislocation accumulation from plastic deformation and dynamic recovery, as expressed by the following:

$$\frac{d\rho_g}{d\varepsilon} = k_1\sqrt{\rho_g} - k_2\rho_g \tag{15}$$

where k_1 is the WH coefficient related to the accumulation of statistically stored dislocations (SSDs) [73], and k_2 is the dynamic recovery coefficient dependent on temperature (T) and strain rate [16]. Within the framework of the proposed phase-field model, the local dislocation density (ρ_g) is treated as a spatially homogeneous variable within the confines of each individual grain constituting the microstructure. Consequently, the model does not incorporate any spatial heterogeneity or distribution of dislocation density within a single grain.

Due to the similarities in composition and processing conditions between the AZ31 Mg alloy used in this study and that investigated by Liu et al. [51], all relevant material parameters, including the Kocks–Mecking coefficients, were adopted from their work. By utilizing the well-established parameters from Liu et al., the model benefits from validated data, thereby reducing uncertainties associated with material property estimation. This approach ensures that the mechanical responses of the AZ31 Mg alloy in the simulations align closely with experimentally observed behavior, enhancing the predictive accuracy of the study.

3.2. Nucleation Process

Experimental observations have shown that during uniform deformation, nucleation events preferentially initiate at triple junctions and grain boundaries, rather than within grain interiors [56]. This preference is attributed to the energetically favorable nucleation sites provided by triple junctions, which possess a higher grain boundary area-to-volume ratio [74,75]. Therefore, these established assumptions are incorporated into the present investigation. Specifically, in this study, nuclei are placed exclusively at triple junctions and grain boundaries, with their spatial distribution randomized.

A circular nucleus, centered at the selected position, is introduced into the simulation domain. The initial order parameter within this nucleus is governed by the following function [22]:

$$\eta_i = \frac{1}{2}\left(1 - \tanh\left(\frac{d - r_{seed}}{2\Delta x}\right)\right) \tag{16}$$

where d represents the radial distance from the center of the nucleus, r_{seed} denotes the radius of the nucleus, and Δx is the grid spacing. Following the introduction of nuclei into the phase-field domain, an initial relaxation period is necessary to allow the microstructure to adjust to the newly formed recrystallized grains [76,77]. This requirement addresses the potential for non-physical artifacts due to overlapping order parameters. It is noteworthy that this methodology is applicable to static recrystallization (SRX) simulations. In such simulations, the holding process, which precedes SRX, can be excluded from the computational domain, thereby not affecting the main microstructure evolution. However, in discontinuous dynamic recrystallization modeling, the coupled processes of deformation and nucleation necessitate a different approach. Incorporating a holding time for nuclei stabilization at each nucleation step is not only computationally inefficient but may also induce microstructural artifacts, such as anomalous grain growth. To circumvent this issue, a two-grain simulation was conducted utilizing identical material and model parameters as the primary simulation. The diffuse interface profile of the order parameters at the grain boundary was then extracted. Subsequently, a fitting procedure was performed on the r_{seed} parameter to ensure the seed order parameter accurately replicated the simulated diffuse interface profile, as depicted in Figure 5. Finally, during the primary simulation, upon insertion of each nucleus into the microstructure, the existing order parameters within the nucleus's spatial domain were replaced with an exact copy (a replica, Grain B in Figure 5) of the nucleus's order parameter profile. This procedure effectively eliminated the requirement for a holding time to establish a symmetrical diffuse interface at grain boundaries.

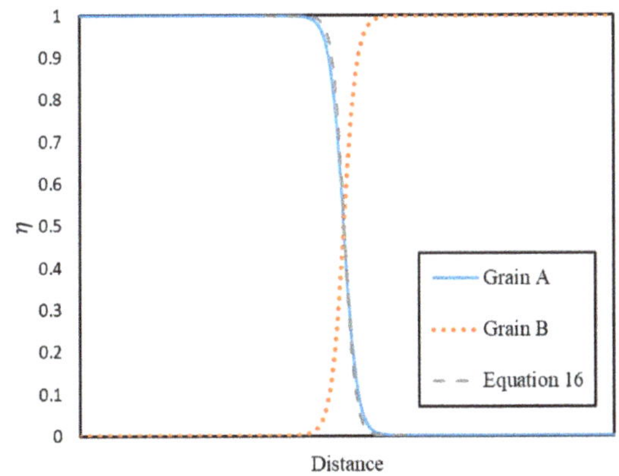

Figure 5. Diffuse interface development in the simulation of two-grain model with fitted r_{seed} in the Equation (16).

Given the determination of nuclei position, size, and order parameter, the quantity of nuclei introduced per time increment into the dynamically evolving microstructure is directly proportional to the nucleation rate. Ding and Guo [16] used the following equation for the nucleation rate:

$$\dot{n} = C\dot{\varepsilon}\exp\left(\frac{-Q_{act}}{RT}\right) \tag{17}$$

where C is a material fitting coefficient. Nucleation is initiated at a discrete grid node when the local dislocation density (ρ) attains the critical value (ρ_c). Upon reaching this threshold, a stochastic nucleation event occurs, introducing a quantity of nuclei (dn), as defined by Equation (18) [78], into the microstructure at each discrete time step. The placement of these nuclei within the microstructure is randomized, simulating the inherent randomness of nucleation phenomena.

$$dn = \frac{\dot{n}\Delta t n_{gb}\Delta x\Delta y}{l_{gb}} \tag{18}$$

where dn represents the nucleation number, and n_{gb} representing the count of discrete grid points defining grain boundaries.

3.3. Grid Deformation Model

The deformation occurring during discontinuous dynamic recrystallization is approximated by altering the simulation grid dimensions. Takaki et al. [27] employed the relationship between nominal strain and true strain to calculate the grid dimension in the y-direction and subsequently determined the grid dimension in the x-direction by maintaining a constant simulation area. While this method is suitable for simulating uniaxial compression straining, it is not directly applicable to modeling the impression creep process. To address this limitation, the results from an upper-bound analysis of the impression creep test [44] are utilized. Specifically, the relationship between strain rate ($\dot{\varepsilon}$) and impression rate ($\dot{U}$), derived assuming a constant friction factor of 0.35, is employed. Based on the value of this constant friction factor, the ratio of impression rate to strain rate ($\dot{U}/\dot{\varepsilon}$) can be calculated and used to determine the impression depth, U, as follows.

$$U = \left(\frac{\dot{U}}{\dot{\varepsilon}}\right) \times \varepsilon \tag{19}$$

The indenter displacement, representing the impression depth, is imposed on the simulation grid along the y-axis. To enforce volume conservation within the two-dimensional simulation domain during plastic deformation, the spatial discretization along the x-axis is dynamically scaled. This scaling factor is determined by the inverse ratio of the current y-dimension to the initial y-dimension, thereby ensuring that the total area of the computational domain remains invariant throughout the deformation process.

4. Dynamic Recrystallization During the Impression Creep Testing

Dynamic recrystallization is a critical phenomenon during high-temperature creep deformation, especially in materials with low stacking fault energy where recovery rates are slow, allowing sufficient stored energy to accumulate for new grain nucleation and growth. As evidenced in pure Pb, Ni, austenite [79], and complex alloys like Nimonic 108 [80,81], DRX induces significant softening through dislocation annihilation within newly formed grains, leading to a rapid increase in creep rate [82], often punctuated by oscillations [34,83]. This softening is manifested as a pronounced stress drop in stress–strain curves, reflecting the influence of DRX kinetics and the resulting microstructure on the rate at which recrystallization occurs. The process commences only when a critical strain, dependent on stress and temperature, is exceeded, with the critical strain generally decreasing with increasing stress and temperature. Impurities can also elevate the strain required for recrystallization [82]. The interplay between the initial microstructure and the evolution of recrystallized grains, coupled with the formation and growth rate of strain-free grains under applied stress, dictates the creep behavior, highlighting the fundamental link between DRX dynamics and the macroscopic mechanical response of materials during high-temperature deformation.

Figure 6a illustrates schematic creep curves depicting the influence of dynamic recrystallization during conventional tensile creep testing. In the absence of DRX, the expected stages of primary, secondary (steady-state), and tertiary creep are clearly observable. Conversely, the occurrence of DRX manifests as a discernible increase in creep rate immediately following the establishment of steady-state conditions [82]. This behavior underscores the softening effect of DRX on the material's creep resistance. Consistent with the observations in tensile creep, dynamic recrystallization is also seen in impression creep tests [47,84]. Analogously, Figure 6b demonstrates a similar phenomenon in impression creep curves, where an elevated impression rate is observed post-DRX initiation. This increase in impression rate reflects the localized softening induced by DRX within the impression zone, mirroring the macroscopic creep rate enhancement observed in conventional creep testing.

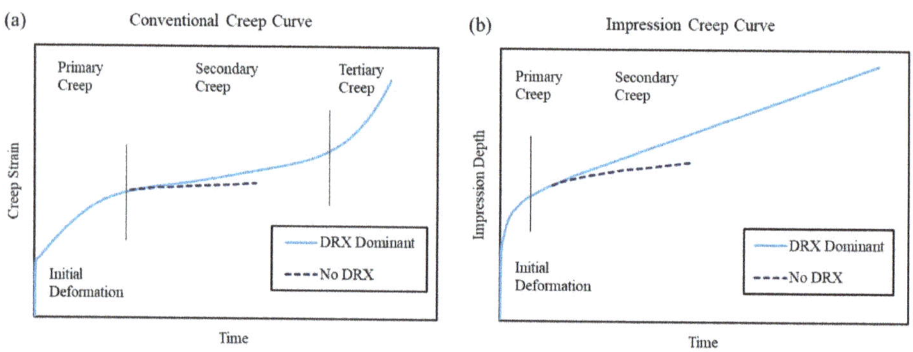

Figure 6. Schematic representation of creep curves illustrating the influence of dynamic recrystallization (DRX): (**a**) conventional tensile creep curve demonstrating the DRX-induced accelerated creep; (**b**) impression creep curve showing the increase in impression rate following the initiation of DRX.

Creep and hot deformation are distinct high-temperature deformation processes, primarily differentiated by their loading conditions and experimental methodologies. In creep, a constant external stress is applied, and the resulting strain is monitored as a function of time, characterizing the material's time-dependent deformation behavior. Conversely, hot deformation involves imposing a constant strain rate, with the corresponding stress response being measured. This fundamental difference in loading protocols—constant stress in creep versus constant strain rate in hot deformation—manifests as divergent macroscopic mechanical responses and testing procedures. However, at the microscopic level, both processes exhibit fundamental similarities in dislocation activity and plastic deformation mechanisms. Thus, despite the macroscopic variations in loading, the underlying microscopic mechanisms of plastic deformation remain essentially consistent [35]. Consequently, due to this shared microscopic basis, the Kocks–Mecking model, which effectively describes dislocation-mediated plastic deformation, can be applied to model both creep and hot deformation processes.

In conventional dynamic recrystallization modeling, such as for hot compression processes [27], a constant strain rate is typically imposed. With a fixed time increment (Δt), the incremental strain ($d\varepsilon$) is calculated. Consequently, given the constancy of both strain rate and Δt, $d\varepsilon$ remains constant throughout the simulation. Utilizing the Kocks–Mecking relationship (Equation (15)), the incremental dislocation density ($d\rho$) is computed within the deformed grains at each time step. This iterative process continues until the dislocation density reaches the critical value (ρ_c), triggering the initiation of nucleation. The stress at each time step can then be determined using the following [27]:

$$\sigma = \alpha G b \sqrt{\rho_{mean}} \tag{20}$$

where α is a dislocation interaction coefficient of approximately 0.5. During nucleation, the dislocation density within the newly formed nuclei is substantially lower than that of the surrounding deformed grains, leading to a reduction in the average dislocation density (ρ_{mean}) and consequently, a decrease in stress. However, in creep processes, stress is maintained constant, and the strain rate is an unknown variable, rendering the incremental strain ($d\varepsilon$) indeterminate. Nevertheless, during nucleation, the incremental dislocation density is determined by the difference between the average dislocation density of the microstructure prior to and following the nucleation event, $d\rho_{mean}$. By rearranging the Kocks–Mecking relationship, the incremental strain can be calculated as follows.

$$d\varepsilon = \frac{d\rho_{mean}}{k_1\sqrt{\rho_{mean}} - k_2\rho_{mean}} \tag{21}$$

From the incremental strain data, the strain rate is calculated using the Δt and $d\varepsilon$. Subsequently, these derived parameters are transformed into impression depth and impression velocity via Equation (19), enabling the construction of the impression depth versus time curve. Nevertheless, the initial instantaneous deformation and strain rate prior to discontinuous dynamic recrystallization remain unknown.

The instantaneous deformation can be estimated using finite element analysis. In this study, the primary creep rate is neglected due to the limited extent of the primary creep region in impression creep processes [39]. The initial steady-state creep rate (before DDRX) can be estimated based on existing literature, and the critical dislocation density for discontinuous dynamic recrystallization is assumed to be consistent with that used in hot compression simulations (Equation (14)). Utilizing this configuration, the dislocation density evolution can be predicted under a constant strain rate, employing a methodology analogous to compression deformation modeling, until the critical dislocation density is attained. Following the onset of nucleation, Equation (21) is used to calculate $d\varepsilon$ and the strain rate. This process continues until complete microstructural recrystallization is achieved. Consequently, the steady-state creep rate in the impression creep process can be predicted.

5. Simulation Stages

The numerical procedure for the developed phase-field discontinuous dynamic recrystallization (PF-DDRX) model is illustrated in the flow chart presented in Figure 7. Dislocation activity is modeled through the dynamic competition between work hardening, dynamic recovery, and DDRX, as described in Section 3.1. Recrystallization grains nucleate according to the mechanisms outlined in Section 3.2. The deformation stored energy, acting as an additional driving force in the PF-DDRX model, interacts with grain boundary energy to govern the evolution of deformed (DEF) and recrystallized (DRX) grains, as detailed in Section 3. These simulations provide insights into microstructural evolution and its influence on macroscopic mechanical responses during impression creep, as well as DRX kinetics.

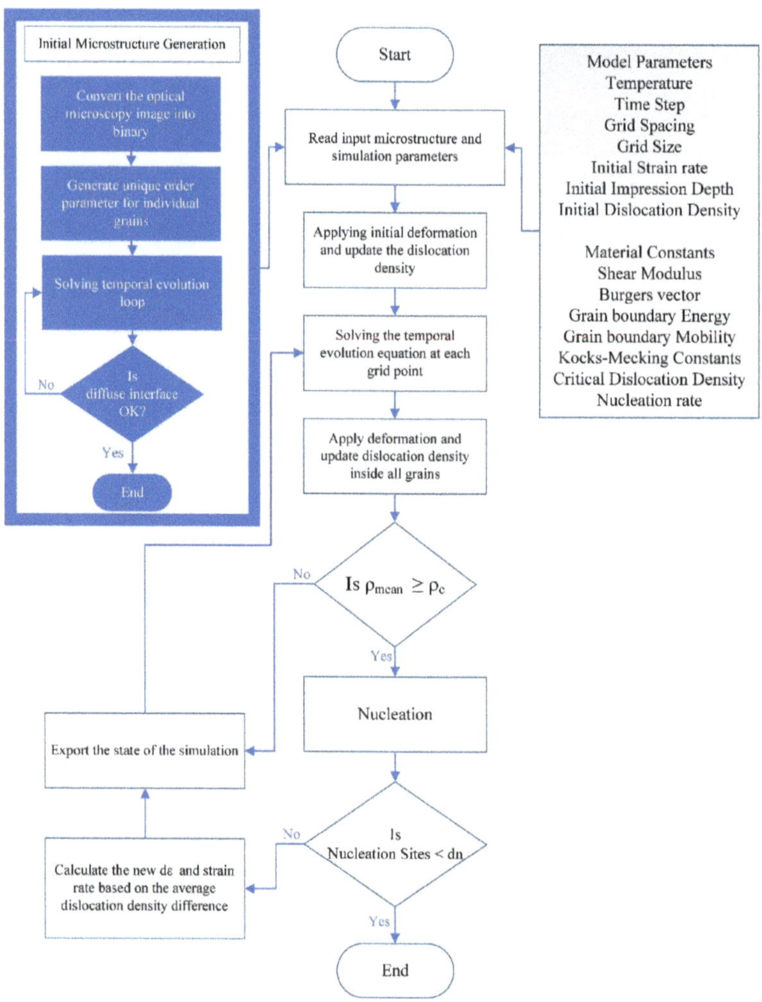

Figure 7. Flow chart of the simulation scheme.

The simulation process comprises the following sequential steps:

1. **Initial Microstructure Generation:** An initial order parameter field is generated from an optical microscopy microstructure image.
2. **Parameter Initialization:** All model and material parameters are input into the simulation.
3. **Initial Deformation and Dislocation Density Calculation:** The initial deformation, obtained from finite element analysis (FEA), is applied to the simulation grid, and the corresponding dislocation density inside each grain in the microstructure is calculated.
4. **Microstructure and Dislocation Evolution Loop:** A loop iterates through the grain dislocation evolution, simulating the interplay of WH, DRV, and DRX.
5. **Stochastic Nucleation:** Nucleation occurs stochastically at eligible grain boundaries.
6. **Incremental Strain and Strain Rate Calculation:** After each nucleation event, the incremental strain and strain rate are calculated, and the values are updated for the next iterative step.
7. **Iteration Continuation:** The simulation monitors the fraction of recrystallized volume at each evolution loop and terminates when complete recrystallization is attained.

6. Results and Discussion

Figure 8 presents the experimental impression creep curve for the AZ31 magnesium alloy, revealing an instantaneous deformation depth of approximately 63 μm at the test's onset. Finite element analysis (FEA), employing material properties and boundary conditions consistent with the experimental setup (250 °C and a constant impression pressure of 200 MPa), yielded an instantaneous impression depth of approximately 70 μm, as illustrated in Figure 9. While precise measurement of the instantaneous deformation depth is challenging in experimental investigations due to factors such as thermal expansion and the micrometer-scale nature of the deformation, the FEA result demonstrates reasonable agreement with the experimental data. In accordance with Section 3.3, the FEA-derived impression depth was first transformed into strain via Equation (19). Subsequently, this calculated strain, together with ρ_0, was applied to Equation (15) to compute the initial average dislocation density, ρ_{ini}, within the grains. The instantaneous impression depth and the calculated average dislocation density were then applied as the initial conditions for the microstructure in the main temporal evolution loop of the phase-field simulation.

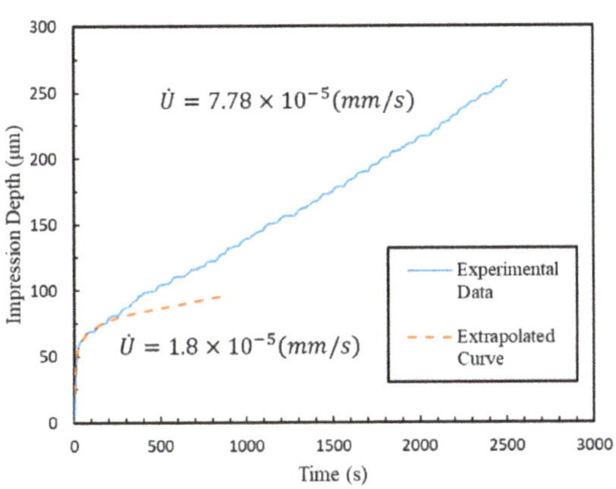

Figure 8. Experimental impression creep data for AZ31 alloy at 250 °C and a constant stress of 200 MPa.

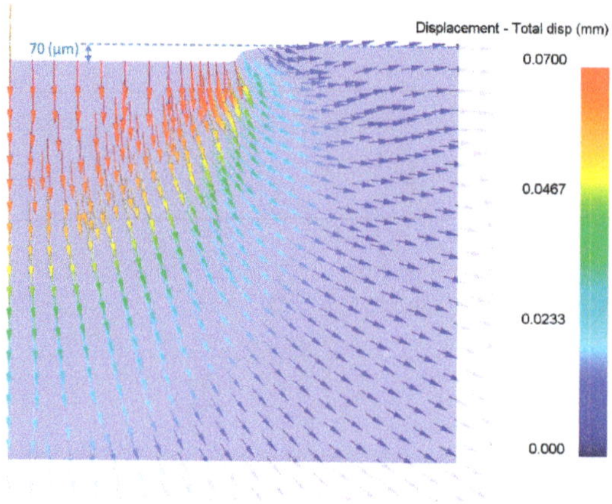

Figure 9. Finite element analysis (FEA) results illustrating the predicted instantaneous indentation depth and total displacement contours (shown as vector plot), with boundary conditions matching the experimental temperature and applied stress.

Figure 8 illustrates that extrapolating a fitted curve from the initial stage of the experimental data, where recrystallization is absent, to later times yields a slope (impression rate) lower than that observed in the steady-state region. This discrepancy in slope is attributed to softening induced by discontinuous dynamic recrystallization (DDRX). The fitted and experimental curves exhibit perfect overlap until approximately 190 s, marking the onset of DDRX.

Using the conversion factors from the upper-bound analysis of the deformation zone in the impression creep test [44], the impression pressure and the impression rate can be converted to the equivalent tensile creep stress and strain rate, respectively, according to

$$\sigma = \eta P \tag{22}$$

$$\dot{\varepsilon} = \beta \dot{U} \tag{23}$$

where η and β are the correlation factors for the stress and strain rate, respectively. The stress conversion factor for the AZ31 alloy was calculated to be 0.26, and the strain rate conversion factor was calculated using a suitable velocity field to be 2.22 according to the instructions detailed in [44]. The impression pressure was converted to stress using Equation (22). Subsequently, the steady-state creep strain rate was determined via the Friedel model [41,85], yielding a value of 3.84×10^{-5} $(1/s)$. This strain rate was then converted to a value of 1.73×10^{-5} (mm/s) using Equation (23) and the declared correlation factor. This calculated strain rate is in close agreement with the fitted impression rate observed in Figure 8. To better optimize the computation of grid size variations during creep deformation, the calculated strain rate value was rounded to 4×10^{-5} $(1/s)$ and adopted as the initial strain rate for the simulation.

Utilizing the established initial simulation configuration, the temporal evolution of deformation in the AZ31 magnesium alloy during impression creep was simulated. Figure 10 presents a comparison between the impression depth versus time curves predicted by the PF-DDRX model and those obtained experimentally. The results indicate that the steady-state impression rate is accurately predicted by the phase-field model. Specifically, the predicted value (8.19×10^{-5} mm/s) demonstrates close agreement (about 5% error) with the experimental steady-state impression rate (7.78×10^{-5} mm/s). Furthermore, the predicted onset of DDRX at 223 s aligns reasonably with the experimental observation of 190 s. These findings confirm a strong correlation between the predicted initial impression depth derived from finite element modeling, the calculated initial impression rate, and empirical observations. Distinctly, the critical indentation depth for the initiation of discontinuous dynamic recrystallization can be identified as the point at which the average dislocation density within the microstructure attains the critical dislocation density, as determined by Equation (14), exhibiting good agreement with experimental observations (73 μm predicted vs. 75 μm measured). The observed deviation between the simulated and experimental creep curves during the initial transient period (t < 100 s) is attributed to the model's inherent assumption of negligible primary creep, which is typically not critical in creep design. The simulation initiates with an idealized, pre-established deformation volume and an instantaneous attainment of a constant creep strain rate, contrasting with the gradual microstructural development characteristic of experimental primary creep. This simplification was implemented to specifically investigate the steady-state creep regime, which constitutes the primary objective of this study. Consequently, while the initial kinetics are not fully captured, this assumption is not anticipated to significantly compromise the accuracy of the predicted steady-state creep rate and the underlying microstructural evolution mechanisms.

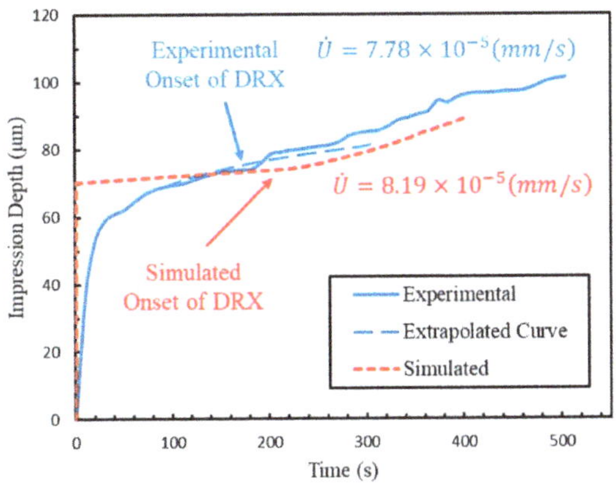

Figure 10. Simulated impression creep curve for the AZ31 alloy at 250 °C and a constant stress of 200 MPa.

Figure 11 illustrates the microstructural evolution during impression creep at 250 °C and 200 MPa, demonstrating the progression of discontinuous dynamic recrystallization with increasing indentation depth. In this figure, white regions represent initial grains, colored regions represent recrystallized grains, and lines delineate grain boundaries. Figure 11a depicts the initial microstructure prior to the onset of nucleation, where the dislocation density within the grains reaches the critical value. Figure 11b–e present the simulated microstructural evolution of discontinuous dynamic recrystallization as it occurs within the impression creep process.

The left column of Figure 11 showcases the evolution of the microstructure, while the right column presents the corresponding grain size distribution histograms. Initially, nuclei formation commences at the original grain boundaries (Figure 11b). Subsequently, new grains nucleate at the boundaries of previously recrystallized grains, leading to the formation of necklace-like structures characteristic of DRX-induced grain refinement (Figure 11c–e).

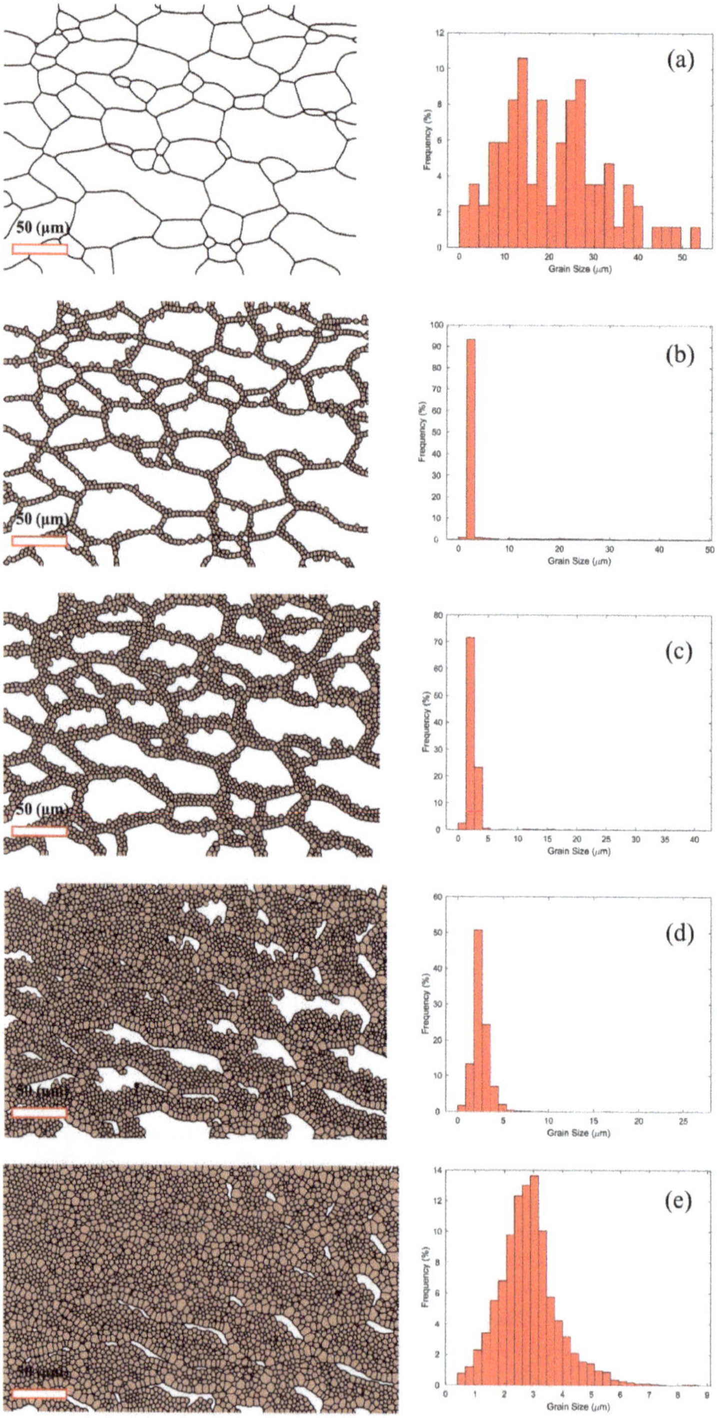

Figure 11. Simulated microstructural and grain size evolution with increasing depth and time during impression creep (250 °C, 200 MPa): (**a**) 73 μm (200 s), (**b**) 76 μm (260 s), (**c**) 79 μm (300 s), (**d**) 85 μm (360 s), and (**e**) 89 μm (400 s).

Analysis of the corresponding grain size distribution histograms (right column of Figure 11) reveals a progressive shift towards smaller grain sizes as DRX progresses. Initially, the grain size distribution exhibits a relatively broad range with a peak in the larger grain size region (Figure 11a). As DRX initiates and new grains form at the original grain boundaries (Figure 11b), a slight shift towards smaller grain sizes is observed. With further strain accumulation and the nucleation of new grains at the boundaries of previously recrystallized grains (Figure 11c–e), the grain size distribution becomes increasingly skewed towards smaller grain sizes, indicating significant grain refinement. This refinement is consistent with the observed necklace-like structures and the gradual expansion of DRX grains into the original grains, as depicted in the left column of Figure 11. The observed refinement of the average grain size during impression creep is a direct consequence of the discontinuous dynamic recrystallization (DDRX) mechanism, wherein the continuous nucleation of fine, strain-free grains leads to a progressive decrease in the overall grain size of the microstructure. The microstructural evolution observed in the left column of Figure 11, which illustrates the formation of the necklace-like structures of recrystallized grains and their subsequent encroachment upon the original, deformed grains, supports this as well.

Figure 12 illustrates the predicted temporal evolution of the DDRX fraction and average grain size. Initially, the DDRX fraction remains at zero until 223 s, corresponding to the point at which the dislocation density inside grains attains the critical value and DDRX commences. Subsequently, the microstructure undergoes complete recrystallization by 400 s. Concurrently, the average grain size exhibits a reciprocal trend, demonstrating a rapid decrease from approximately 32.5 μm to 5.5 μm within the first 130 s following DDRX initiation, before stabilizing at 3.7 μm. This inverse correlation between DDRX fraction and grain size, consistent with observations in other studies [17,20], underscores the microstructural refinement resulting from the formation of new, smaller grains during discontinuous dynamic recrystallization.

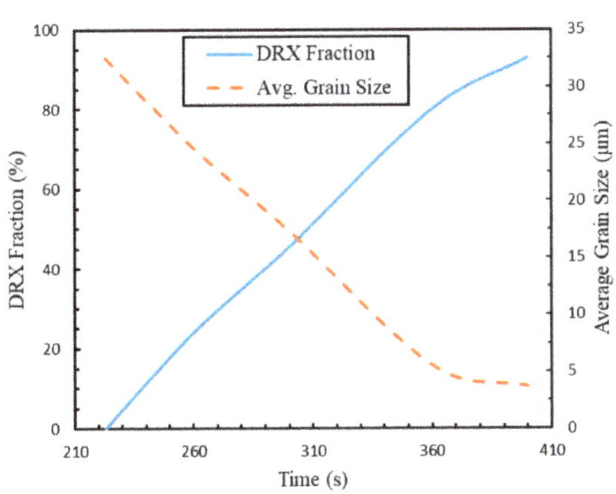

Figure 12. Discontinuous dynamic recrystallization fraction and average grain size evolution during impression creep simulation.

For the purpose of validating the simulated microstructural evolution, experimental analysis of the AZ31 magnesium alloy during impression creep testing was conducted using optical microscopy. The resulting experimental microstructures, illustrated in Figure 13, exhibit a comparable trend to those predicted by the simulation, as presented in Figure 11.

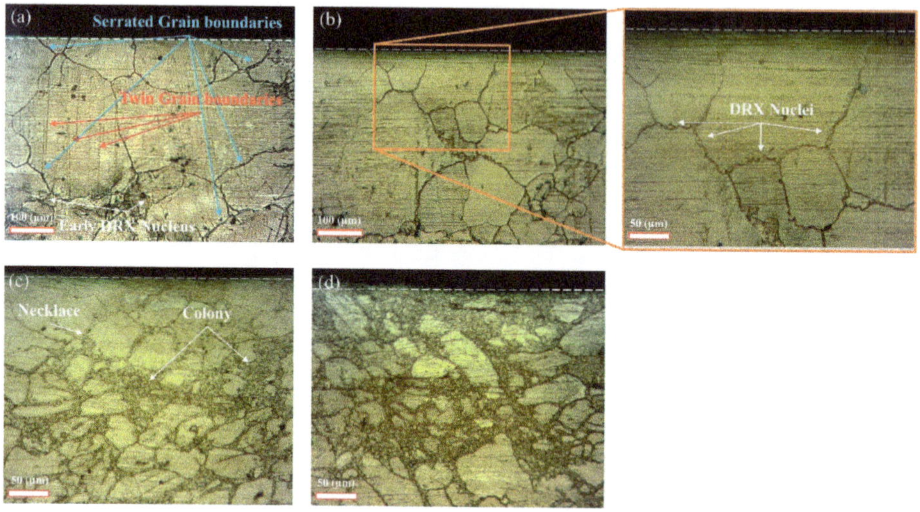

Figure 13. Experimental microstructural evolution with increasing indenter depth during impression creep (250 °C, 200 MPa) at impression depths of (**a**) 70 µm (90 s), (**b**) 75 µm (180 s), (**c**) 80 µm (270 s), and (**d**) 90 µm (380 s). The dashed white line indicates the specimen's surface.

The acquisition of the metallographic images presented herein posed significant experimental challenges. Precise measurement of impression depth was difficult, and sample preparation was a protracted process necessitating the utilization of multiple specimens. As depicted in Figure 13a, the serrated grain boundaries, which are known to be susceptible to dynamic recrystallization (DRX) [86,87], exhibit no discernible evidence of DRX at an impression depth of 80 µm. However, a limited number of DRX nuclei are observed at triple junctions, where the elevated grain boundary area-to-volume ratio appears to promote recrystallization at dislocation densities below the critical threshold [74, 75]. Subsequently, the DRX process is initiated, with nuclei formation occurring along grain boundaries, resulting in the characteristic necklace structure observed in Figure 13b. Further progression of the DRX process, as illustrated in Figure 13c,d, is characterized by the gradual development of nucleation into the grain interiors, culminating in the formation of DRX grain colonies [88]. Figure 14a,b present the fully recrystallized microstructure of the AZ31 magnesium alloy, along with an enlarged selected area for detailed microstructural analysis, respectively. The grain size distribution presented in Figure 14c exhibits a close correlation between experimental and simulated data. Furthermore, the measured average grain size of 3.16 µm demonstrates good agreement with a predicted value of 3.7 µm. This concordance validates the selected material parameters and model parameters, including the seed radius (r_{seed}) and grid spacing and size, confirming their consistency with experimental observations.

Figure 16 presents a series of assembled microstructure images, obtained from a specimen subjected to an impression depth of approximately 450 µm, to investigate a larger area in the vicinity of the indenter.

The region immediately adjacent to the indenter exhibits the earliest complete recrystallization during discontinuous dynamic recrystallization, a consequence of the elevated strain and stress concentrations prevalent in this area. With increasing impression depth, the strain field expands, and the DRX zone propagates to encompass the larger surrounding regions of the initially recrystallized microstructure.

Areas characterized by smaller grain sizes demonstrate accelerated recrystallization kinetics compared to larger grains. This phenomenon is attributed to the higher dislocation density within smaller grains at a given strain, coupled with their increased grain boundary-to-volume ratio, which promotes rapid nucleation and growth.

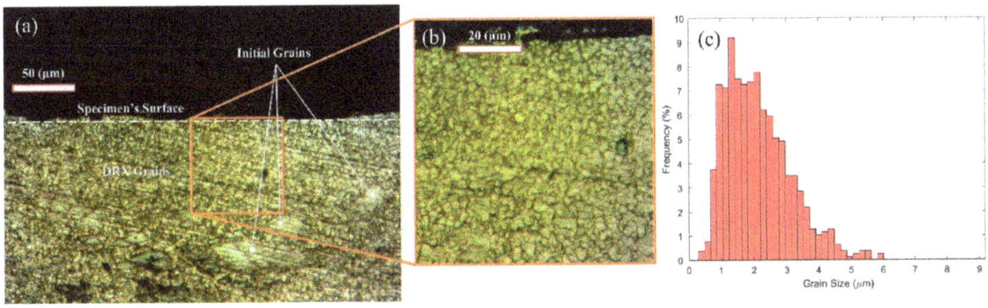

Figure 14. Microstructure of AZ31 showing complete recrystallization at 100 μm impression depth: (**a**) microstructure image, (**b**) magnified view of the selected region in (**a**), and (**c**) grain size distribution histogram.This study has demonstrated that the developed phase-field model accurately predicts both the steady-state impression rate and the discontinuous dynamic recrystallization microstructural evolution during impression creep testing of the AZ31 magnesium alloy. To further elucidate the DRX phenomenon within this context, microstructural analysis was conducted at progressively increasing impression depths. As the indenter penetrates deeper into the specimen during the impression creep process, the DRX regime extends to greater depths and encompasses wider spatial areas, as illustrated in Figures 15 and 16. The microstructure beneath the DRX colony, indicative of the incipient stage of discontinuous dynamic recrystallization, exhibits a necklace-like structure.

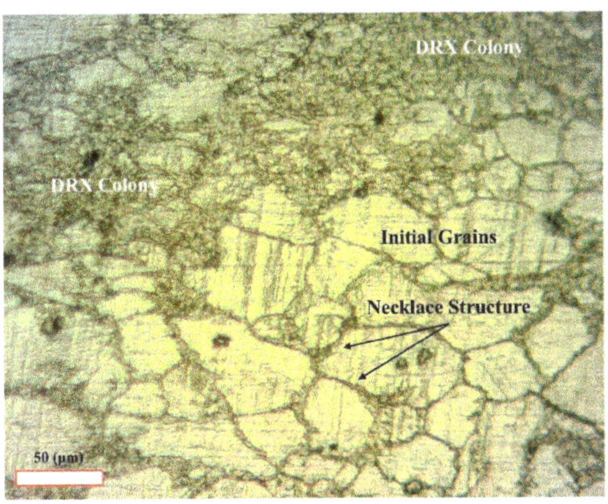

Figure 15. Experimental microstructural evolution of AZ31 after impression creep testing, focusing on lower observation depths.

Conversely, larger grains exhibit elevated dislocation densities near their grain boundaries due to the accumulation of geometrically necessary dislocations (GNDs), facilitating the rapid formation of necklace structures. However, their lower intragranular dislocation density and larger grain volume result in delayed recrystallization, necessitating higher strains for completion. The increased grain volume and lower average dislocation density in larger grains contribute to this delayed recrystallization. As observed in the enlarged region depicted in Figure 16, the DRX colony forms beneath larger grains, despite the average strain in the upper zone exceeding that of the colony zone.

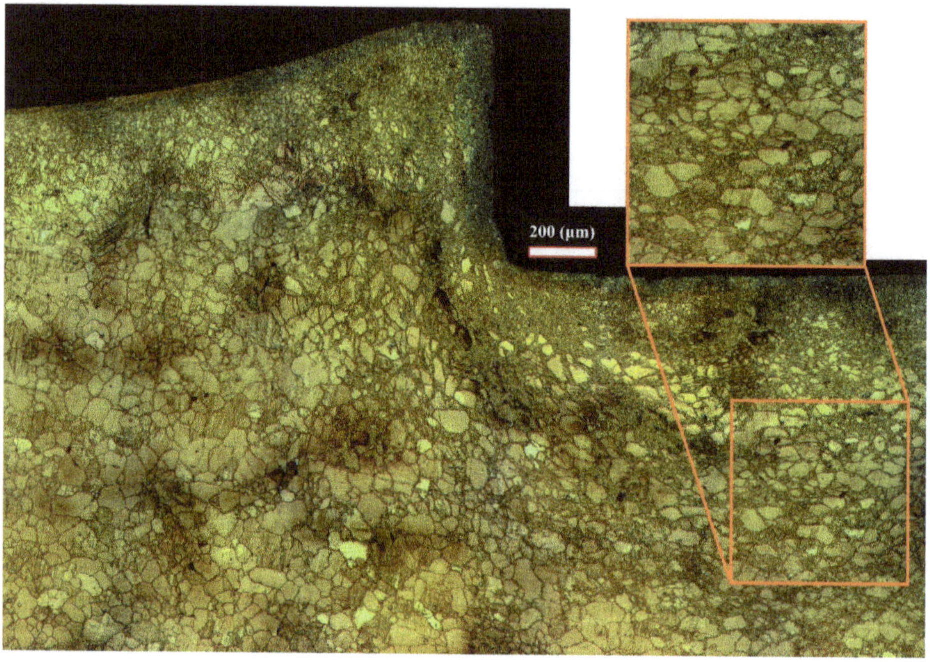

Figure 16. Assembled microstructure images of AZ31 after impression creep at an impression depth of 450 μm.

7. Conclusions

This study developed and validated a phase-field model, coupled with the Kocks–Mecking (KM) dislocation density evolution theory, to predict the steady-state creep rate and microstructural evolution of the AZ31 magnesium alloy during discontinuous dynamic recrystallization (DDRX) in impression creep testing. Key findings and contributions are summarized below:

1. An algorithm was implemented to extract the initial grain structure directly from optical microscopy images of the homogenized microstructure. Finite element analysis (FEA) was then employed to determine the instantaneous deformation, and the resulting impression depth was applied as the initial condition for the phase-field model.
2. Model parameters, including hardening and softening coefficients related to dislocation activity, grain boundary energy, and mobility, were obtained from existing literature.
3. The predicted impression creep rate of 8.19×10^{-5} mm/s demonstrated good agreement with the experimentally measured rate of 7.78×10^{-5} mm/s.
4. The predicted average grain size of the fully recrystallized microstructure was 3.7 μm, which also showed good agreement with the experimental observation of 3.16 μm.
5. The predicted recrystallization kinetics and grain topology evolution demonstrated excellent agreement with experimental observations, validating the model's predictive capability for microstructure evolution.

The proposed phase-field model offers a reliable and accurate approach for predicting the steady-state creep rate, grain size distribution, and microstructural evolution during DDRX in the AZ31 magnesium alloy. This methodology provides valuable insights for lifetime prediction and microstructure development in this material under impression creep conditions.

Building upon the insights gained from this two-dimensional phase-field model, future research endeavors could significantly benefit from extending the current framework to three-dimensional simulations. Such an expansion would enable a more comprehensive and spatially resolved understanding of the complex microstructural evolution during discontinuous dynamic recrystallization.

Author Contributions: Conceptualization, R.E.; methodology, A.R. and R.E.; software, A.R.; validation, A.R. and R.E.; formal analysis, A.R.; investigation, A.R.; resources, R.E. and E.B.; data curation, A.R.; writing—original draft preparation, A.R.; writing—review and editing, R.E. and E.B.; visualization, A.R.; supervision, R.E.; project administration, R.E.; funding acquisition, R.E. and E.B. All authors have read and agreed to the published version of the manuscript.

Funding: This research was funded by the Research Council of Shiraz University, grant number 99-GR-ENG 15. The APC was funded by Brunel University of London.

Data Availability Statement: The data presented in this study are not publicly available at this time as they form part of an ongoing research project. Data may be made available in the future upon reasonable request and after the completion of the related study.

Conflicts of Interest: The authors declare no conflicts of interest.

References

1. Fatemi-Varzaneh, S.M.; Zarei-Hanzaki, A.; Beladi, H. Dynamic recrystallization in AZ31 magnesium alloy. *Mater. Sci. Eng. A* **2007**, *456*, 52–57. [CrossRef]
2. Yang, X.-Y.; Ji, Z.-S.; Miura, H.; Sakai, T. Dynamic recrystallization and texture development during hot deformation of magnesium alloy AZ31. *Trans. Nonferrous Met. Soc. China* **2009**, *19*, 55–60. [CrossRef]
3. He, D.-G.; Lin, Y.C.; Chen, M.-S.; Li, L. Kinetics equations and microstructural evolution during metadynamic recrystallization in a nickel-based superalloy with δ phase. *J. Alloys Compd.* **2017**, *690*, 971–978. [CrossRef]
4. Maghsoudi, M.H.; Zarei-Hanzaki, A.; Changizian, P.; Marandi, A. Metadynamic recrystallization behavior of AZ61 magnesium alloy. *Mater. Des.* **2014**, *57*, 487–493. [CrossRef]
5. Ullmann, M.; Graf, M.; Schmidtchen, M.; Kawalla, R. Metadynamic Recrystallization Kinetics of Twin Roll Cast AZ31 Alloy during Hot Deformation. *Procedia Eng.* **2014**, *81*, 1559–1564. [CrossRef]
6. Chao, H.Y.; Sun, H.F.; Chen, W.Z.; Wang, E.D. Static recrystallization kinetics of a heavily cold drawn AZ31 magnesium alloy under annealing treatment. *Mater. Charact.* **2011**, *62*, 312–320. [CrossRef]
7. Seyed Salehi, M.; Serajzadeh, S. Simulation of static recrystallization in non-isothermal annealing using a coupled cellular automata and finite element model. *Comput. Mater. Sci.* **2012**, *53*, 145–152. [CrossRef]
8. Yang, X.-Y.; Zhu, Y.-K.; Miura, H.; Sakai, T. Static recrystallization behavior of hot-deformed magnesium alloy AZ31 during isothermal annealing. *Trans. Nonferrous Met. Soc. China* **2010**, *20*, 1269–1274. [CrossRef]
9. Humphreys, F.J.; Hatherly, M. Chapter 13—Hot Deformation and Dynamic Restoration. In *Recrystallization and Related Annealing Phenomena*, 2nd ed.; Humphreys, F.J., Hatherly, M., Eds.; Elsevier: Oxford, UK, 2004; pp. 415–450.
10. Mirzadeh, H. Grain refinement of magnesium alloys by dynamic recrystallization (DRX): A review. *J. Mater. Res. Technol.* **2023**, *25*, 7050–7077. [CrossRef]
11. Estrin, Y. Dislocation theory based constitutive modelling: Foundations and applications. *J. Mater. Process. Technol.* **1998**, *80–81*, 33–39. [CrossRef]
12. Roters, F.; Raabe, D.; Gottstein, G. Work hardening in heterogeneous alloys—A microstructural approach based on three internal state variables. *Acta Mater.* **2000**, *48*, 4181–4189. [CrossRef]
13. Peczak, P.; Luton, M.J. A Monte Carlo study of the influence of dynamic recovery on dynamic recrystallization. *Acta Metall. Mater.* **1993**, *41*, 59–71. [CrossRef]
14. Rollett, A.D.; Luton, M.J.; Srolovitz, D.J. Microstructural simulation of dynamic recrystallization. *Acta Metall. Mater.* **1992**, *40*, 43–55. [CrossRef]
15. Goetz, R.L.; Seetharaman, V. Modeling Dynamic Recrystallization Using Cellular Automata. *Scr. Mater.* **1998**, *38*, 405–413. [CrossRef]

16. Ding, R.; Guo, Z.X. Coupled quantitative simulation of microstructural evolution and plastic flow during dynamic recrystallization. *Acta Mater.* **2001**, *49*, 3163–3175. [CrossRef]
17. Chen, M.-S.; Yuan, W.-Q.; Li, H.-B.; Zou, Z.-H. Modeling and simulation of dynamic recrystallization behaviors of magnesium alloy AZ31B using cellular automaton method. *Comput. Mater. Sci.* **2017**, *136*, 163–172. [CrossRef]
18. Kugler, G.; Turk, R. Modeling the dynamic recrystallization under multi-stage hot deformation. *Acta Mater.* **2004**, *52*, 4659–4668. [CrossRef]
19. Wang, L.; Fang, G.; Qian, L. Modeling of dynamic recrystallization of magnesium alloy using cellular automata considering initial topology of grains. *Mater. Sci. Eng. A* **2018**, *711*, 268–283. [CrossRef]
20. Cai, Y.; Sun, C.Y.; Li, Y.L.; Hu, S.Y.; Zhu, N.Y.; Barker, E.I.; Qian, L.Y. Phase field modeling of discontinuous dynamic recrystallization in hot deformation of magnesium alloys. *Int. J. Plast.* **2020**, *133*, 102773. [CrossRef]
21. Gentry, S.P.; Thornton, K. Simulating recrystallization in titanium using the phase field method. *IOP Conf. Ser. Mater. Sci. Eng.* **2015**, *89*, 012024. [CrossRef]
22. Gentry, S.P.; Thornton, K. Sensitivity analysis of a phase field model for static recrystallization of deformed microstructures. *Model. Simul. Mater. Sci. Eng.* **2020**, *28*, 065002. [CrossRef]
23. Takaki, T.; Hirouchi, T.; Hisakuni, Y.; Yamanaka, A.; Tomita, Y. Multi-Phase-Field Model to Simulate Microstructure Evolutions during Dynamic Recrystallization. *Mater. Trans.* **2008**, *49*, 2559–2565. [CrossRef]
24. Takaki, T.; Tomita, Y. Static recrystallization simulations starting from predicted deformation microstructure by coupling multi-phase-field method and finite element method based on crystal plasticity. *Int. J. Mech. Sci.* **2010**, *52*, 320–328. [CrossRef]
25. Zhang, J.; Zheng, C.-W.; Li, D.-Z. A Multi-phase Field Model for Static Recrystallization of Hot Deformed Austenite in a C–Mn Steel. *Acta Metall. Sin. Engl. Lett.* **2018**, *31*, 208–215. [CrossRef]
26. Zhu, B.; Militzer, M. 3D phase field modelling of recrystallization in a low-carbon steel. *Model. Simul. Mater. Sci. Eng.* **2012**, *20*, 085011. [CrossRef]
27. Takaki, T.; Hisakuni, Y.; Hirouchi, T.; Yamanaka, A.; Tomita, Y. Multi-phase-field simulations for dynamic recrystallization. *Comput. Mater. Sci.* **2009**, *45*, 881–888. [CrossRef]
28. Takaki, T.; Yamanaka, A.; Tomita, Y. Phase-Field Modeling for Dynamic Recrystallization. In *From Creep Damage Mechanics to Homogenization Methods: A Liber Amicorum to Celebrate the Birthday of Nobutada Ohno*; Altenbach, H., Matsuda, T., Okumura, D., Eds.; Springer International Publishing: Cham, Switzerland, 2015; pp. 441–459.
29. Zhao, P.; Low, T.S.E.; Wang, Y.; Niezgoda, S.R. An integrated full-field model of concurrent plastic deformation and microstructure evolution: Application to 3D simulation of dynamic recrystallization in polycrystalline copper. *Int. J. Plast.* **2016**, *80*, 38–55. [CrossRef]
30. Zhao, P.; Wang, Y.; Niezgoda, S.R. Microstructural and micromechanical evolution during dynamic recrystallization. *Int. J. Plast.* **2018**, *100*, 52–68. [CrossRef]
31. Chen, L.; Chen, J.; Lebensohn, R.A.; Ji, Y.Z.; Heo, T.W.; Bhattacharyya, S.; Chang, K.; Mathaudhu, S.; Liu, Z.K.; Chen, L.Q. An integrated fast Fourier transform-based phase-field and crystal plasticity approach to model recrystallization of three dimensional polycrystals. *Comput. Methods Appl. Mech. Eng.* **2015**, *285*, 829–848. [CrossRef]
32. Creep polygonization and dynamic recrystallization. In *Creep of Crystals: High-Temperature Deformation Processes in Metals, Ceramics and Minerals*; Poirier, J.-P., Ed.; Cambridge University Press: Cambridge, UK, 1985; pp. 169–193.
33. Andrade, E.N.D.C. Creep of Metals and Recrystallization. *Nature* **1948**, *162*, 410. [CrossRef]
34. Stüwe, H.P.; Ortner, B. Recrystallization in Hot Working and Creep. *Met. Sci.* **1974**, *8*, 161–167. [CrossRef]
35. Spigarelli, S.; El Mehtedi, M. Creep as an extension of hot working: A unified approach to high temperature deformation of AZ31 alloy. *Mater. Sci. Eng. A* **2010**, *527*, 5708–5714. [CrossRef]
36. Estrin, Y.; Mecking, H. A unified phenomenological description of work hardening and creep based on one-parameter models. *Acta Metall.* **1984**, *32*, 57–70. [CrossRef]
37. Mecking, H.; Kocks, U.F. Kinetics of flow and strain-hardening. *Acta Metall.* **1981**, *29*, 1865–1875. [CrossRef]
38. Chu, S.N.G.; Li, J.C.M. Impression creep; a new creep test. *J. Mater. Sci.* **1977**, *12*, 2200–2208. [CrossRef]
39. Sastry, D.H. Impression creep technique—An overview. *Mater. Sci. Eng. A* **2005**, *409*, 67–75. [CrossRef]
40. Yang, Y.; Xiong, X.; Chen, J.; Peng, X.; Chen, D.; Pan, F. Research advances of magnesium and magnesium alloys worldwide in 2022. *J. Magnes. Alloys* **2023**, *11*, 2611–2654. [CrossRef]
41. Kim, H.-K.; Kim, W.-J. Creep behavior of AZ31 magnesium alloy in low temperature range between 423 K and 473 K. *J. Mater. Sci.* **2007**, *42*, 6171–6176. [CrossRef]
42. Luo, A.A. Recent magnesium alloy development for elevated temperature applications. *Int. Mater. Rev.* **2004**, *49*, 13–30. [CrossRef]
43. Kassner, M.E. Chapter 1—Fundamentals of Creep in Materials. In *Fundamentals of Creep in Metals and Alloys*, 3rd ed.; Kassner, M.E., Ed.; Butterworth-Heinemann: Boston, MA, USA, 2015; pp. 1–6.
44. Rezvani, A.; Ebrahimi, R. Mathematical Modeling of the Deformation Zone Under an Impression Creep Indenter Using Upper-Bound Theory. *J. Mater. Eng. Perform.* **2025**. [CrossRef]

45. Zhang, Y.; Yang, L.; Ge, C.; Pang, S.; Wang, X.; Zhang, Z.; Han, Z. Influence of Ca and Sr Addition on Impression Creep Behavior of Mg-4Al-RE Alloy. *J. Mater. Eng. Perform.* **2019**, *28*, 394–403. [CrossRef]
46. Kim, H.-K. The correlation between the impression and double shear creep of AZ31 magnesium alloy. *Mater. Sci. Eng. A* **2012**, *551*, 1–6. [CrossRef]
47. Naveena, J.; Kumar, G.; Mathew, M.D. Finite Element Analysis of Plastic Deformation During Impression Creep. *J. Mater. Eng. Perform.* **2015**, *24*, 1741–1753. [CrossRef]
48. Yang, F.; Li, J.C.M.; Shih, C.W. Computer simulation of impression creep using the hyperbolic sine stress law. *Mater. Sci. Eng. A* **1995**, *201*, 50–57. [CrossRef]
49. Nabariya, R.J.; Goyal, S.; Vasudevan, M.; Arivazhagan, N. Finite Element Analysis of Impression Creep. *Mater. Today Proc.* **2018**, *5 Pt 2*, 12320–12329. [CrossRef]
50. Teja, V.; Bedi, R.; Kumar, M. Finite Element Analysis of Impression Creep on P91. *IOP Conf. Ser. Mater. Sci. Eng.* **2022**, *1248*, 012023. [CrossRef]
51. Liu, X.; Li, L.-X.; He, F.-Y.; Zhou, J.; Zhu, B.-W.; Zhang, L.-Q. Simulation on dynamic recrystallization behavior of AZ31 magnesium alloy using cellular automaton method coupling Laasraoui–Jonas model. *Trans. Nonferrous Met. Soc. China* **2013**, *23*, 2692–2699. [CrossRef]
52. Chen, L.-Q. Phase-Field Models for Microstructure Evolution. *Annu. Rev. Mater. Res.* **2002**, *32*, 113–140. [CrossRef]
53. Moelans, N.; Blanpain, B.; Wollants, P. An introduction to phase-field modeling of microstructure evolution. *Calphad* **2008**, *32*, 268–294. [CrossRef]
54. Moelans, N.; Blanpain, B.; Wollants, P. Quantitative analysis of grain boundary properties in a generalized phase field model for grain growth in anisotropic systems. *Phys. Rev. B* **2008**, *78*, 024113. [CrossRef]
55. Moelans, N.; Blanpain, B.; Wollants, P. Quantitative Phase-Field Approach for Simulating Grain Growth in Anisotropic Systems with Arbitrary Inclination and Misorientation Dependence. *Phys. Rev. Lett.* **2008**, *101*, 025502. [CrossRef] [PubMed]
56. Humphreys, J.; Rohrer, G.S.; Rollett, A. Chapter 5—Mobility and Migration of Boundaries. In *Recrystallization and Related Annealing Phenomena*, 3rd ed.; Humphreys, J., Rohrer, G.S., Rollett, A., Eds.; Elsevier: Oxford, UK, 2017; pp. 145–197.
57. Moelans, N. A quantitative and thermodynamically consistent phase-field interpolation function for multi-phase systems. *Acta Mater.* **2011**, *59*, 1077–1086. [CrossRef]
58. Moelans, N.; Godfrey, A.; Zhang, Y.; Jensen, D.J. Phase-field simulation study of the migration of recrystallization boundaries. *Phys. Rev. B* **2013**, *88*, 054103. [CrossRef]
59. Allen, S.M.; Cahn, J.W. A microscopic theory for antiphase boundary motion and its application to antiphase domain coarsening. *Acta Metall.* **1979**, *27*, 1085–1095. [CrossRef]
60. Gruber, J.; Ma, N.; Wang, Y.; Rollett, A.D.; Rohrer, G.S. Sparse data structure and algorithm for the phase field method. *Model. Simul. Mater. Sci. Eng.* **2006**, *14*, 1189–1195. [CrossRef]
61. Vedantam, S.; Patnaik, B.S.V. Efficient numerical algorithm for multiphase field simulations. *Phys. Rev. E* **2006**, *73*, 016703. [CrossRef]
62. Turnbull, D. Theory of Grain Boundary Migration Rates. *JOM* **1951**, *3*, 661–665. [CrossRef]
63. Liu, Y.-X.; Lin, Y.C.; Li, H.-B.; Wen, D.-X.; Chen, X.-M.; Chen, M.-S. Study of dynamic recrystallization in a Ni-based superalloy by experiments and cellular automaton model. *Mater. Sci. Eng. A* **2015**, *626*, 432–440. [CrossRef]
64. Prosgolitis, C.G.; Lambrakos, S.G.; Zervaki, A.D. Phase-Field Modeling of Nugget Zone for a AZ31-Mg-Alloy Friction Stir Weld. *J. Mater. Eng. Perform.* **2018**, *27*, 5102–5113. [CrossRef]
65. Groh, S.; Marin, E.B.; Horstemeyer, M.F.; Bammann, D.J. Dislocation motion in magnesium: A study by molecular statics and molecular dynamics. *Model. Simul. Mater. Sci. Eng.* **2009**, *17*, 075009. [CrossRef]
66. Fan, D.; Chen, L.Q. Computer simulation of grain growth using a continuum field model. *Acta Mater.* **1997**, *45*, 611–622. [CrossRef]
67. Ning, Y.Q.; Wang, T.; Fu, M.W.; Li, M.Z.; Wang, L.; Zhao, C.D. Competition between work-hardening effect and dynamic-softening behavior for processing as-cast GH4720Li superalloys with original dendrite microstructure during moderate-speed hot compression. *Mater. Sci. Eng. A* **2015**, *642*, 187–193. [CrossRef]
68. Puchi-Cabrera, E.S.; Staia, M.H.; Guérin, J.D.; Lesage, J.; Dubar, M.; Chicot, D. An experimental analysis and modeling of the work-softening transient due to dynamic recrystallization. *Int. J. Plast.* **2014**, *54*, 113–131. [CrossRef]
69. Luo, F.; Peng, H.; Chen, H.; Xiao, X.; Xie, W.; Wang, H.; Yang, B. Dislocation substructure-controlled softening of Cu-20Ni-20Mn alloy. *Mater. Charact.* **2019**, *147*, 253–261. [CrossRef]
70. Frost, H.J.; Ashby, M.F. Deformation-Mechanism Maps: The Plasticity and Creep of Metals and Ceramics. Franklin Book Company, Incorporated: Elkins Park, PA, USA, 1982.
71. Honeycombe, R.W.K.; Pethen, R.W. Dynamic recrystallization. *J. Less Common Met.* **1972**, *28*, 201–212. [CrossRef]
72. Sakai, T.; Ohashi, M. Dislocation substructures developed during dynamic recrystallisation in polycrystalline nickel. *Mater. Sci. Technol.* **1990**, *6*, 1251–1257. [CrossRef]

73. Arsenlis, A.; Parks, D.M. Crystallographic aspects of geometrically-necessary and statistically-stored dislocation density. *Acta Mater.* **1999**, *47*, 1597–1611. [CrossRef]
74. Clemm, J.; Fisher, J.C. The influence of grain boundaries on the nucleation of secondary phases. *Acta Metall.* **1955**, *3*, 70–73. [CrossRef]
75. Wu, G.L.; Jensen, D.J. Orientations of recrystallization nuclei developed in columnar-grained Ni at triple junctions and a high-angle grain boundary. *Acta Mater.* **2007**, *55*, 4955–4964. [CrossRef]
76. Jokisaari, A.M.; Permann, C.; Thornton, K. A nucleation algorithm for the coupled conserved–nonconserved phase field model. *Comput. Mater. Sci.* **2016**, *112*, 128–138. [CrossRef]
77. Rezvani, A.; Ebrahimi, R.; Bagherpour, E. Static Recrystallization Simulation of Interstitial Free-Steel by Coupling Multi-Phase-Field and Crystal Plasticity Model Considering Dislocation Density Distribution. *Adv. Eng. Mater.* **2025**, 2500117. [CrossRef]
78. He, F.; Wu, C.; Shi, L. Phase-field simulation of dynamic recrystallization in friction stir weld nugget zone of dissimilar Al/Mg alloys. *J. Mater. Res. Technol.* **2023**, *27*, 2670–2683. [CrossRef]
79. Gladman, T. Creep Strength in Steel and High-Temperature Alloys. *Steel Times* **1972**, *200*, 910.
80. Mayes, P.; Hancock, P. Grain-boundary sliding and recrystallization of nimonic 108 during creep. *Met. Sci. J.* **1973**, *7*, 69–75. [CrossRef]
81. Mayes, F.; Hancock, P. Recrystallization of Nimonic 108 during Creep. *Met. Sci.* **1975**, *9*, 145–148. [CrossRef]
82. Koul, A.K.; Immarigeon, J.P.A. Dynamic recrystallization during creep in a 45 Pct Ni-35 pct Fe-20 pct Cr alloy system. *Metall. Trans. A* **1985**, *16*, 51–57. [CrossRef]
83. Gottstein, G. Dynamic recrystallization of Cu single crystals during tensile deformation in creep. *Met. Sci.* **1983**, *17*, 497–502. [CrossRef]
84. Larsen-Badse, J. Indentation creep of soft metals. In Proceedings of the Annual Meeting of AIME, Los Angeles, CA, USA, 19–23 February 1967.
85. Hyun, C.Y.; Kim, H.K. Creep deformation behaviour of AZ31 magnesium alloy over wide range of temperature and stress. *Mater. High Temp.* **2014**, *31*, 34–40. [CrossRef]
86. Korla, R.; Chokshi, A.H. A Constitutive Equation for Grain Boundary Sliding: An Experimental Approach. *Metall. Mater. Trans. A* **2014**, *45*, 698–708. [CrossRef]
87. Roodposhti, P.S.; Sarkar, A.; Murty, K.L. Microstructural development of high temperature deformed AZ31 magnesium alloys. *Mater. Sci. Eng. A* **2015**, *626*, 195–202. [CrossRef]
88. Essadiqi, E.; Liu, W.J.; Kao, V.; Yue, S.L.; Verma, R. Recrystallization in AZ31 Magnesium Alloy during Hot Deformation. *Mater. Sci. Forum* **2005**, *475–479*, 559–562. [CrossRef]

crystals

MDPI

Article

Process Optimization Simulation of Residual Stress in Martensitic Steel Considering Phase Transformation

Yuzheng Cui [1,2] and Guang Yang [3,4,*]

1 School of Management, Northeastern University, Qinhuangdao 066004, China; cuiyuzheng@neuq.edu.cn
2 Valley of Data Technology Group, Hangzhou VOD Intelligent Technology Co., Ltd., Hangzhou 311100, China
3 Department of Engineering Mechanics, Center for Nano and Micro Mechanics, and Applied Mechanics Lab, Tsinghua University, Beijing 100084, China
4 Key Laboratory of Mechanical Reliability for Heavy Equipments and Large Structures of Hebei Province, Yanshan University, Qinhuangdao 066004, China
* Correspondence: yangg99845@163.com

Abstract

The solid phase transformation of martensitic steel during heat treatment will affect the stress and temperature. Previous residual stress prediction models ignore the effect of phase transition on residual stress. In order to predict residual stress accurately, a residual stress calculation method considering solid phase transition was presented. The measures to reduce residual stress in quenching medium, cooling rate, and the starting temperature and tempering temperature of the martensitic transformation were studied. The experimental results show that residual stress decreases after air cooling. In a certain range, residual stress can be reduced during heat treatment by decreasing the cooling rate and the martensite start temperature. The recommended tempering temperature is 380 °C.

Keywords: martensitic steel; martensitic transformation; residual stress; optimization of heat treatment process; finite element simulation

1. Introduction

Martensitic steel is extensively utilized in transportation and mechanical components because of its favorable characteristics, including excellent plasticity, extended service life, and high strength [1–5]. Martensitic steel undergoes the martensitic transformation during heat treatment. Residual stress is affected by the martensitic transformation [6–9]. The martensitic transformation can lead to volume and yield strength changes, which will affect the residual stress of martensitic steel after heat treatment [10,11]. The residual stress produced in the heat treatment process affects the dimensional stability, fatigue strength, stress corrosion resistance and crack resistance of the member [12,13]. Therefore, the accurate prediction of residual stress is very important. The experimental method can only measure the local residual stress of martensitic steel, which has limitations. The residual stress field of martensitic steel can be obtained by finite element simulation, and the residual stress of martensitic steel components can be predicted. A residual stress model that takes into account the martensitic transformation is crucial for accurately forecasting residual stress, which in turn is essential for optimizing the heat treatment process to control the residual stress in martensitic steel.

By optimizing the heat treatment process (such as quenching and tempering), the residual stress during the martensitic transformation can be effectively controlled, thereby improving the fatigue strength and toughness of the material. By reducing residual stress or providing a favorable residual stress state, the service life of structural components can be significantly extended and the risk of failure reduced. Economic aspects can be improved by optimizing processes to reduce later repair and replacement costs. In areas such as the aviation, automotive, and energy sectors, there are extremely high demands on material properties, and optimizing residual stress is critical to meeting these demands. By controlling residual stress, material characteristics can be customized to meet performance requirements in specific applications.

Some studies have measured residual stress in martensitic steels by experimental tests. The effects of different quenching rates on residual stress were studied by X-ray diffraction [14]. Although these experiments provide valuable information for understanding the distribution of residual stress, it is often difficult to fully reflect the residual stress state under actual process conditions due to the complexity of experimental conditions and samples. The finite element analysis method has been used to simulate stress evolution during the martensitic transformation [15]. Although these numerical models can predict the distribution of residual stress fairly well, they often rely on empirical parameters and simplified assumptions, and the results may not be accurate enough. In some studies, the multi-scale model method is used to try to combine the relationship between microstructure and macroscopic properties to investigate the effect of the martensitic transformation on residual stress more comprehensively. However, these methods are often computationally complex and require large amounts of experimental data to verify their accuracy. Most of the current research focuses on martensitic steels with specific alloy composition, such as some low-alloy steels, while the research on high-alloy martensitic steels is relatively rare. The effects of different alloy compositions on the phase transformation process and residual stress are significantly different, but systematic comparative studies are insufficient. There are few studies on the effect of new process on the martensitic transformation and residual stress in the current literature, and many studies are still located in the traditional process conditions, lacking in-depth discussion under modern manufacturing technology. In this study, an improved finite element method, combined with the phase field model and the thermo-mechanical coupling method, is used to simulate the residual stress generation in the martensitic transformation process more accurately. This method can better consider the effect of local microstructure change on stress state. This will help optimize the heat treatment process for specific materials. This study will explore the effects of advanced heat treatment processes on the martensitic transformation and residual stress, which will not only enrich theoretical research but also help guide practical industrial applications.

The prediction of residual stress involves thermodynamics, phase transformation dynamics, and elastoplastic mechanics [16,17]. The residual stress model should take into account various factors, including heat transfer, metallurgy, and mechanics [18,19]. For martensitic steel, the martensitic transformation has only been considered in a few studies on welding [20–22]. Wang et al. [23] utilized ABAQUS 2022 to determine the residual stress distribution in welded joints of martensitic steel. Similarly, Zubairuddin et al. [24] employed SYSWELD to analyze the stress and deformation in martensitic steel. Although the martensitic transformation was considered, latent heat was also ignored. Deng et al. [25] obtained the stress field of martensitic steel through simulation but did not consider the effect of martensitic phase transition. In their research, the effects of volume change and yield strength variation resulting from the martensitic transformation were taken into account, whereas plasticity and latent heat were disregarded. The martensitic transformation can cause large volume expansion [26]. Chen et al. [27] found in their research that large volume strain during heat treatment has a significant impact on residual stress. The effect of phase transition on residual stress was neglected in previous studies of residual stress models. In order to predict the residual stress more accurately, a residual stress prediction model considering phase transition is proposed [28,29]. Therefore, the effect of phase transition should be considered in finite element simulation to accurately predict the residual stress of martensitic steel.

A residual stress prediction method considering phase change is proposed. The measures to control the residual stress are discussed in terms of quenching medium, cooling rate, and the starting temperature and tempering temperature of the martensitic transition.

2. Calculation Method of Residual Stress Considering Martensitic Transformation

2.1. Microstructure Field Calculation

The sample analyzed in this paper is low-carbon martensitic steel after quenching. Consequently, the martensitic transformation is incorporated into the microstructure calculations with the martensite volume fraction following the Koistinen–Marburger equation [10]:

$$\varphi_{\mathrm{M}} = 1 - \exp[-\alpha(M_s - T)] \tag{1}$$

φ_{M} is the martensite volume fraction; M_s is the martensitic transformation point; T is temperature; α is a constant which is 0.011.

2.2. Temperature Field Calculation

The calculation equation of the whole temperature field is written as [30]

$$\rho c\ (\varphi_{\mathrm{M}}, T)\ \dot{T}\ = \nabla[k(\varphi_{\mathrm{M}}, T) \cdot \nabla T] + \dot{Q} \tag{2}$$

ρ is material density; c is heat capacity; k is the heat transfer coefficient; T is temperature; $\dot{T}$ is the partial of temperature with respect to time; ∇ is the gradient operator; $\dot{Q}$ is the latent heat of phase transformation per unit time.

Q is the latent heat of phase transformation, calculated by the formula

$$Q = \Delta H \cdot \dot{\varphi}_{\mathrm{M}} \tag{3}$$

ΔH is the heat released during the decomposition of austenite per unit volume; $\dot{\varphi}_{\mathrm{M}}$ is the martensitic phase variable per unit time.

The boundary condition is defined as follows [11]:

$$k\left[\frac{\partial T}{\partial x}l_x + \frac{\partial T}{\partial y}l_y + \frac{\partial T}{\partial z}l_z\right] = h(T - T_\infty) \tag{4}$$

l_x, l_y, and l_z are direction cosines in the x, y, and z directions, respectively; h is the surface heat transfer coefficient; T and T_∞ are the surface and ambient temperatures, respectively.

2.3. Stress Field Calculation

The thermal elastoplastic behavior of martensitic steel during the quenching process is a nonlinear problem addressed using incremental theory. In this analysis, the Von-Mises yield criterion, isotropic hardening conditions, and the Prandtl–Reuss flow rule are implemented. The volume change of martensitic steel is illustrated in Figure 1. The total strain increment is defined in the following manner [31]:

$$\Delta\varepsilon_{ij} = \Delta\varepsilon_{ij}^{\mathrm{e}} + \Delta\varepsilon_{ij}^{\mathrm{p}} + \Delta\varepsilon_{ij}^{\mathrm{th}} + \Delta\varepsilon_{ij}^{\mathrm{tr}} + \Delta\varepsilon_{ij}^{\mathrm{tp}} \tag{5}$$

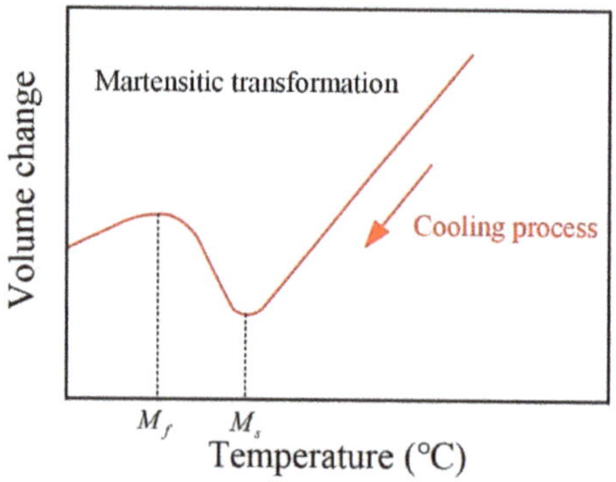

Figure 1. Schematic diagram of volume variation during quenching.

$\Delta\varepsilon_{ij}^{e}$, $\Delta\varepsilon_{ij}^{P}$, $\Delta\varepsilon_{ij}^{th}$, $\Delta\varepsilon_{ij}^{tr}$, and $\Delta\varepsilon_{ij}^{tp}$ are elastic strain increment, plastic strain increment, thermal strain increment, phase change plastic increment, and phase change plastic strain increment, respectively.

The material exhibits uniform mechanical and physical properties in all directions, allowing the increments of microstructure strain and thermal strain to be expressed as follows [32]:

$$\Delta\varepsilon_{ij}^{tr} = \varphi_{M}[\varepsilon_{M}(T+\Delta T) - \varepsilon_{A}(T+\Delta T)] \tag{6}$$

$$\Delta\varepsilon_{ij}^{th} = \varphi_{M}[\varepsilon_{M}(T+\Delta T) - \varepsilon_{M}(T)] \tag{7}$$

ε_{M} is the relative strain of the martensitic microstructure; ε_{A} is the relative strain of the austenitic microstructure.

In line with the GJ theory, the Lebond formula is employed to compute transformation plasticity [33,34]:

$$d\varepsilon^{tr} = \frac{3}{2}KSf(\varepsilon)\varepsilon \tag{8}$$

S is the stress deviation; K is the coefficient of transformation plasticity; ε is the rate of transformation plasticity; $f(\varepsilon)$ is the kinetic formula of transformation plasticity.

2.4. Simulation Route

In simulations that take into account the martensitic transformation, the influence of phase transitions on stress is integrated into the constitutive equation and implemented through an ABAQUS user subroutine. The simulated route is shown in Figure 2. First, ABAQUS 2022 is utilized to create a 3D model, wherein heat-related material properties are entered into the material parameter editing interface. Next, during the simulation, the martensite fraction is determined based on temperature and the Koistinen–Marburger formula [34]. Subsequently, the latent heat generated by the phase transition is incorporated into the HETVAL subroutine as an internal heat source [35]. The changes in phase transformation plasticity and yield strength are addressed by the UHARD subroutine, while the impact of volume change on stress is managed by the UEXPAN subroutine. This model is applicable to metals undergoing martensitic phase transitions. The finite element model constructed in this study is mainly applicable to metal materials with uniform material properties. The material parameters used in the model are based on references, and its physical and mechanical properties have been verified in a number of experiments. However, for heterogeneous materials or composites, models need to be adjusted accordingly to account for the effects of material interfaces and polyphase structures. The finite element model in the present study is mainly designed for regular circular geometries and is suitable for structures with symmetry and uniform loading conditions. However, for complex asymmetries or geometries with sharp corners, the model requires mesh optimization and parameter adjustment to ensure the accuracy of the analysis results.

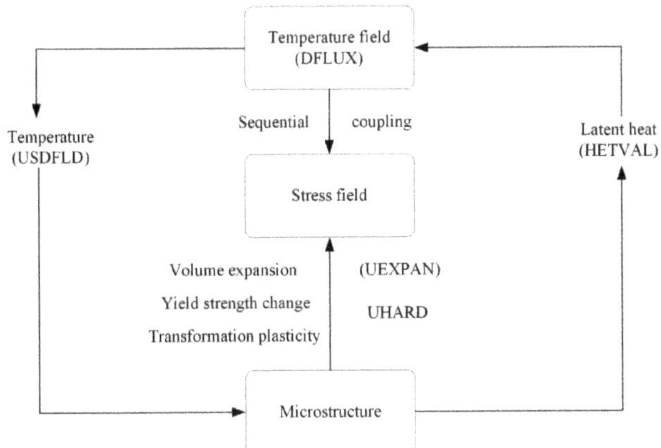

Figure 2. Simulation route.

2.5. Calculation Model and Parameters

The thermal and physical properties of martensitic steel involved in the model and mechanical property parameters are shown in Figure 3. In metallurgical and mechanical calculations, the temperature-dependent properties of materials should be considered first. The corresponding thermal property data (such as thermal conductivity λ, specific heat capacity C) and mechanical properties parameters (such as yield stress σ_y, elastic modulus E, Poisson ratio ν, linear thermal expansion coefficient α). The density of martensitic steel changes little with temperature. It is considered that the density of martensitic steel during quenching is 7800 kg·m^{-3}. The heat transfer coefficient used in the model is based on martensitic steel as outlined in reference [36]. The thermal conductivity of air cooling is 0.025 W(m·K), the thermal conductivity of oil cooling is 0.2 W(m·K), and the thermal conductivity of water cooling is 0.6 W(m·K). The martensitic steel model is constructed using finite element analysis software ABAQUS 2022, as depicted in Figure 4. The thermodynamic coupling finite element method is used to simulate the interaction between temperature and stress fields, especially the stresses caused by phase transitions and temperature gradients during heat treatment. The dimensions of the ring are 100 × 100 × 50 mm^3, with a cell size of 2 mm, resulting in a total of 62,500 cells. In finite element analysis, the mesh size is the key factor that affects the calculation accuracy and convergence. Smaller mesh sizes often provide more accurate results but significantly increase computational costs. Therefore, it is necessary to conduct grid sensitivity analysis to determine the appropriate mesh size and ensure the reliability and computational efficiency of the model results. In order to ensure the accuracy of the model results, a grid sensitivity analysis was carried out. Three different mesh sizes (coarse mesh, medium mesh, and fine mesh) were selected for simulation calculation. The key parameters include maximum stress, maximum displacement, etc. The results show that with the decrease in mesh size, the changes in key parameters tend to be stable, indicating that the model results converge gradually. Through the grid sensitivity analysis, the middle grid was selected as the final model grid to balance the calculation accuracy and efficiency. The boundary conditions of the temperature field were convection boundary conditions, and the initial temperature was set at room temperature (20 °C). To validate the accuracy of the model, data from the literature were employed for comparison. The model validation in this study mainly relies on experimental data in the existing literature. Specifically, we selected the key parameters of stress and strain in the literature for comparison. During the verification process, the model simulation results were compared with experimental results in the literature under the same conditions, such as temperature, pressure, and other environmental variables. In order to evaluate the accuracy of the model, we use the mean square error as an evaluation index. The results show that the mean square error between the predicted value of the model and the experimental data is 0.05, which indicates that the model has good fitting performance. Although there is no independent experimental verification in this study, the validity of the model is preliminarily verified by comparing it with a number of data from the literature. Future studies will include further experimental validation to consolidate the reliability of the model. In addition, we plan to expand the data source to cover more experimental conditions to fully evaluate the suitability and stability of the model. Current models have indeed been developed primarily for specific types of martensitic steel materials and their specific component geometry. The ring workpiece has symmetry and uniformity, which makes it possible to simplify the calculation when analyzing the stress distribution. The symmetrical geometry helps to understand and predict the distribution of residual stress, especially on the inner and outer surfaces of the ring.

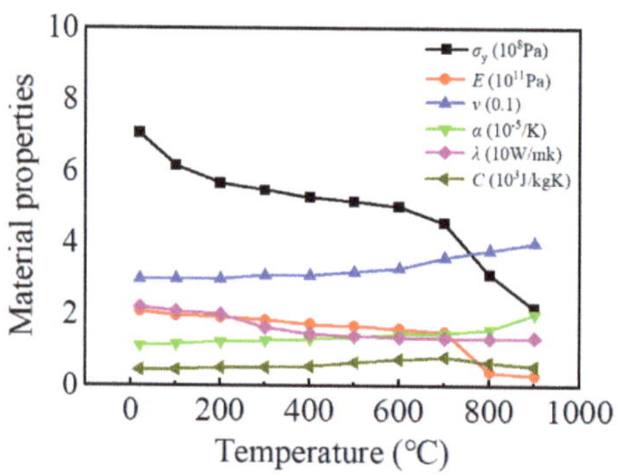

Figure 3. Material properties of martensitic steel.

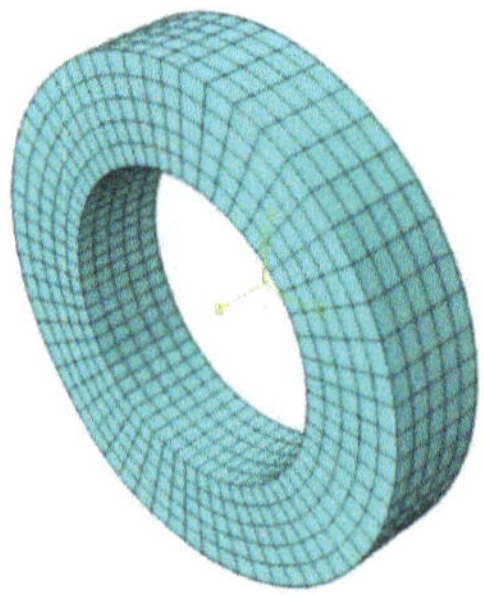

Figure 4. Grid division of martensitic steel model.

3. Heat Treatment Process Regulates Residual Stress

3.1. Microstructure

The microstructure of low-carbon martensitic steel is composed of martensitic lath and residual austenite as shown in Figure 5. After tempering, the lath martensite structure is tempered martensite structure, and the structure is the smallest, and there is a small amount of dispersed carbide.

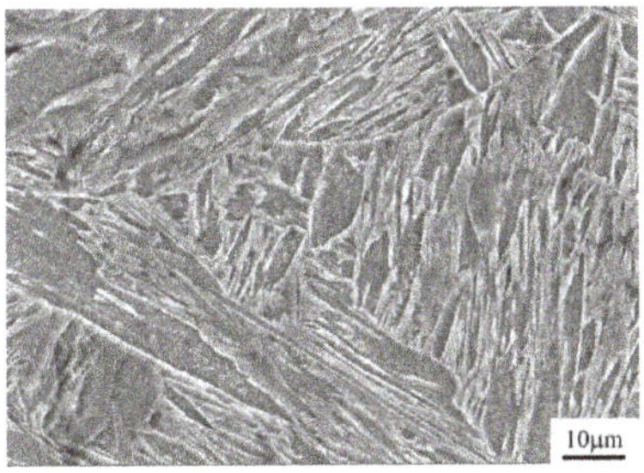

Figure 5. Microstructure of martensitic steel.

3.2. Temperature and Stress During Quenching

Figure 6 shows the temperature field variation of martensitic steel during water quenching at 20 °C. When the sample is water-quenched, the temperature of sample edge drops the fastest followed by the surface. The core of the sample is not in direct contact with the medium and relies on heat conduction. Therefore, the sample remains at a very high temperature. With the quenching, the core temperature of the sample decreases gradually. Due to the different cooling rates of the core and surface of the sample, an uneven temperature field is generated. When the quenching time is 300 s, the temperature of the core and surface gradually tends to be consistent with that of the medium.

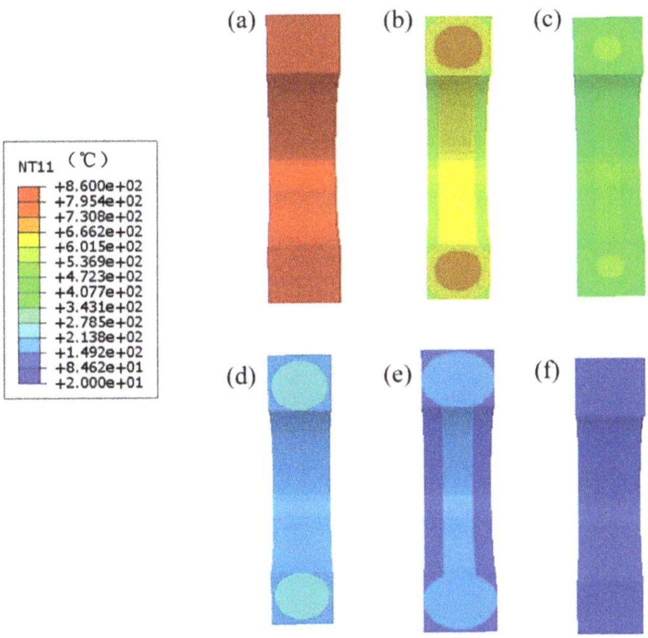

Figure 6. Temperature field of martensitic steel during quenching at 20 °C: (**a**) 0 s; (**b**) 10 s; (**c**) 15 s; (**d**) 25 s; (**e**) 50 s; (**f**) 300 s.

Figure 7 shows the change in the stress field of martensitic steel during water quenching at 20 °C. When the sample just enters the water, the surface cools rapidly and shrinks, resulting in tensile residual stress. Under conditions of synchronous shrinkage, compressive residual stress is generated at the surface. As quenching time progresses, the temperature of the core gradually decreases, and its cooling rate exceeds that of the surface.

Consequently, the core begins to shrink. At this point, the surface temperature is low, and its strength is high, leading to an increased restraining effect from the core. This results in a gradual increase in residual stress within the core, while the residual stress at the surface decreases. As the core continues to shrink, it transitions from compressive residual stress to tensile residual stress at 25 s, while the surface shifts from an initial state of tensile residual stress to compressive residual stress. Ultimately, a stress state characterized by inner tension and outer compression is established. Throughout the quenching process, the stress values for both the core and surface increase progressively. Upon the completion of quenching, the tensile residual stress in the core can reach 440 MPa, while the compressive residual stress at the surface can reach −442 MPa.

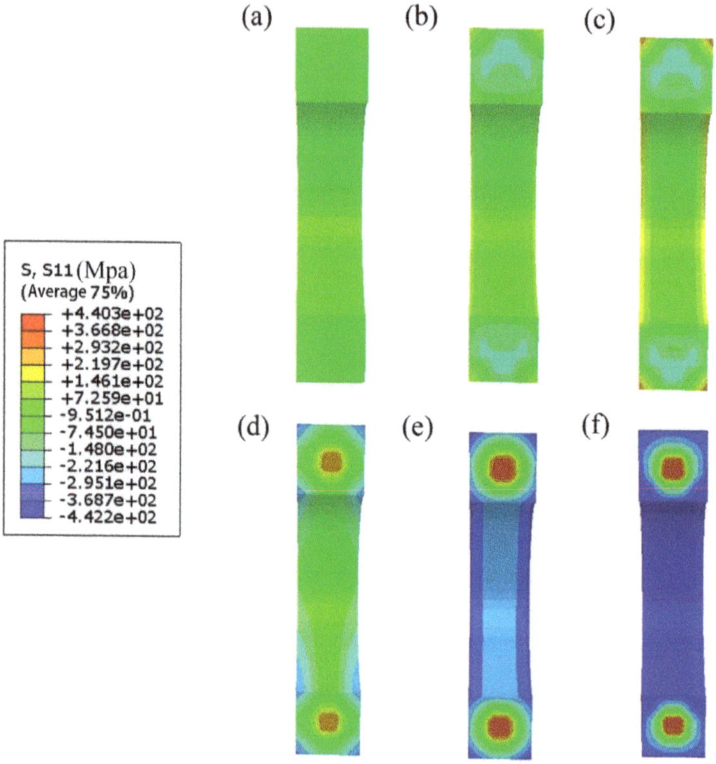

Figure 7. Axial stress field of martensitic steel during quenching at 20 °C: (**a**) 0 s; (**b**) 10 s; (**c**) 15 s; (**d**) 25 s; (**e**) 50 s; (**f**) 300 s.

The analysis above indicates that the evolution paths of the temperature field and stress field in martensitic steel during quenching differ significantly. To monitor the changes in temperature and stress at key locations during the quenching process, nodes are selected at the surface and at the center of the core. As shown in Figure 8a, the exothermic effect of phase transformation is distinctly observable around 385 °C. Over time, the temperature difference between the surface and the core gradually reduces. Although the temperature of the sample continues to decrease, there are no significant changes in residual stress during the later stages of quenching, as shown in Figure 8b.

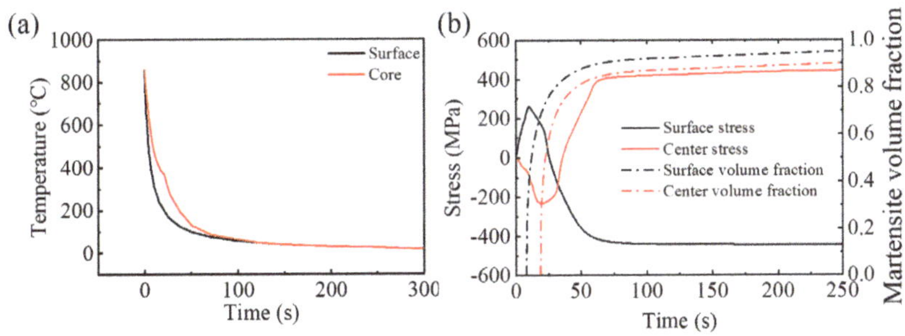

Figure 8. (**a**) Variation curves of temperature field during surface and core quenching; (**b**) Variation curves of residual stress field during surface and core quenching.

3.3. *Influence of Quenching Medium on Residual Stress*

Different quenching media have different cooling capacity for the sample and the corresponding quenching stress is also different. In this excerpt, three different cooling media are used for calculation and comparative study. Assuming the hardenability of the sample is large enough, the whole martensitic microstructure can be obtained even if it is air-cooled. The distribution of residual stress during the cooling process of martensitic steel is analyzed, revealing that the stress distribution after quenching varies with different cooling media, as illustrated in Figure 9. In the case of air cooling, which is relatively gentle, the temperature difference between the surface and the core remains small. For oil cooling, the temperature difference is also modest due to the slower cooling rates, resulting in tensile residual stress at the surface and compressive residual stress in the core. Conversely, water quenching involves a rapid cooling rate, producing a substantial temperature gradient within the sample. As a result, the residual stress levels are markedly higher following water quenching.

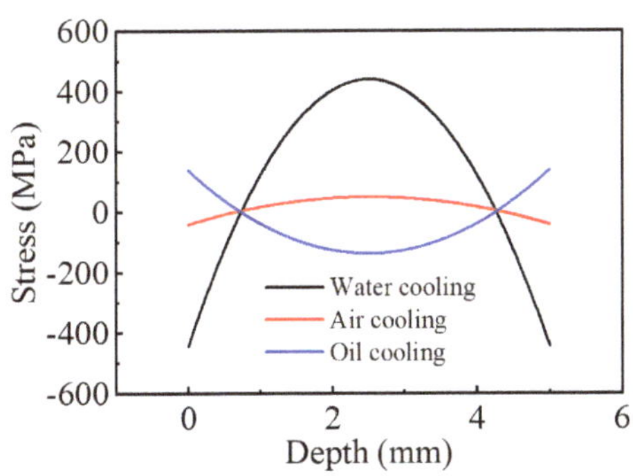

Figure 9. Influence of different quenching media on residual stress.

3.4. *Influence of Cooling Rate on Residual Stress*

The residual stress distribution after quenching in water at 20 °C, 40 °C, 60 °C, and 80 °C, respectively, are shown in Figure 10 and Table 1. In Table 1, quenching temperature refers to the temperature of the water in which the quenching is carried out. The temperature selected is the commonly used quenching temperature of martensitic steel. The quenching residual stress at the four temperatures presents a symmetrical distribution. The residual stress value is higher when the quenching temperature is 20 °C. The surface stress is 440 MPa and the heart stress is −442 MPa. When quenching temperature is 60 °C, surface and core residual stresses are 335 MPa and −340 MPa. When the quenching temperature rises to 80 °C, the residual stress of the sample does not exceed 260 MPa. The results indicate that as the quenching temperature decreases, the temperature difference between the specimen and the cooling medium increases. Consequently, lowering the cooling rate can lead to a reduction in the residual stress of martensitic steel.

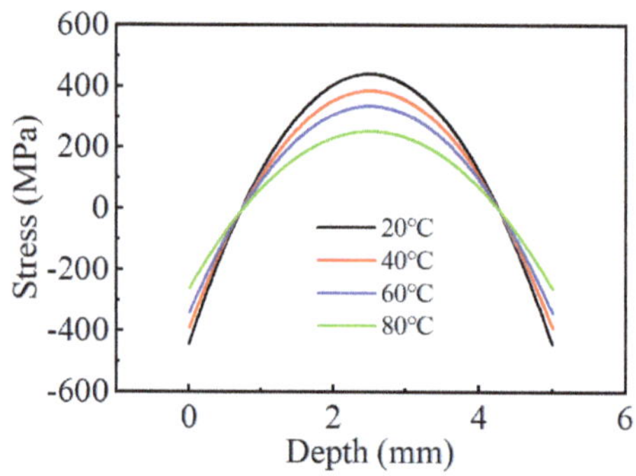

Figure 10. Influence of quenching temperature on residual stress.

Table 1. The influence of quenching temperatures on residual stress.

Quenching Temperatures (°C)	20	40	60	80
Surface residual stresses (MPa)	−442	−390	−340	−260
Core residual stresses (MPa)	439	384	334	251

Figure 11 illustrates the evolution of temperature and stress at the surface and core of the sample following water quenching at 20 °C and 60 °C. When the sample is initially submerged in water, the cooling rate at 60 °C is lower than that at 20 °C. According to the stress evolution curves, both quenching temperatures result in tensile residual stress at the surface and compressive residual stress at the core upon entering the water. As quenching time progresses, the stress at the surface transitions from tensile to compressive, while the core stress shifts from compressive to tensile.

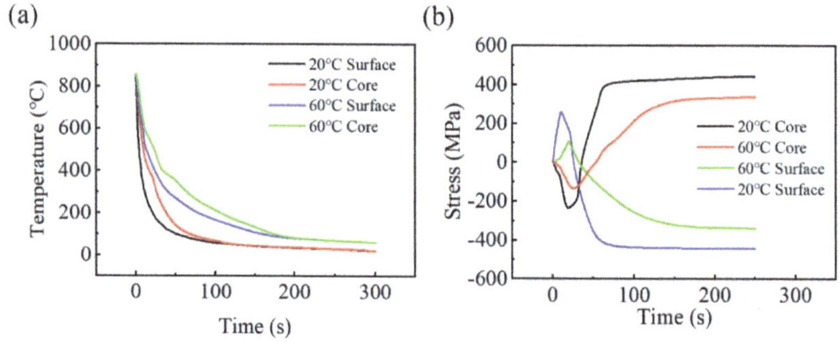

Figure 11. Evolution of surface and core temperature and stress fields during quenching at different temperatures: (**a**) temperature field; (**b**) stress field.

Figure 12 and Table 2 present the impact of varying transition starting temperatures on residual stress. The peaks of tensile and compressive stresses are influenced by the transition starting temperature. As the transition starting temperature rises, the stress levels gradually increase. Consequently, the peak of tensile stress shifts towards the center, while the phase change stress decreases. The combination of thermal stress and phase change stress contributes to an increase in surface compressive stress. Thus, lowering the starting temperature for the martensitic transformation is advantageous for reducing residual stress.

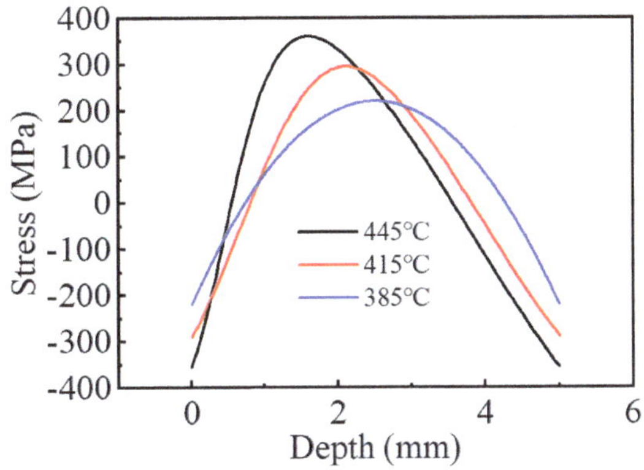

Figure 12. Effect of martensitic transformation starting temperature on residual stress.

Table 2. The influence of Ms on the residual stress.

Starting temperature of martensitic transformation (°C)	445	415	385
Finishing temperature of martensitic transformation (°C)	270	240	210

Residual stress is inevitably generated during phase transition. Tempering can effectively reduce the residual stress. Tempering temperature refers to the temperature at which the metal is heated after tempering. The residual stress at 320 °C, 350 °C, and 380 °C without tempering is calculated. The evolution of residual stress under four conditions is shown in Figure 13 and Table 3. In Table 3, the tempering temperature is understood as the heating temperature of the material. Although the overall stress distribution trend is not obvious after tempering, the stress after tempering is lower than that before tempering. Tempering reduces the internal stress of martensitic steel, especially the residual stress with uneven distribution. The residual stress decreases with the increasing tempering temperature.

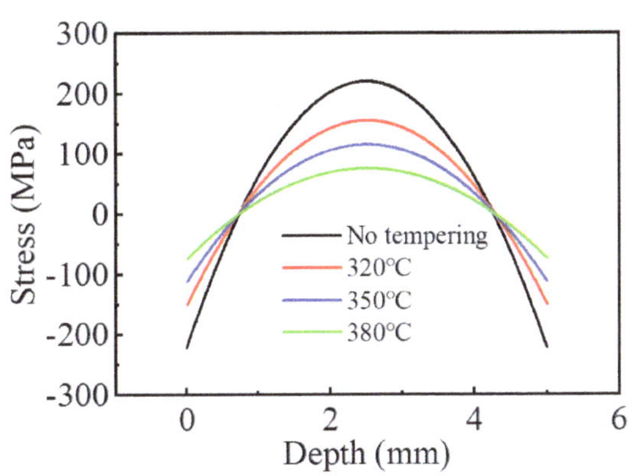

Figure 13. Residual stress at different tempering temperatures.

Table 3. The influence of tempering temperatures on the residual stress.

Tempering Temperatures (°C)	No Tempering	320	350	380
Surface residual stresses (MPa)	−221	−150	−112	−74
Core residual stresses (MPa)	220	154	114	74

4. Conclusions

A method of residual stress prediction considering phase transition is proposed. The measures to reduce residual stress are discussed in terms of quenching medium, cooling rate, and the initial temperature and tempering temperature of the martensitic transformation by finite element simulation. The residual stress generated after quenching varies depending on the type of cooling medium used. In the case of air cooling, the residual stress in martensitic steel is lower, and the distribution of this stress is more uniform. An investigation into the residual stress of martensitic steel following tempering at 320 °C, 350 °C, and 380 °C reveals that increasing the tempering temperature leads to a reduction in residual stress. The optimal tempering temperature suggested is 380 °C. The findings indicate that lowering the initial temperature of phase transition helps to decrease residual stress. It is recommended to set the initial phase transition temperature at 385 °C. The reliability of the model is highly dependent on the material constitutive relationship and experimental data, but in some cases, the relevant parameters are difficult to obtain or inaccurate, which affects the prediction results. A combination of macro-scale and micro-scale modeling methods is used to more accurately capture stress distribution and phase transition processes inside the material. In order to further promote the application of martensitic steel and other types of steel, the residual stress forms of different alloy components and heat treatment processes on other steels such as low-alloy steel and high-strength steel and their effects on material properties were studied, and optimized treatment schemes were sought. The effects of different cooling rates, temperatures, and media on the residual stress and microstructure of martensitic steel and other steels during phase transformation were systematically studied in order to explore the best heat treatment process. Long-term fatigue tests and environmental stress tests are carried out to evaluate the performance and potential failure mode of martensitic steel under actual working conditions, providing a more reliable reference for industrial applications.

Author Contributions: Y.C. and G.Y. contributed to the study conception and design. Finite element modeling and data collection and analysis were performed by G.Y. The first draft of the manuscript was written by Y.C. and all authors commented on previous versions of the manuscript. All authors have read and agreed to the published version of the manuscript.

Funding: This work was supported by the Fundamental Research Funds for the Central Universities (N2423003).

Data Availability Statement: The raw data required to reproduce these findings cannot be shared at this time as the data also form part of an ongoing study.

Conflicts of Interest: Author Yuzheng Cui was employed by the company Hangzhou VOD Intelligent Technology Co., Ltd. The remaining authors declare that the research was conducted in the absence of any commercial or financial relationships that could be construed as a potential conflict of interest.

References

1. Pant, B.; Sundar, R.; Kumar, H. Studies towards development of laser peening technology for martensitic stainless steel and titanium alloys for steam turbine applications. *Mater. Sci. Eng. A* **2013**, *587*, 352–358. [CrossRef]
2. Jiang, T.; Sun, J.; Wang, Y.; Liu, H.; Guo, S.; Liu, W.; Liu, Y. Strong grain-size effect on martensitic transformation in high-carbon steels made by powder metallurgy. *Powder Technol.* **2020**, *363*, 652–656.

3. Zubairuddin, M.; Albert, S.; Vasudevan, M. Thermomechanical analysis of preheat effect on grade P91 steel during GTA welding. *Mater. Manuf. Process.* **2016**, *31*, 366–371.
4. B'eres, L.; Irmer, W.; Balogh, A. Inconsistency of classification of creep resistant steels in European standard EN 288–3. *Sci. Technol. Weld. Join.* **2013**, *2*, 236–238.
5. Kim, H.; Lee, M.; Yoon, S.; Vucko, F.; Lee, C.; Theirry, D. Diffusible hydrogen behavior and delayed fracture of cold rolled martensitic steel in consideration of automotive manufacturing process and vehicle service environment. *J. Mater. Res. Technol.* **2020**, *9*, 13483–13501.
6. Boyle, E.; Northwood, D.; Bowers, R.; Sun, X.; Bauerle, P. Microstructural effects on residual stress, retained austenite, and case depth of carburized automotive steels. *SAE Int. J. Mater. Manuf.* **2009**, *1*, 697–708.
7. Denis, S.; Archambault, P.; Gautier, E.; Simon, A.; Beck, G. Prediction of residual stress and distortion of ferrous and non-ferrous metals: Current status and future developments. *J. Mater. Eng. Perform.* **2002**, *11*, 92–102.
8. Julien, R.; Velay, V.; Vidal, V.; Dahan, Y.; Forestier, R.; R'ezaï-Aria, F. Characterization and modeling of forged Ti-6Al-4V Titanium alloy with microstructural considerations during quenching process. *Int. J. Mech. Sci.* **2018**, *142*, 456–467.
9. Denis, S.; Farias, D.; Simon, A. Mathematical model coupling phase transformations and temperature evolutions in steels. *ISIJ Int.* **1992**, *32*, 316–325.
10. Koistinen, D.; Marburger, R. A general equation prescribing the extent of the austenite-martensite transformation in pure iron-carbon alloys and plain carbon steels. *Acta Metall.* **1959**, *7*, 59–60.
11. Abdollahpoor, A.; Chen, X.; Pereira, M. Sensitivity of the final properties of tailored hot stamping components to the process and material parameters. *J. Mater. Process. Technol.* **2016**, *228*, 125–136. [CrossRef]
12. Teixeira, J.; Marechal, D.; Wimpory, R.; Denis, S.; Lefebvre, F.; Frappier, R. Formation of residual stresses during quenching of Ti17 and Ti-6Al-4V alloys: Influence of phase transformations. *Mater. Sci. Eng. A.* **2022**, *832*, 142456. [CrossRef]
13. Denis, S.; Sjostr, S.; Simon, A. Coupled temperature, stress, phase transformation calculation. *Metall. Mater. Trans. A* **1987**, *18*, 1203–1212. [CrossRef]
14. Bhadeshia, H. Developments in martensitic and bainitic steels: Role of the shape deformation. *Mater. Sci. Eng. A* **2004**, *378*, 34–39. [CrossRef]
15. Zhang, Y.; Zhan, D.; Qi, X.; Jiang, Z. Effect of tempering temperature on the microstructure and properties of ultrahigh-strength stainless steel. *J. Mater. Sci. Technol.* **2019**, *35*, 1240–1249. [CrossRef]
16. Emanuelli, A.; Molinari, L.; Facchini, E.; Sbettega, S.; Carmignato, M.; Bandini, M. Effect of heat treatment temperature and turning residual stresses on the plain and notch fatigue strength of Ti-6Al-4V additively manufactured via laser powder bed fusion. *Int. J. Fatigue* **2022**, *162*, 107009. [CrossRef]
17. Denis, S.; Gautier, E.; Sjostr, S.; Simon, A. Influence of stresses on the kinetics of pearlitic transformation during continuous cooling. *Acta Metall.* **1987**, *35*, 1621–1632. [CrossRef]
18. Chobaut, N.; Carron, D.; Saelzle, P.; Drezet, J. Measurements and modeling of stress in precipitation-hardened aluminum alloy AA2618 during gleeble interrupted quenching and constrained cooling. *Metall. Mater. Trans. A* **2016**, *47*, 5641–5649. [CrossRef]
19. Gaurav, D.; Chandan, P. Experimental investigation on microstructure, mechanical properties, and residual stresses of dissimilar welded joint of martensitic P92 and AISI 304L austenitic stainless steel. *Int. J. Press. Vessel. Pip.* **2021**, *194*, 104536.
20. Rae, W. Thermo-metallo-mechanical modelling of heat treatment induced residual stress in Ti–6Al–4V alloy. *Mater. Sci. Technol.* **2019**, *35*, 747–766. [CrossRef]
21. Tan, P.; Shen, F.; Li, B.; Zhou, K. A thermo-metallurgical-mechanical model for selective laser melting of Ti6Al4V. *Mater. Des.* **2019**, *168*, 107642. [CrossRef]
22. Teixeira, J.; Denand, B.; Aeby, E.; Denis, S. Simulation of coupled temperature, microstructure and internal stresses evolutions during quenching of a β-metastable titanium alloy. *Mater. Sci. Eng. A* **2016**, *651*, 615–625. [CrossRef]
23. Kumar, S.; Yadav, V.; Sharma, S.; Pandey, C.; Goyal, A.; Kumar, P. Role of dissimilar Ni-based ERNiCrMo-3 filler on the microstructure, mechanical properties and weld induced residual stresses of the ferritic/martensitic P91 steel welds joint. *Int. J. Press. Vessel. Pip.* **2021**, *193*, 104443. [CrossRef]
24. Ahn, J.; He, E.; Chen, L.; Wimpory, R.; Dear, J.; Davies, C. Prediction and measurement of residual stresses and distortions in fibre laser welded Ti-6Al-4V considering phase transformation. *Mater. Des.* **2017**, *115*, 441–457. [CrossRef]
25. Wang, X.; Hu, L.; Chen, D. Effect of martensitic transformation on stress evolution in multi-pass butt-welded 9% Cr heat-resistant steel pipes. *Acta Metall. Sin.* **2017**, *53*, 888–896.
26. Zubairuddin, M.; Albert, S.; Mahadevan, S. Experimental and finite element analysis of residual stress and distortion in GTA welding of modified 9Cr-1Mo steel. *J. Mech. Sci. Technol.* **2014**, *28*, 5095–5105.
27. Deng, D.A.; Murakawa, H. Prediction of welding residual stress in multi-pass buttwelded modified 9Cr–1Mo steel pipe considering phase transformation effects. *Comput. Mater. Sci.* **2006**, *37*, 209–219. [CrossRef]
28. Ahmed, T.; Rack, H. Phase transformations during cooling in α+β titanium alloys. *Mater. Sci. Eng. A* **1998**, *243*, 206–211. [CrossRef]

29. Chen, S.; Zhang, Y.; Wu, Q.; Gao, H.; Gao, Z.; Li, X. Effect of solid-state phase transformation on residual stress of selective laser melting Ti6Al4V. *Mater. Sci. Eng. A* **2021**, *819*, 141299. [CrossRef]
30. Beres, L.; Balogh, A.; Irmer, W. Welding of martensitic creep-resistant steels. *Weld. J.* **2001**, *80*, 191–195.
31. Goerres, G.; Revesz, T.; Duncan, J. The deformation of metals under small stresses during phase transformations. *Proc. R. Soc. Lond.* **1965**, *283*, 403–422.
32. Leblond, J. Mathematical modelling of transformation plasticity in steels II: Coupling with strain hardening phenomena. *Int. J. Plast.* **1989**, *5*, 573–591.
33. Leblond, J.; Devaux, J.; Devaux, J. Mathematical modelling of transformation plasticity in steels I: Case of ideal-plastic phases. *Int. J. Plast.* **1989**, *5*, 551–572.
34. Liu, Y. *Determination and Finite Element Simulation of Quench Stress Distribution in Medium Carbon Steel*; Shanghai Jiao Tong University: Shanghai, China, 2017.
35. Liu, Y.; Qin, S.; Zhang, J. Influence of Transformation Plasticity on the Distribution of Internal Stress in Three Water-Quenched Cylinders. *Metall. Mater. Trans. A* **2017**, *48*, 4943–4956. [CrossRef]
36. Yang, G.; Xia, S.L.; Zhang, F.C.; Branco, R.; Long, X.Y.; Li, Y.G.; Li, J.H. Effect of tempering temperature on monotonic and low-cycle fatigue properties of a new low-carbon martensitic steel. *Mater. Sci. Eng. A* **2021**, *826*, 141939. [CrossRef]

crystals

MDPI

Article

Constitutive Equation and Heat Distortion Behavior of TA4 Titanium Alloy

Lifeng Ma [1], Wenshuai Liu [1], Yanchun Zhu [1,*], Ling Qin [2] and Jingfeng Zou [1]

1 School of Mechanical Engineering, Taiyuan University of Science and Technology, Taiyuan 030024, China; s202212210119@stu.tyust.edu.cn (W.L.); zjf_rftyust@163.com (J.Z.)
2 Department of Petroleum Engineering, University of Wyoming, Laramie, WY 82071, USA; lqin1@uwyo.edu
* Correspondence: lzlzyc@163.com

Abstract

In this study, the high-temperature thermal deformation behavior of the TA4 alloy was investigated by thermal compression experiments. The effects of deformation temperature and strain rate on the rheological stress are described by analyzing the variation of stress–strain curves with different parameters and establishing the constitutive equation based on the dynamic material theory model. Thermal processing diagrams were established and plotted to analyze the optimal processing zone and the destabilization zone under different strains. From the thermal machining diagram, it can be concluded that the optimum machining zone at a strain of 0.9 is 1040~1133 K/0.01~0.7 s^{-1}. The optimum machining zone at a strain of 0.6 is 940~1000 K/0.01~0.04 s^{-1}. The optimum machining zone at a strain of 0.3 is 940~1000 K/0.01~0.08 s^{-1}. The effects of different deformation conditions on the thermal deformation mechanism were analyzed in conjunction with EBSD characterization. The results showed that dynamic recrystallization (DRX) was the main deformation softening mechanism when at low strain rate ($\leq$0.1 s^{-1}). At higher strain rates (>0.1 s^{-1}) and lower temperatures (<1083 K and $\geq$933 K), the main deformation softening mechanism was DRV; at higher temperatures ($\geq$1083 K and $\leq$1133 K), the main deformation softening mechanism was DRX.

Keywords: TA4 titanium alloy; hot deformation behavior; hot working diagram; recrystallize

1. Introduction

Titanium, known as the "third metal" after iron and aluminum due to its abundant reserves in the Earth's crust, has garnered significant attention. Titanium alloys, owing to their excellent comprehensive properties such as high specific strength, low density, low thermal expansion coefficient, high biocompatibility, and high corrosion resistance, are widely used in aerospace, petrochemical, automotive, shipbuilding, and biomedical fields [1–4]. In recent years, titanium alloys have gained strong momentum in the civil market, particularly as materials for consumer electronics. They have already been commercialized in electronic fabrication and aerospace industrial applications [5].

During the hot deformation process, the deformability of titanium alloys is influenced by both deformation parameters and material microstructure. Therefore, selecting appropriate deformation parameters such as strain rate, strain, and deformation temperature is crucial for the processing and mechanical performance of the final product [6,7]. These parameters interact with each other and jointly influence the hot mechanical processing of titanium alloys [8]. A constitutive model can be developed to represent the relationship

 https://doi.org/10.3390/cryst15040290

between stress and strain under different loads, as well as to predict the hot processing performance under various deformation parameters. Additionally, the use of hot working diagrams allows for a more accurate determination of the material's processing window, providing guidance for the optimization of deformation parameters during hot processing. Hao et al. [9] conducted hot compression experiments on near-α titanium alloys, studying their hot deformation behavior, microstructural evolution, and hot working diagrams; the flow behavior and flow softening mechanism evolution of the deformation under different phase fields are obtained. Wang et al. [10] used isothermal compression testing on the Ti-55511 titanium alloy to investigate the effects of various hot deformation parameters on the alloy's properties and microstructural evolution; the heat deformation behavior and microstructure evolution of the Ti-55511 titanium alloy in the range of experimental parameters were obtained and the peak stress could be predicted. Lu et al. [11] analyzed the flow stress–strain curves of the TA10 titanium alloy under different strain rates and temperatures, developed a hot plasticity constitutive equation for pure titanium alloy based on a hyperbolic sine Arrhenius equation, and elucidated the flow-softening mechanism of this alloy. The evolution of the microstructure during the processing, such as the solidification process, subsequent heat treatment [12], thermal deformation behavior, dynamic recrystallization, and dynamic recovery [13], inevitably affects the machinability of the final product. Therefore, studying the influence of different deformation parameters on microstructural evolution during hot processing is essential for further exploring suitable manufacturing processes for the TA4 titanium alloy [14–16].

The selection of thermal processing parameters for the TA4 titanium alloy is investigated in this study based on real production requirements, with a focus on industrially relevant temperature ranges (933–1133 K) and strain rates (0.01–10 s^{-1}). Hot compression tests were conducted on the alloy, and a hot plasticity constitutive equation was developed by integrating the hyperbolic sinusoidal Arrhenius equation with the dynamic material theory model. Thermal processing diagrams were then generated for different strain levels. Subsequently, optimal processing zones for the TA4 alloy were identified based on these diagrams. The effects of various parameters on the microstructural evolution, including dynamic recrystallization and dynamic recovery, were analyzed using Electron Backscatter Diffraction (EBSD). By examining the thermal deformation mechanisms, the optimal processing parameters for achieving the desired microstructure and properties were determined. This study provides valuable theoretical and practical insights into the manufacturing process, microstructure customization, and enhanced thermal processing techniques for the TA4 titanium alloy.

2. Materials and Methods

2.1. Experimental Materials

The experimental material used in this study is commercial-grade TA4 titanium wire (GB/T 3620.1) [17]. The titanium wire was processed through multiple vacuum remelting, followed by forging and high-line rolling. The TA4 titanium alloy is an α-type titanium alloy. Figure 1 is the metallographic microstructure of the alloy after full annealing, it can be seen that its micro structure after annealing is almost entirely equiaxial α-phase, the intergranular position will have a small amount of β-phase (dark), the distribution of grains is uniform, and the average grain size is about 10 μm. Table 1 also provides the elemental composition and content of the TA4 titanium alloy used in the experiment.

Table 1. Standard chemical composition (wt%) of the TA4 titanium alloy according to GB/T 3620.1.

Fe	C	N	H	O	Ti
0.1500	0.0500	0.0070	0.0021	0.3600	Bal.

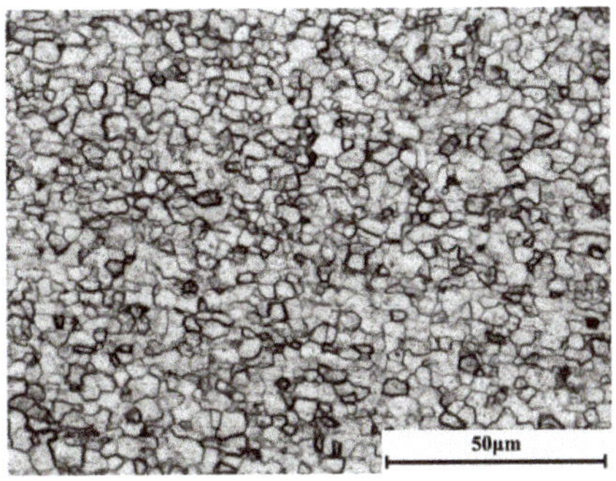

Figure 1. Microstructure of the TA4 titanium alloy (LM).

2.2. High-Temperature Thermal Compression Test

Hot compression tests were conducted using a Gleeble-3800 thermomechanical simulator (Dynamic Systems Inc., Poughkeepsie, NY, USA). The samples used in the experiments were TA4 alloy rods with dimensions of ϕ8.0 mm × 12.0 mm. The experimental procedure is shown in Figure 2a. The deformation temperatures of the samples were set at 933, 983, 1033, 1083, and 1133 K, with a deformation degree of 60%. The strain rates were 0.01, 0.1, 1, and 10 s^{-1}. During the experiment, the samples were heated on the Gleeble-3800 simulation tester at a rate of 10 K/s to the specified compression deformation temperature and held at that temperature for 3 min to ensure uniform temperature distribution within the sample. The sample was then compressed to 60% strain at the specified constant strain rate, and water cooling was applied after the compression. At the end of the experiment, the Gleeble-3800 analog testing machine recorded the load and displacement during compression by means of a force transducer and a displacement transducer and generated a stress–strain curve.

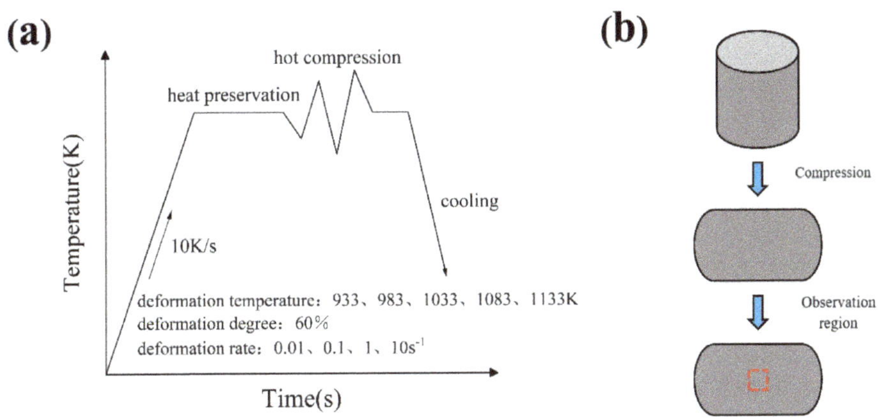

Figure 2. Schematic diagram of thermal compression test of TA4 titanium alloy: (**a**) processing parameters; (**b**) compression diagram.

To minimize the influence of friction on the experiment, lubricant was applied to both ends of the sample in contact with the Gleeble-3800 simulator, and graphite sheets were attached. Additionally, the chamber was vacuumed and protected with gas before the experiment. To maintain uniform temperature during the entire hot compression process, a pair of S-type Pt/Rh thermocouples were spot-welded at the midpoint along the height of the sample. The force and displacement sensors in the simulator actively collected the load and displacement during deformation, and the internal program automatically converted these measurements into true stress–strain curves.

2.3. Microstructural Characterization

To examine the microstructure of the specimen after thermal deformation, the compressed specimen was cut along its axis. The central region of the cross-section, as shown in Figure 2b, was selected for observation. The cut specimens were ground with sandpaper and polished using silica abrasive and suspension. The specimens used for metallographic observation were etched with Kroll's reagent and the prepared specimens were subsequently analyzed under a metallurgical light microscope (LM; Eclipse MA200, Nikon, Hong Kong, China) to observe the microstructure. To achieve the surface quality required for the EBSD examinations, the samples for EBSD research were electrolytically polished in a solution of 10% perchloric acid and 90% methanol, with a voltage of 20 V, a temperature of 30 °C, and a time of 30 s. The EBSD research was conducted using a Gemini SEM 500 field-emission scanning electron microscope equipped with EBSD Aztec Symmetry. The instrument operating voltage was 20 kV, the scanning step size was 0.15 μm. Finally, the subgrain boundary structure, grain morphology and local orientation were analyzed by Aztec Crystal software (version 6.1, Oxford Instruments, Abingdon, UK).

3. Results

3.1. Flow Stress Behavior

Stress–strain curves are fundamental in materials science and engineering, as they offer crucial insights into a material's mechanical behavior under various loading conditions [18].

The hot deformation behavior of titanium alloys is commonly analyzed using true stress–strain curves. The true stress–strain curves of the TA4 titanium alloy under different deformation conditions are shown in Figure 3. From the figure, it can be observed that the flow stress of the TA4 titanium alloy generally increases rapidly at first and then stabilizes under different stress–strain curves. This is because, in the initial stage of compression, the alloy undergoes intense work hardening, leading to a rapid increase in flow stress, which is reflected in the continuous rapid increase in flow stress on the stress–strain curve. As compression time increases, softening mechanisms such as dynamic recovery and dynamic recrystallization begin to take effect. When a stable state is reached, these two effects counteract each other, resulting in the flow stress stabilizing [19–21].

Figure 4 shows the peak stress values of the TA4 titanium alloy at different deformation parameters, in which it can be seen that the peak stress of the material decreases significantly with increasing deformation temperature and decreasing strain rate. From the analysis, it was found that firstly, the strain rate is proportional to the peak stress at the same deformation temperature. This is because the compression time of the material becomes shorter when the strain rate increases. When the deformation time of the material is not enough to allow the softening mechanisms such as dynamic restitution and dynamic recrystallization within the alloy to be fully activated, this is manifested by an increase in peak stress. Secondly, at lower temperatures (below 1033 K), the increase in peak stress with increase in strain rate is relatively modest, 113% versus 190%, respectively. This indicates that the change in strain rate at low temperatures has less effect on the heat distortion process of the TA4 titanium alloy. At higher temperatures (above 1033 K), the increase in peak stress rises to 304%, 296%, and 340%. This indicates that the thermal deformation process of the material becomes more sensitive to the strain rate with increasing temperatures. This is due to the fact that the higher the deformation temperature of the alloy, the higher the dislocation activity of the alloy, resulting in a higher deformation storage energy. At the same time, the nucleation and growth of dynamically recrystallized grains are more rapid with increasing temperature, leading to an increase in the softening mechanism of the alloy and a consequent decrease in its resistance to deformation.

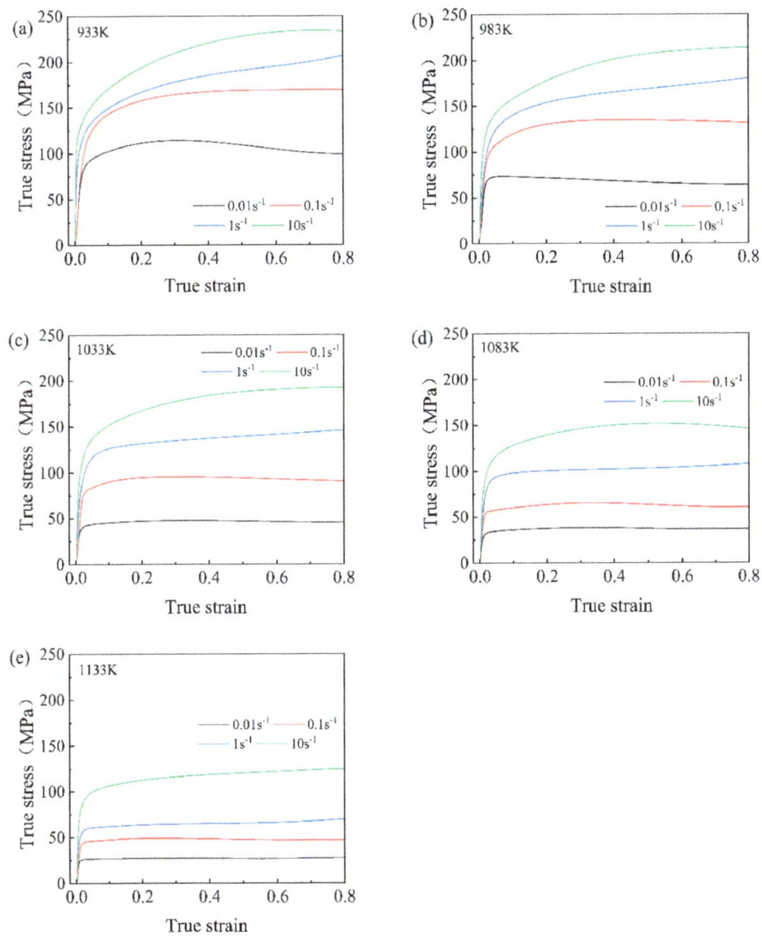

Figure 3. When the strain rate is between 0.01 s−1 and 10 s−1, the true stress–strain curves of the TA4 alloy at different deformation temperatures are as follows: (**a**) 933 K; (**b**) 983 K; (**c**) 1033 K; (**d**) 1083 K; (**e**) 1133 K.

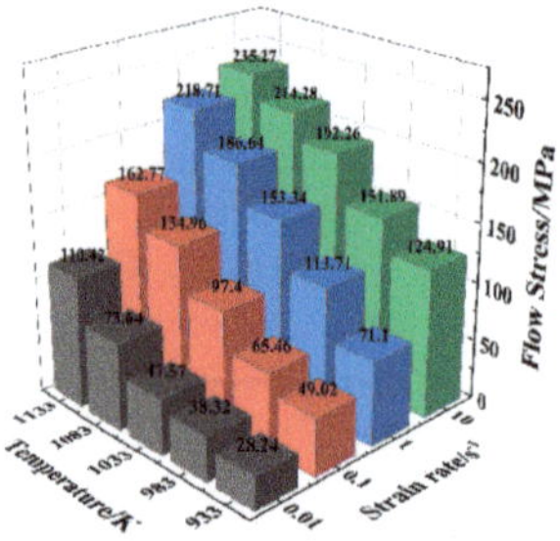

Figure 4. Peak stresses of alloys in different deformation conditions.

3.2. *Derivation of the Constitutive Equation*

To study the relationship between stress, strain, and temperature during the deformation of metallic materials, a constitutive equation must be developed for the material. The Arrhenius model is commonly used to construct the constitutive equation for the hot deformation behavior of titanium alloys. This hyperbolic sine constitutive model is not only widely applicable but also offers high predictive accuracy. Considering these factors, this model was selected to construct the constitutive equation for the TA4 alloy at a strain of 0.5. The mathematical expression of this equation consists of three equations, as shown in Equation (1) [22,23]:

$$\dot{\varepsilon} = \begin{cases} A_1\sigma^{n_1}\exp(-\frac{Q}{RT}) & (\alpha\sigma < 0.8) \\ A_2\exp(\beta\sigma)\exp(-\frac{Q}{RT}) & (\alpha\sigma > 1.2) \\ A_3[\sinh(\alpha\sigma)]^{n_2}\exp\left(-\frac{Q}{RT}\right) & (for\ all) \end{cases} \tag{1}$$

In Equation (1), $\dot{\varepsilon}$ is the strain rate; σ can represent the peak stress, steady state variable rheological stress, or the rheological stress of the specified strain; n_1, n_2 are temperature-independent stress indices; A_1, A_2, A_3, and β are material constants that are also temperature-independent; Q is the apparent activation energy of thermal deformation at different temperatures; T is the thermodynamic temperature at absolute temperature; R is the gas constant in the ideal state; α is a parameter characterizing the pressure level, and the relationship with other parameters is $\alpha = \beta/n$.

It has been shown that the third equation is usually used to construct the constitutive equations for different stress conditions. The hyperbolic sine form was proposed by Sellars and Tegart, which can express the constitutive equations more accurately under different stress conditions. It is well adapted for use and widely used, and is expressed as follows:

$$\dot{\varepsilon} = A_3[\sinh(\alpha\sigma)]^{n_2}\exp\left(-\frac{Q}{RT}\right) \tag{2}$$

The Zener–Hollomon parameter Z [24], which expresses the relationship between the strain rate $\dot{\varepsilon}$ and the absolute temperature T, was used as follows:

$$Z = \dot{\varepsilon}exp\left(\frac{Q}{RT}\right) \tag{3}$$

According to Equations (2) and (3):

$$Z = A_3[\sinh(\alpha\sigma)]^{n} \tag{4}$$

Equation (4), where Z is the temperature-compensated strain rate factor. The meanings of the other letters have been pointed out above.

According to the definition of hyperbolic sine function, the following is true:

$$\sinh^{-1}(\alpha\sigma) = \left[\ln(\alpha\sigma) + \left(\left(\alpha\sigma\right)^2 + 1\right)^{\frac{1}{2}}\right] \tag{5}$$

According to Equations (3)–(5), the following can be obtained:

$$\sigma = \frac{1}{\alpha}\ln\left\{\left(\frac{z}{A_3}\right)^{\frac{1}{n_2}} + \left[\left(\frac{z}{A_3}\right)^{\frac{2}{n_2}} + 1\right]^{\frac{1}{2}}\right\} \tag{6}$$

The natural logarithm can be obtained by solving for each side of Equation (1):

$$ln\dot{\varepsilon} = A_4 + n_1 ln\sigma \tag{7}$$

$$ln\dot{\varepsilon} = A_5 + \beta\sigma \tag{8}$$

$$ln\dot{\varepsilon} = n_2 ln[sinh(\alpha\sigma)] + A_6 \tag{9}$$

The steady-state stress and strain rates during deformation were taken, and according to Equations (7) and (8), the least-squares method was adopted for the linear regression process, and the relationship between $ln\sigma - ln\dot{\varepsilon}$ and $\sigma - ln\dot{\varepsilon}$ was plotted with the help of Origin data analysis software (Version 2023b, OriginLab Corporation, USA), and the results are shown in Figure 5 [25]. The fitting results show that the linear correlation coefficients of $ln\sigma - ln\dot{\varepsilon}$ curves at 933 to 1133 K and $\sigma - ln\dot{\varepsilon}$ curves at 933 to 1133 K are all greater than 0.99, indicating a good fitting reliability.

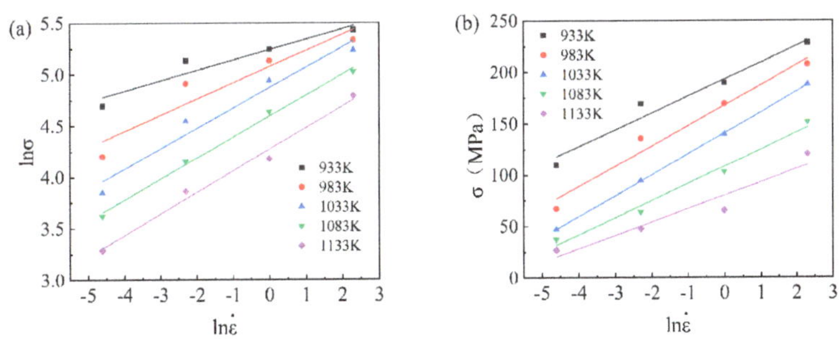

Figure 5. The relationship between rheological stress and strain rate of TA4 alloy: (**a**) $\ln\sigma - \ln\dot{\varepsilon}$; (**b**) $\sigma - \ln\dot{\varepsilon}$.

Taking the average value of the inverse of the slope of the fitted straight line in Figure 5a for 933 to 1133 K as n_1, there is n_1 = 4.9046; taking the average value of the inverse of the slope of the fitted straight line in Figure 5b for 933 to 1133 K as β, there is β = 0.0492 MPa^{-1}; and according to the definition of α, we can obtain $\alpha = \beta/n = 0.0101$ MPa^{-1}.

The strain stress and strain rates at different deformation temperatures were substituted according to Equation (9), the linear regression was fitted by the least-squares method, and the $\ln[\sinh(\alpha\sigma)] - ln\dot{\varepsilon}$ relationship was plotted with the help of Origin data analysis software, as shown in Figure 6. The fitting results show that the linear correlation coefficients of the $\ln[\sinh(\alpha\sigma)] - ln\dot{\varepsilon}$ curves at 933 to 1133 K are all greater than 0.99, indicating that the double logarithmic relationship between the hot deformation rheological stresses and the strain rate satisfies the linear relationship better. The results show that the relationship between the rheological stress and strain rates of the TA4 alloy at a constant deformation temperature can be described by a hyperbolic sine relationship.

A certain temperature range, assuming constant Q at the same strain rate, is obtained according to Equation (9):

$$\ln[\sinh(\alpha\sigma)] = A_7 + \frac{A_8}{T} \tag{10}$$

The strain rate $\dot{\varepsilon}$, the stress σ, and the calculated α value were brought into Equation (10), while the least squares method was taken for the linear regression treatment to plot the $\ln[\sinh(\alpha\sigma)] - \frac{1}{T}$ relationship, as shown in Figure 6. The fitting results show that the linear correlation coefficients of the $\ln[\sinh(\alpha\sigma)] - \frac{1}{T}$ curves are greater than 0.99 at all strain rates, indicating that the relationship between the rheological stress and the temperature of the TA4 alloy at a high temperature and constant strain rate can be described by the hyperbolic sine relationship.

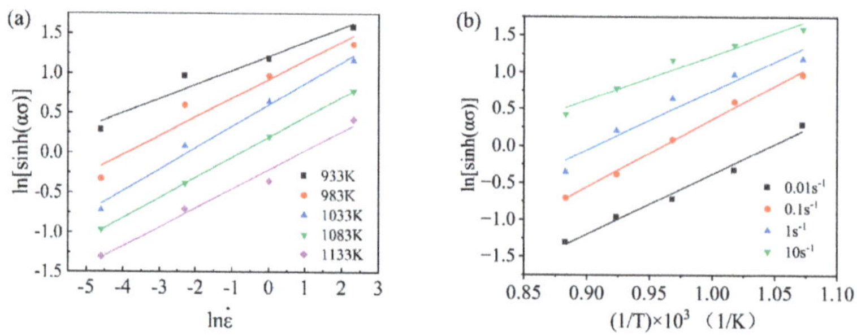

Figure 6. (**a**) $\ln[\sinh(\alpha\sigma)] - ln\dot{\varepsilon}$ fit curve; (**b**) $\ln[\sinh(\alpha\sigma)] - \frac{1}{T}$ fit curve.

Taking the natural logarithm of the derivative on both sides of the third equation of Equation (1), the hot deformation activation energy equation can be obtained as follows:

$$Q = R\left\{\frac{\partial(\ln\dot{\varepsilon})}{\partial\ln[\sinh(\alpha\sigma)]}\right\}_T \left\{\frac{\partial\ln[\sinh(\alpha\sigma)]}{\partial\left(\frac{1}{T}\right)}\right\}_{\varepsilon} \tag{11}$$

According to the previous calculations, the three terms on the right side of Equation (11) are the gas constant in the ideal state, the inverse of the slope of the $\ln[\sinh(\alpha\sigma)] - ln\dot{\varepsilon}$ curve at a certain temperature, and the slope of the $\ln[\sinh(\alpha\sigma)] - \frac{1}{T}$ curve at a certain strain rate, and the values of the latter two are 3.9634 and 9.1436, respectively. Substituting into Equation (11) yields the value of the deformation activation energy Q as 301 KJ/mol.

Taking the logarithm of both sides of Equation (4), the following can be obtained:

$$lnZ = lnA + n\ln[\sinh(\alpha\sigma)] \tag{12}$$

The obtained average deformation activation energy Q value, strain rate, and deformation temperature were substituted into Equation (12), and the plot of the $lnZ - \ln[\sinh(\alpha\sigma)]$ relationship plotted by linear regression using the least squares method is shown in Figure 7. Since the linear correlation coefficient between the variables of the relationship plot is greater than 0.98, the stress index n can be expressed by the slope of this fitted curve, and lnA can be expressed by the intercept. The data processing can obtain $n = 4.2924$ and $A = 1.9653 \times 10^{14}$.

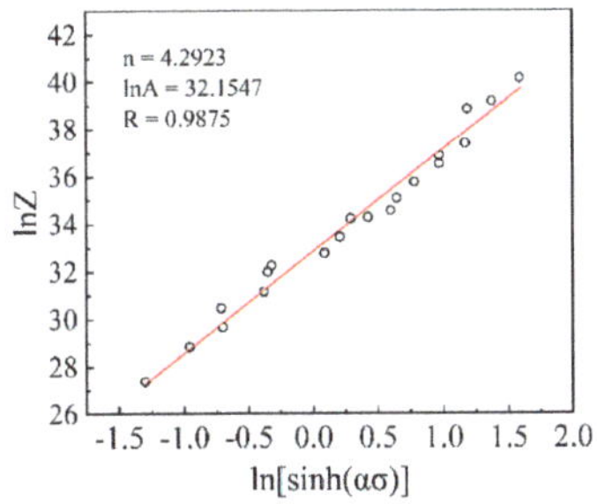

Figure 7. Relationship diagram of lnZ − ln[sinh(ασ)] TA4 alloy.

Combined with the above calculation results, all the material parameters for constructing the constitutive equation of the TA4 alloy have been found, which are as follows: stress level factor $\alpha = 0.0101$ MPa^{-1}, activation energy of deformation $Q = 301$ KJ/mol, stress exponent $n = 4.2923$, and structure factor $A = 1.9654 \times 10^{14}$, respectively. Substituting the above parameters into Equation (2), the eigenstructural equation for a strain of 0.5 temperature range of 933–1133 K can be found as follows:

$$\dot{\varepsilon} = 1.9654 \times 10^{14}[sinh(0.0101\sigma)]^{4.2923}\exp\left(-\frac{301000}{RT}\right) \tag{13}$$

Similarly, the parameter data of Arrhenius model for different strain ranges and different deformation conditions can be calculated by the above method, as shown in Table 2.

Table 2. Arrhenius model parameter.

Material Constant	True Strain							
	0.1	0.2	0.3	0.4	0.5	0.6	0.7	0.8
α	0.0119	0.0112	0.0107	0.0105	0.0105	0.0105	0.0105	0.0105
Q	317.1401	320.5820	315.5531	304.7553	295.4912	283.9743	276.3184	272.3312
n	5.5782	5.1983	4.9145	4.6582	4.4234	4.2371	4.0892	4.0433
lnA	34.2532	34.7864	34.3312	33.1813	32.1540	30.8460	29.9552	29.4720

However, Bani [26] argued that the important effect of strain on this process should be taken into account when analyzing the high temperature deformation of titanium alloys. The results showed that strain has a significant effect on the deformation activation energy and material constants (α, Q, n and lnA). Therefore, in order to investigate the intrinsic modeling of titanium alloys under different strains, the relationship of strain on deformation activation energy and material constants is constructed as a polynomial function as shown in Equation (14). The effect of strain on the constitutive equation is reflected by the application of the polynomial function in the constitutive equation.

$$\begin{cases} \alpha = C_0 + C_1\varepsilon + C_2\varepsilon^2 + C_3\varepsilon^3 + C_4\varepsilon^4 + C_5\varepsilon^5 \\ Q = D_0 + D_1\varepsilon + D_2\varepsilon^2 + D_3\varepsilon^3 + D_4\varepsilon^4 + D_5\varepsilon^5 + D_6\varepsilon^6 \\ n = E_0 + E_1\varepsilon + E_2\varepsilon^2 + E_3\varepsilon^3 + E_4\varepsilon^4 + E_5\varepsilon^5 + E_6\varepsilon^6 \\ lnA = F_0 + F_1\varepsilon + F_2\varepsilon^2 + F_3\varepsilon^3 + F_4\varepsilon^4 + F_5\varepsilon^5 + F_6\varepsilon^6 \end{cases} \tag{14}$$

According to Equation (14), the correlations of $\alpha - \varepsilon$, $n - \varepsilon$, $Q - \varepsilon$, and $lnA - \varepsilon$ are plotted, and the fitted curves are shown in Figure 8. From the figure, it can be found that the correlation coefficients of each fitted curve of strain versus material constants are greater than 0.99, which has a good reliability. It indicates that this polynomial function can be used to describe the strain-to-material constant relationship during high-temperature deformation of titanium alloys.

The data obtained from the experiments were substituted into Equation (14) and the equations were fitted using a nonlinear fitting algorithm to calculate the model material constants, the results of which are shown in Table 3.

As a result, the Arrhenius model for strain compensation can be obtained, and the comparison between the predicted and experimental values of this model is shown in Figure 9.

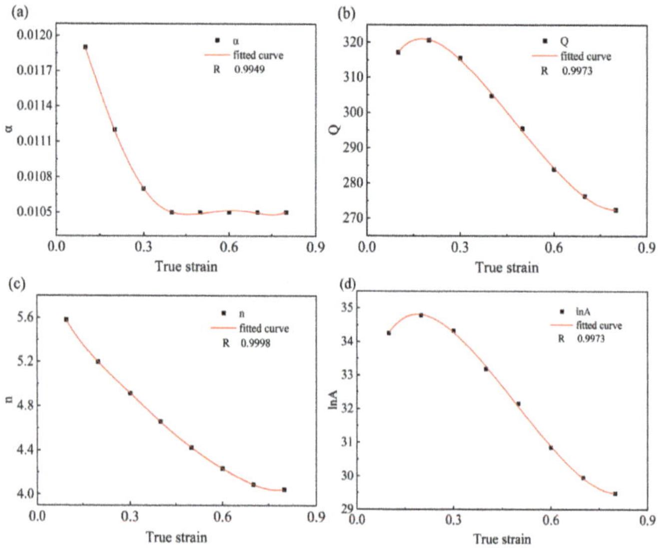

Figure 8. Fitting the variation curve with (**a**) α, (**b**) Q, (**c**) n, and (**d**) ln A polynomials.

Table 3. Polynomial coefficient results for α, Q, n, ln A.

	Polynomial Coefficient						
	C_0	C_1	C_2	C_3	C_4	C_5	
$\alpha = f(\varepsilon)$	0.0124	−0.0017	−0.0493	0.1819	−0.2362	0.1058	
	D_0	D_1	D_2	D_3	D_4	D_5	D_6
$Q = f(\varepsilon)$	303.546	168.836	−132.412	−2689.08	8011.976	−9079.71	3772.389
	E_0	E_1	E_2	E_3	E_4	E_5	E_6
$n = f(\varepsilon)$	6.1858	−7.9749	23.8815	−56.1106	73.0744	−48.186	13.5021
	F_0	F_1	F_2	F_3	F_4	F_5	F_6
$lnA = f(\varepsilon)$	32.5454	21.0176	−17.966	−292.158	869.1344	−981.078	406.9907

3.3. *Evaluation of the Ontological Model*

In order to be able to accurately evaluate the reliability of the predicted results of these constitutive equations, it is necessary to analyze the error between the experimental values and the predicted values of the constitutive equations. Here, the correlation coefficient (R), mean absolute percentage error (MAPE), and mean root mean square error (RMSE) are introduced to analyze the error numerically with the following expressions [27,28]:

$$R = \frac{\sum_{i=1}^{N}(E_i - \overline{E})(P_i - \overline{P})}{\sqrt{\sum_{i=1}^{N}(E_i - \overline{E})^2 \sum_{i=1}^{N}(P_i - \overline{P})^2}} \quad (15)$$

$$MAPE = \frac{l}{N}\sum_{i=1}^{N}\left|\frac{E_i - P_i}{E_i}\right| \times 100\% \tag{16}$$

$$RMSE = \sqrt{\frac{I}{N}\sum_{i=1}^{N}(E_i - P_i)^2} \tag{17}$$

In the above equation, E_i is the experimental stress value; P_i is the predicted stress value; $\overline{E}$ is the average of experimental and predicted values; N is the total amount of data. In general, the strength of the linear correlation between the model predicted data and the experimental measured data can be characterized by the correlation coefficient (R). However, for this intrinsic model, it is not always valid to rely solely on the correlation coefficient to characterize the correlation between the model data and the experimental data. Therefore, the *MAPE* and the *RMSE* need to be calculated here to accurately characterize the accuracy of the constitutive equations with unbiased statistical parameters. In order to explore the generalization ability of the proposed model, the corresponding *MAPE* and *RMSE* values were calculated and the linear relationship between the experimental values of the flow stresses and the predicted values of the model were plotted as shown in Figure 10.

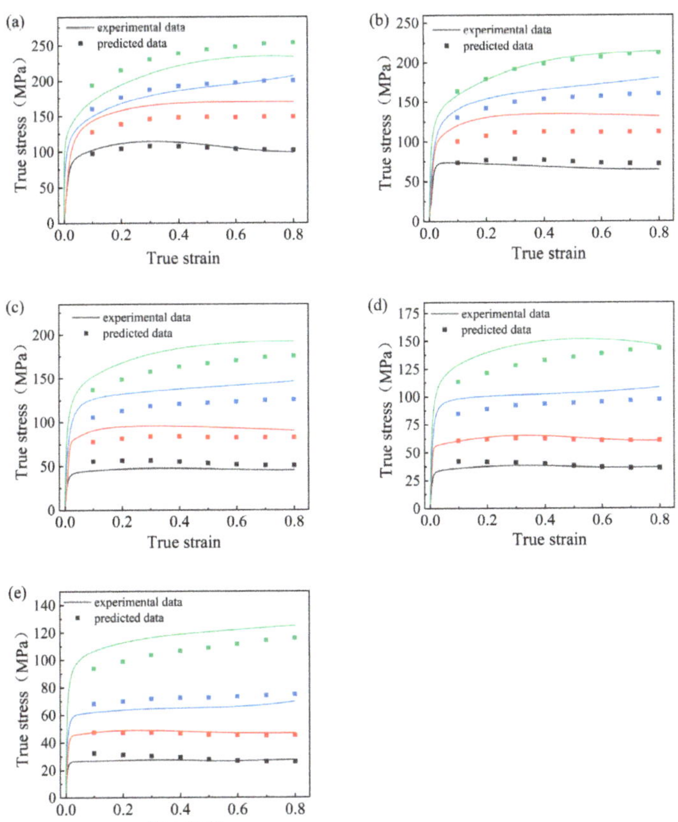

Figure 9. Comparison between experimental stress values and predicted stress values: (**a**) 933 K; (**b**) 983 K; (**c**) 1033 K; (**d**) 1083 K; (**e**) 1133 K.

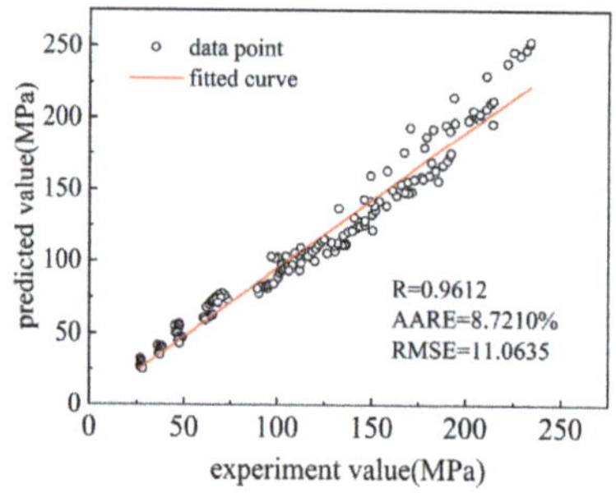

Figure 10. The correlation between experimental and predicted values.

From the figure, it can be seen that the *MAPE* and *RMSE* values of the model are 8.7210% and 11.0635 MPa, respectively. The data values at the whole temperature are around the optimal regression line, which can indicate that the intrinsic model has a high accuracy. This indicates that the developed TA4 Arrhenius eigen structural equation has good accuracy and its correlation is high enough to be used in the calculation of deformation resistance for the actual rolling process [29].

3.4. *Creation and Analysis of Thermal Processing Diagrams*

In actual production, selecting the appropriate thermal processing parameters for the TA4 alloy requires not only constructing the constitutive equations but also establishing the corresponding thermal processing diagram to visualize the material's processing and forming properties. The hot working diagram can delineate the safety zone and rheological instability zone of plastic processing, which can provide guidance for the selection of hot working process parameters and the control of microstructure and properties of metal materials.

Based on Prasad's [30] instability criterion and its dynamic material model, the external energy P is mainly consumed in the dissipation G required for plastic deformation and the co-dissipation J required for microstructure evolution, which is given by Equation (18) as follows:

$$P = \sigma\dot{\varepsilon} = G + J = \int_0^{\varepsilon} \sigma d\dot{\varepsilon} + \int_0^{\sigma} \dot{\varepsilon} d\sigma \tag{18}$$

m is the strain rate sensitivity factor with the following Equation (19); η is the power dissipation factor of the processed material with the following Equation (20):

$$m = \frac{\partial \ln \sigma}{\partial \ln \dot{\varepsilon}} \tag{19}$$

$$\eta = 2\left(1 - \frac{1}{1+\mathrm{m}}\right) = \frac{2\mathrm{m}}{1+\mathrm{m}} \tag{20}$$

When the material presents an ideal linear dissipative state, the value of the strain rate sensitivity index m is 1, at which time the dissipation coefficient J reaches its maximum value J_{max}, expressed as follows:

$$J_{max} = \frac{\sigma\dot{\varepsilon}}{2} \tag{21}$$

Therefore, the energy dissipation rate η can be expressed as follows:

$$\eta = \frac{J}{J_{max}} = \frac{2m}{m+1} \tag{22}$$

In the above equation, m is the strain rate sensitivity index; η is the energy dissipation rate.

ξ is the destabilization parameter; when the destabilization parameter is less than 0, to judge the occurrence of rheological destabilization, Equation (21) is as follows:

$$\xi(\dot{\varepsilon}) = \frac{\partial \ln\left(\frac{m}{m+1}\right)}{\partial \ln \dot{\varepsilon}} + m < 0 \tag{23}$$

Different strains of the rheological instability parameter ξ value can be plotted as a contour map of the hot working diagram—that is, the instability map—and superimposed with the energy dissipation map. The contour lines in the map represent the values of the energy dissipation efficiency factor; the shaded area in the map represents the part where the instability parameter ξ is less than zero, i.e., the plastic rheological instability region. The thermal processing diagram of the TA4 titanium alloy at different real strains is shown in Figure 11. The destabilization zone at strain 0.3 is 933~1023 K/0.01~4 s^{-1}. The destabilization zone at strain 0.6 is 933~1023 K/0.02~7 s^{-1}. In this strain interval, the destabilization zone gradually moves to the high strain rate interval with the increase of strain, but the overall change of the destabilization zone is small. Among them, the destabilization zone at strain 0.9 is 933~1023 K/0.04~10 s^{-1}, and the area of the destabilization zone at this time increases compared with that at low strain. With the increase in strain, the area of instability is gradually shifted from the location of low strain rate to the area of high strain rate and is gradually extended from the low-temperature region to the high-temperature region. Overall, the area of the destabilized region increases with the increase in strain.

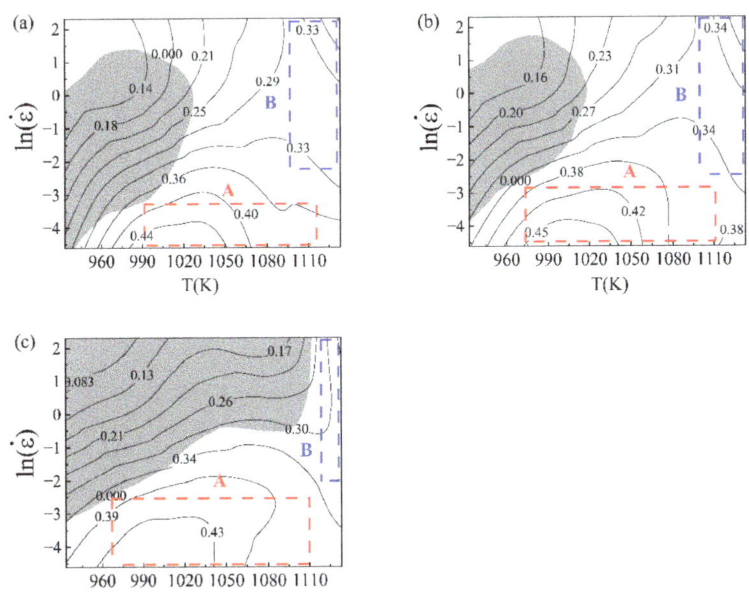

Figure 11. Hot working diagram of TA4 alloy under different strains: (**a**) ε = 0.3; (**b**) ε = 0.6; (**c**) ε = 0.9.

The optimum processing region is divided into two regions: the medium-temperature low-strain region A and the high-temperature high-strain region B. Region B should be avoided when designing the thermal processing parameters due to the high temperature and strain rate required. It can be seen that the energy dissipation efficiency factor is closely related to the deformation temperature and strain rate, which increases and then decreases with the rise in temperature and strain rate. The maximum region of energy dissipation efficiency factor is located in the region of medium temperature and low strain rate, and the maximum value is near 0.44. The maximum value is around 0.44. The optimal processing region is at a strain of 0.9, 1040–1133 K/0.01–0.08 s^{-1}. The optimal processing region is at a strain of 0.6, 940–1000 K/0.01–0.04 s^{-1}. The optimal processing region is at a strain of 0.3, 940–1000 K/0.01–0.08 s^{-1}.

3.5. EBSD Analyses

It is difficult to elucidate the specific deformation mechanism of the TA4 titanium alloy based on the stress–strain curves and thermal processing diagrams alone. In order to further investigate the effects of different thermal deformation parameters on the thermal deformation mechanism of the TA4 titanium alloy, its compression samples at different temperatures at the same strain rate (1 s^{-1}) and at the same temperatures (1033 K) at different strain rates were subjected to EBSD analysis.

Figure 12 shows the grain size statistics of titanium alloys under different deformation parameters. From the Figure 12a,b, it can be seen that the average grain area of the alloy increases significantly with the decrease in the strain rate, from 50.23 to 89.64 μm^2, and the grain shape tends to be more isometric. The orientation difference distribution graph indicates the proportion of the grain boundary orientation angle, and Figure 13 shows the orientation difference angle distribution of titanium alloys under different deformation parameters. The blue line in the figure represents the large-angle grain boundaries (HAGBs) with dislocation angles greater than 15°, and the red line represents the small-angle grain boundaries (LAGBs) with dislocation angle angles ranging from 2 to 15°. It can be obtained in Figure 13a,b,f that the percentage of LAGB decreases significantly from 42.5 to 29.8% as the strain rate decreases. The original small-angle grain boundaries formed by dislocations are transformed into large-angle grain boundaries due to the process of new grain generation and growth caused by the prolongation of deformation time with dynamic recrystallization (DRX). It is shown that the main thermal deformation mechanism of the TA4 titanium alloy at a low strain rate is DRX [31].

In Figure 12c–e, it can be seen that the average grain area of titanium alloys increases with the increase in deformation temperature. The average grain area increases slightly from 37.17 to 37.77 μm^2 when rising from a low temperature of 933 K to a high temperature of 1083 K, and the grain shape does not change much. When the heating temperature rises from 1083 to 1113 K, the average grain area increases significantly from 37.17 to 49.49 μm^2, and the elongated grains to the equiaxial transformation trend. DRX is usually accompanied by significant changes in grain area, especially at high temperatures accompanied by grain recrystallization and the grain growth process [32]. Dynamic recovery (DRV) is the recovery process of the original grains, in which the grains are elongated and flattened, and the grain boundaries are widened, but the change of grain size is not obvious [33]. In Figure 13c–e,g, it is found that when the heating temperature increases from the low temperature 933 K to high temperature 1083 K, the titanium alloy LAGBs account for a slight increase from 47.8 to 48.4%. When the temperature increases from 1083 to 1113 K, the percentage of LAGBs starts to decrease, from 48.4 to 40.6%. In the DRX process, nascent recrystallized grains only partially exist, or even no LAGBs exist, and the boundaries between small grains will be reorganized by atomic diffusion during grain growth, resulting in the reduction in LAGBs, and a large portion of the LAGBs are generated in the DRV process. This means that the main thermal deformation mechanism during the heating of the TA4 titanium alloy from a low temperature (933 K) to high temperature (1083 K) is DRV, which leads to little change in grain size and a slight increase in the percentage of LAGBs. When the temperature is above 1083 K, the main thermal deformation mechanism is DRX, resulting in a significant increase in grain size and a decrease in the percentage of LAGBs [34].

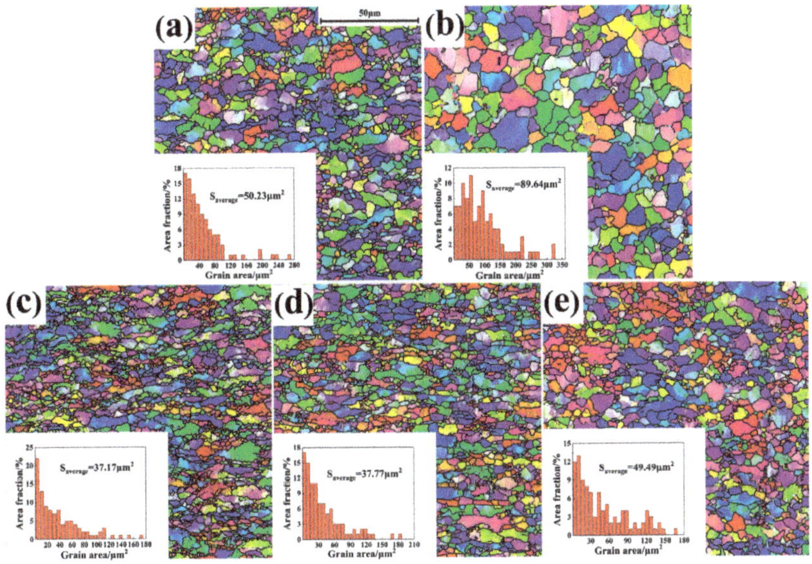

Figure 12. Deformed alloy grain sizes: (**a**) 1033 K/0.1 s^{-1}; (**b**) 1033 K/0.01 s^{-1}; (**c**) 933 K/1 s^{-1}; (**d**) 1083 K/1 s^{-1}; (**e**) 1133 K/1 s^{-1}.

To further discuss the thermal deformation mechanism of the TA4 titanium alloy, the percentage of dynamically recrystallized grains, substructures, and deformed grains in the typical characteristic regions of the TA4 titanium alloy were counted using the grain orientation spread (GOS) parameter as shown in Figure 14. In these microstructures, grains with GOS $\leq 2°$ are considered as recrystallized grains and are marked in a blue color. Grains with GOS > 7° are deformed grains and are marked in a red color. Grains with $2° <$ GOS $\leq 7°$ are substructured and are marked in a yellow color. In Figure 15, some equiaxed grains surrounded by HAGBs appear in the low dislocation region, which are recrystallized grains. There are also some irregularly shaped grains surrounded by LAGBs, which are substructures, and these grains are usually caused by the DRV process [35,36].

It can be found in Figure 14a,b,g that the percentage of recrystallized grains increases significantly from 18.4 to 36.6% with the decrease in strain rate, and the percentage of substructures decreases from 66.5 to 60.5%. This indicates that a considerable portion of the substructure is transformed into recrystallized grains at this time, and this phenomenon is a hallmark feature of dynamic recrystallization. From Figure 15a,b, it can be obtained that the dislocation density decreases significantly with the decrease in strain rate, and the average value changes from 0.55 to 0.34. This indicates that with the decrease in strain rate, the recrystallization process has enough time to consume dislocations, form new grains, and reduce the dislocation density. It is also noted that 1033 K/0.01 s^{-1} is in the optimal processing zone A of Figure 11, when the grain size is the largest, the percentage of recrystallization is the largest, and the average orientation difference is the lowest value of only 0.49, which indicates that it has a better processing performance and is in agreement with the results obtained from the thermal processing diagram. When the deformation temperature is gradually increased, the percentage of both recrystallization and substructure increases [37]. However, it is worth noting that when the temperature increases from 933 to 1083 K, the percentage of recrystallization only increases from 14.1 to

14.4%, and the percentage of substructure increases significantly from 45.9 to 61.8% as shown in Figure 14c,d,f. From Figure 15c,d, we see that the dislocation density decreases significantly at this time, the average value changes from 0.95 to 0.63, and the kernel average misorientation (KAM) value near the grain boundary is larger than that in the grain. This indicates that a large number of dislocations generated inside the grains during the deformation process are rearranged and reorganized under the influence of DRV to form small-angle grain boundaries, which are also known as subgranular boundaries, and the most substructures surrounded by LAGBs are formed. At the same time, DRV is more likely to cause the dislocations inside the grains to slip, climb, and annihilate, which leads to a relatively low KAM value inside the grains. When the temperature increases from 1083 to 1133 K, the percentage of recrystallization increases from 14.4 to 22%, the percentage of substructure increases from 61.8 to 67%, and the percentage of deformed grains decreases from 23.8 to 11%. At this time, it can be seen from Figure 15d,e that its overall dislocation density increases, the average value changes from 0.63 to 0.79. The increase in the percentage of recrystallized grains in this temperature interval implies that the dynamic recrystallization process occurs further. However, the rise of the overall dislocations indicates that it occurs at this point in the process of initial recrystallization nucleation generation in DRX [38].

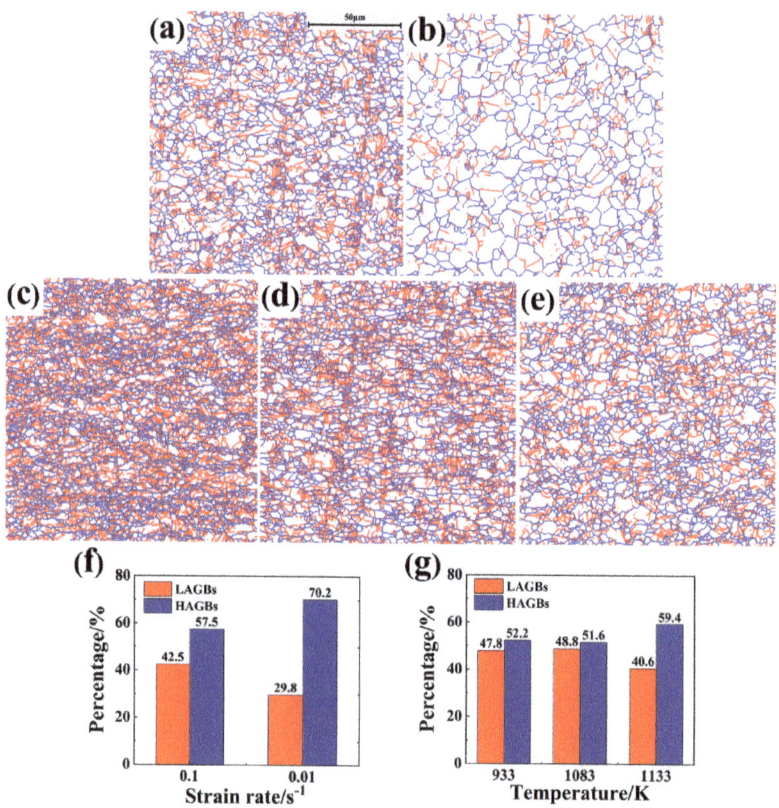

Figure 13. Grain boundary ratio of titanium alloys with different parameters: (**a**) 1033 K/0.1 s^{-1}; (**b**) 1033 K/0.01 s^{-1}; (**c**) 933 K/1 s^{-1}; (**d**) 1083 K/1 s^{-1}; (**e**) 1133 K/1 s^{-1}; (**f,g**) size angle grain boundary ratio.

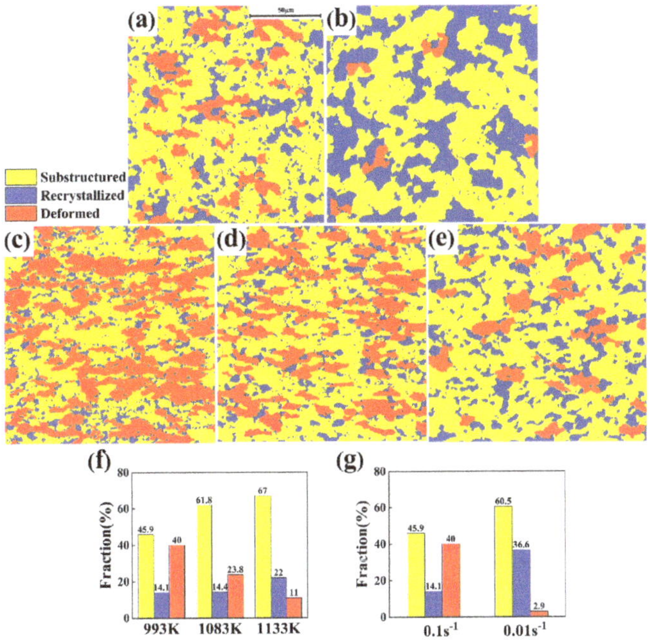

Figure 14. Recrystallization, substructure and deformation grain of the alloys with different parameters: (**a**) 1033 K/0.1 s^{-1}; (**b**) 1033 K/0.01 s^{-1}; (**c**) 933 K/1 s^{-1}; (**d**) 1083 K/1 s^{-1}; (**e**) 1133 K/1 s^{-1}; (**f**,**g**) the proportion of different grain types.

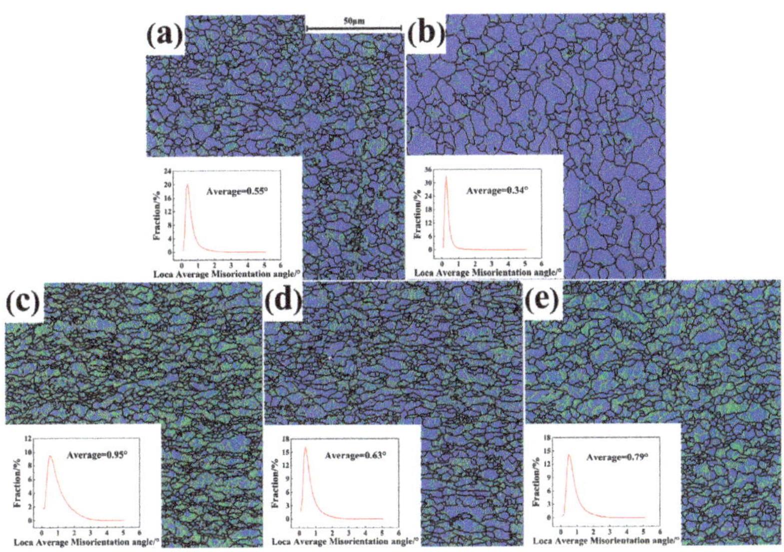

Figure 15. KAM distributions: (**a**) 1033 K/0.1 s^{-1}; (**b**) 1033 K/0.01 s^{-1}; (**c**) 933 K/1 s^{-1}, (**d**) 1083 K/1 s^{-1}; (**e**) 1133 K/1 s^{-1}.

4. Discussion

Recent studies on the thermal deformation of α-titanium alloys have provided valuable insights into their behavior under various processing conditions. Ma et al. [39]

conducted uniaxial tensile tests over a temperature range of 1173–1248 K and strain rates of 0.0005–0.01 s^{-1}, establishing a theoretical basis for parameter selection during the superplastic deformation phase. Similarly, Wang et al. [40] examined the flow characteristics and dynamic recrystallization (DRX) mechanisms within a temperature range of 1073–1243 K and strain rates from 0.001 to 5 s^{-1}. Their research resulted in a process map with three optimized processing windows and a destabilization zone, contributing further to the understanding of superplastic deformation. Yu et al. [41] explored the relationship between deformation mechanisms and microstructural evolution in the 1093–1193 K/0.01–10 s^{-1} parameter domain, developing an evolutionary model for thermal deformation mechanisms and defining a processing exclusion zone.

These studies share two key features: first, they focus primarily on temperatures near the critical phase transition temperature, mainly investigating phase-transition-induced microstructural changes; second, they lack a thorough analysis of broader temperature ranges, particularly with respect to dynamic recrystallization, microstructural evolution, and the quantitative relationships between process parameters in these regions.

Our study addresses these limitations by extending the investigation beyond the traditional phase transition temperature region, focusing on a processing window (933–1133 K/0.01–10 s^{-1}) commonly used in industrial applications. We systematically examine the thermal deformation mechanisms and microstructural evolution of near-α titanium alloys within this parameter range, employing thermal simulation experiments and various characterization techniques. This research not only establishes a foundation for selecting processing parameters but also fills a critical gap in the theoretical framework for processing in this temperature range. Ultimately, it provides valuable insights to guide the precise shape control of titanium alloy components.

In this work, the influence of processing parameters on microstructural evolution is analyzed by modeling the competing mechanisms of dynamic recrystallization (DRX) and dynamic recovery (DRV) during the hot deformation of the TA4 titanium alloy. Experimental results reveal a strong sensitivity to strain rate: at a fixed temperature of 1033 K, the DRX volume fraction is only 18.4% at a strain rate of 0.1 s^{-1} but increases to 36.6% when the strain rate is reduced to 0.01 s^{-1}. Statistical analysis shows that when the strain rate is $\leq$0.1 s^{-1}, recrystallized grains are broadly distributed in the microstructure (with an average orientation difference of <0.55°), confirming DRX as the dominant softening mechanism. However, when the strain rate exceeds 0.1 s^{-1}, a temperature-sensitive behavior emerges in the deformation mechanism. Further analysis reveals that at lower temperatures (933–1083 K), DRV, characterized by dislocation cell wall remicrostructure, is the primary softening mechanism, with a larger average orientation difference and a combined ratio of subcrystalline and deformed grains exceeding 78%. At higher temperatures (1083–1133 K), while DRX remains the dominant softening mechanism, the average orientation difference remains >0.63°, indicating that DRX is still inhibited by high strain rates. Based on these findings, the optimal processing window is determined to be a heating temperature of 1033 $\pm$ 20 K combined with a strain rate of 0.01–0.05 s^{-1}.

5. Conclusions

In this paper, isothermal compression tests were carried out on the TA4 titanium alloy specimens using the Gleeble-3800 thermal simulation testing machine to study the thermal deformation behavior of the TA4 titanium alloy in the processing range of 933–1133 K/0.01–10 s^{-1}, and the following conclusions were drawn:

(1) The constitutive equation based on the strain-compensated Arrhenius model was established for the TA4 titanium alloy. After analysis, the correlation coefficient between the predicted and experimental values of this constitutive model is 0.9612, the

average absolute percentage error is 8.7210%, and the average root mean square error is 11.0635 MPa, which indicates that this constitutive equation has a better prediction ability for the rheological stresses in the temperature interval of 933–1133 K and the interval of strain rate of 0.01–10 s^{-1} for the TA4 titanium alloy.

(2) The optimal processing region of the TA4 titanium alloy is mainly divided into two regions, in which the region with the largest energy dissipation efficiency factor is located in the region of medium temperature and low strain rate, with the maximum value of 0.44. The optimal processing region with the larger energy dissipation efficiency factor is as follows at different strains: the optimum machining zone at a strain of 0.9 is 1040~1133 K/0.01~0.7 s^{-1}. The optimum machining zone at a strain of 0.6 is 940~1000 K/0.01~0.04 s^{-1}. The optimum machining zone at a strain of 0.3 is 940~1000 K/0.01~0.08 s^{-1}.

(3) With the decrease of strain rate and the increase of deformation temperature, the grain area increases, the proportion of large-angle grain boundaries increases, and the proportion of dynamic recrystallization increases. The dislocation density increases with decreasing strain rate, and shows a tendency of decreasing and then increasing with increasing deformation temperature.

(4) The main deformation softening mechanism of the TA4 titanium alloy during thermal deformation is DRV at a low temperature and high strain rate, at a high temperature and high strain rate, and at a low strain rate. The optimal process window is determined to be a 1033 ± 20 K heating temperature matched to 0.01–0.05 s^{-1} strain rate combination.

Author Contributions: Conceptualization, L.M. and W.L.; methodology, Y.Z.; software, L.Q; validation, W.L. and Y.Z.; formal analysis, W.L.; investigation, J.Z.; resources, Y.Z.; data curation, J.Z.; writing—original draft preparation, W.L.; writing—review and editing, Y.Z.; supervision, L.M.; project administration, L.M.; funding acquisition, L.Q. All authors have read and agreed to the published version of the manuscript.

Funding: This work was supported by the National Natural Science Foundation of China (No. 52375363), (No. B2408100.W12), (No. 52205400), and (No. 52375362), the Applied Basic Research Project of Shanxi Province (No. 202203021211208), and the Technological Innovation Talent Team Special Plan of Shanxi Province (202204051002002).

Data Availability Statement: Data will be made available on request.

Conflicts of Interest: The authors declare no conflict of interest.

References

1. Wang, Z.B.; Hu, H.X.; Zheng, Y.G.; Ke, W.; Qiao, Y.X. Comparison of the corrosion behavior of pure titanium and its alloys in fluoride-containing sulfuric acid. *Corros. Sci.* **2016**, *103*, 50–65. [CrossRef]
2. Shahmir, H.; Pereira, P.H.R.; Huang, Y.; Langdon, T.G. Mechanical properties and microstructural evolution of nanocrystalline titanium at elevated temperatures. *Mater. Sci. Eng. A* **2016**, *669*, 358–366. [CrossRef]
3. Yuan, B.; Zheng, Y.; Wang, Y.; Chen, Q.; Gong, L.; Lv, M. Effect of temperature on hydrogen absorption characteristic and microstructural evolution of TC21 alloy. *J. Alloys Compd. Interdiscip. J. Mater. Sci. Solid-State Chem. Phys.* **2015**, *648*, 794–802.
4. Zhao, H.; Shan, X.; Tang, Y.; Wang, B.; Tian, Y. Hot deformation behavior and dynamic recrystallization mechanism of Ti2ZrTa0.75 refractory complex concentrated alloy. *Mater. Charact.* **2023**, *203*, 113061.
5. Després, A.; Mithieux, J.D.; Sinclair, C.W. Modelling the relationship between deformed microstructures and static recrystallization textures: Application to ferritic stainless steels. *Acta Mater.* **2021**, *219*, 117226.
6. Fan, X.G.; Yang, H.; Gao, P.F. Prediction of constitutive behavior and microstructure evolution in hot deformation of TA15 titanium alloy. *Mater. Des.* **2013**, *51*, 34–42. [CrossRef]
7. Momeni, A.; Abbasi, S.M. Effect of hot working on flow behavior of Ti–6Al–4V alloy in single phase and two phase regions. *Mater. Des.* **2010**, *31*, 3599–3604. [CrossRef]
8. Yang, Y.; Xu, D.; Cao, S.; Wu, S.; Zhu, Z.; Wang, H.; Li, L.; Xin, S.; Qu, L.; Huang, A. Effect of strain rate and temperature on the deformation behavior in a Ti-23.1Nb-2.0Zr-1.0O titanium alloy. *J. Mater. Sci. Technol.* **2021**, *73*, 52–60. [CrossRef]

9. Hao, F.; Xiao, J.; Feng, Y.; Wang, Y.; Ju, J.; Du, Y.; Wang, K.; Nie, Z.; Tan, C. Tensile deformation behavior of a near-α titanium alloy Ti-6Al-2Zr-1Mo-1V under a wide temperature range. *J. Mater. Res. Technol.* **2020**, *9*, 2818–2831.
10. Wang, X.; Liu, P.; Liang, C.; Lu, T.; Feng, T.; Niu, H.; Dong, Y.; Liu, X. Investigation on the thermal deformation mechanisms and constitutive model of Ti-55511 titanium alloy. *J. Mater. Res. Technol.* **2024**, *33*, 6780–6797. [CrossRef]
11. Li, L.; Ma, G.; Huang, H.; Xiao, H.; Yuan, Z.; Li, Y.; Zhou, R. Flow behavior analysis and prediction of flow instability of a lamellar TA10 titanium alloy. *Mater. Charact.* **2022**, *194*, 112403. [CrossRef]
12. Zhang, Y.; Huang, T.; Zhou, Z.; Li, M.; Tan, L.; Gan, B.; Jie, Z.; Qin, L.; Zhang, J.; Liu, L. Variation of Homogenization Pores during Homogenization for Nickel-Based Single-Crystal Superalloys. *Adv. Eng. Mater.* **2021**, *23*, 2001547. [CrossRef]
13. Ji, R.; Zhu, K.; Zhang, H.; Luo, H.; Mao, J. Microstructure evolution, mechanical response and strengthening models for TA15 titanium alloy during thermal processes: A brief review. *J. Mater. Res. Technol.* **2024**, *28*, 1644–1656. [CrossRef]
14. Abbasi, S.M.; Momeni, A.; Lin, Y.C.; Jafarian, H.R. Dynamic softening mechanism in Ti-13V-11Cr-3Al beta Ti alloy during hot compressive deformation. *Mater. Sci. Eng. A* **2016**, *665*, 154–160. [CrossRef]
15. Xiang, Y.; Xiang, W.; Yuan, W. Flow softening and microstructural evolution of near β titanium alloy Ti-55531 during hot compression deformation in the α + β region. *J. Alloys Compd.* **2023**, *955*, 170165. [CrossRef]
16. Wang, Y.; Yang, G.; Zhang, S.; Xiu, S. Effect of crystal orientation on micro-stress distribution in a damage-tolerant titanium alloy TC21. *J. Alloys Compd. Interdiscip. J. Mater. Sci. Solid-State Chem. Phys.* **2022**, *924*, 166637.
17. Si-Wei, D.; Xiang-Long, D. Interpretation of the Latest Edition of ≪GB/T 3620.1-2016 Titanium and Titanium Alloy Grades and Chemical Compositions≫. Science & Technology Vision 2019.
18. Xie, H.; Wang, J.; Li, Y.; Song, J.; Hu, H.; Qin, L.; Zhao, H.; Li, C.; Cui, Y.; Tan, Z. Fast shot speed induced microstructure and mechanical property evolution of high pressure die casting Mg-Al-Zn-RE alloys. *J. Mater. Process. Technol.* **2024**, *331*, 118523. [CrossRef]
19. Han, L.; Zhang, H.; Cheng, J.; Zhou, G.; Wang, C.; Chen, L. Thermal Deformation Behavior of Ti-6Mo-5V-3Al-2Fe Alloy. *Crystals* **2021**, *11*, 1245. [CrossRef]
20. Chai, Y.; Zhu, Y.; Qin, L.; Luo, Y.; Niu, Y.; Shao, Z. High-temperature hot deformation behavior and processing map of Ti-22Al-25Nb alloy. *Mater. Today Commun.* **2024**, *41*, 110599. [CrossRef]
21. Prasad, Y.V.R.K.; Gegel, H.L.; Doraivelu, S.M.; Malas, J.C.; Morgan, J.T.; Lark, K.A.; Barker, D.R. Modeling of dynamic material behavior in hot deformation: Forging of Ti-6242. *Metall. Trans. A* **1984**, *15*, 1883–1892.
22. Shokry, A.; Gowid, S.; Kharmanda, G. An improved generic Johnson-Cook model for the flow prediction of different categories of alloys at elevated temperatures and dynamic loading conditions. *Mater. Today Commun.* **2021**, *27*, 102296. [CrossRef]
23. Zhao, Y.; Sun, J.; Li, J.; Yan, Y.; Wang, P. A comparative study on Johnson-Cook and modified Johnson-Cook constitutive material model to predict the dynamic behavior laser additive manufacturing FeCr alloy. *J. Alloys Compd.* **2017**, *723*, 179–187. [CrossRef]
24. Chen, W.; Guan, Y.; Wang, Z. Modeling of Flow Stress of High Titanium Content 6061 Aluminum Alloy Under Hot Compression. *J. Mater. Eng. Perform.* **2016**, *25*, 4081–4088. [CrossRef]
25. Liao, H.; Wu, Y.; Zhou, K.; Yang, J. Hot deformation behavior and processing map of Al–Si–Mg alloys containing different amount of silicon based on Gleebe-3500 hot compression simulation. *Mater. Des.* **2015**, *65*, 1091–1099. [CrossRef]
26. Abbasi-Bani, A.; Zarei-Hanzaki, A.; Pishbin, M.H.; Haghdadi, N. A comparative study on the capability of Johnson–Cook and Arrhenius-type constitutive equations to describe the flow behavior of Mg–6Al–1Zn alloy. *Mech. Mater.* **2014**, *71*, 52–61. [CrossRef]
27. Wang, Y.; Peng, J.; Zhong, L.; Pan, F. Modeling and application of constitutive model considering the compensation of strain during hot deformation. *J. Alloys Compd.* **2016**, *681*, 455–470. [CrossRef]
28. Zhou, Z.; Fan, Q.; Xia, Z.; Hao, A.; Yang, W.; Ji, W.; Cao, H. Constitutive Relationship and Hot Processing Maps of Mg-Gd-Y-Nb-Zr Alloy. *J. Mater. Sci. Technol.* **2017**, *33*, 637–644. [CrossRef]
29. Mandal, S.; Rakesh, V.; Sivaprasad, P.V.; Venugopal, S.; Kasiviswanathan, K.V. Constitutive equations to predict high temperature flow stress in a Ti-modified austenitic stainless steel. *Mater. Sci. Eng. A* **2009**, *500*, 114–121. [CrossRef]
30. Robi, P.S.; Dixit, U.S. Application of neural networks in generating processing map for hot working. *J. Mater. Process. Technol.* **2003**, *142*, 289–294. [CrossRef]
31. Long, F.-W.; Jiang, Q.-W.; Xiao, L.; Li, X.-W. Compressive Deformation Behaviors of Coarse- and Ultrafine-Grained Pure Titanium at Different Temperatures: A Comparative Study. *Mater. Trans.* **2011**, *52*, 1617–1622. [CrossRef]
32. Li, L.; Li, M.Q.; Luo, J. Flow softening mechanism of Ti–5Al–2Sn–2Zr–4Mo–4Cr with different initial microstructures at elevated temperature deformation. *Mater. Sci. Eng. A* **2015**, *628*, 11–20. [CrossRef]
33. Hou, Z.; Liu, H.; Lu, L.; Li, W.; Zhang, Z.; Lu, W.; Feng, Q.; Jia, B.; Song, K. Effect of rare earth yttrium and the deformation process on the thermal deformation behavior and microstructure of pure titanium for cathode rolls. *J. Mater. Res. Technol.* **2024**, *33*, 4192–4205. [CrossRef]
34. Xu, G.; Tao, J.; Deng, Y.; Zheng, B.; Zhang, Y.; Jiang, Y. Multi-stage hot deformation and dynamic recrystallization behavior of low-cost Ti–Al–V–Fe alloy via electron beam cold hearth melting. *J. Mater. Res. Technol.* **2022**, *20*, 1186–1203. [CrossRef]

35. Al-Samman, T. Modification of texture and microstructure of magnesium alloy extrusions by particle-stimulated recrystallization. *Mater. Sci. Eng. A* **2013**, *560*, 561–566. [CrossRef]
36. Qu, F.S.; Zhou, Y.H.; Zhang, L.Y.; Wang, Z.H.; Zhou, J. Research on hot deformation behavior of Ti-5Al-5Mo-5V-1Cr-1Fe alloy. *Mater. Des.* **2015**, *69*, 153–162. [CrossRef]
37. Xu, J.; Zeng, W.; Zhang, X.; Zhou, D. Analysis of globularization modeling and mechanisms of alpha/beta titanium alloy. *J. Alloys Compd.* **2019**, *788*, 110–117. [CrossRef]
38. Yang, Y.; Wang, H.; Feng, Z.; Shi, Q.; Yang, B.; Chen, M.; Qi, H.; Wang, X. Study on the warm deformation behavior and microstructure evolution of the MDIFed Ti–6Al–4V titanium alloy. *J. Mater. Res. Technol.* **2024**, *33*, 8929–8943. [CrossRef]
39. Ma, L.; Wan, M.; Li, W.; Shao, J.; Bai, X.; Zhang, J. Superplastic deformation mechanical behavior and constitutive modelling of a near-α titanium alloy TNW700 sheet. *Mater. Sci. Eng. A* **2021**, *817*, 141419. [CrossRef]
40. Wang, Y.; Li, Z.; Wang, H.; Hou, M.; Yu, K.; Xu, Y.; Xiao, H. Flow behavior and dynamic recrystallization mechanism of a new near-alpha titanium alloy Ti-0.3 Mo-0.8 Ni-2Al-1.5 Zr. *J. Mater. Res. Technol.* **2024**, *30*, 3863–3876. [CrossRef]
41. Yu, J.; Li, Z.; Qian, C.; Huang, S.; Xiao, H. Investigation of deformation behavior, microstructure evolution, and hot processing map of a new near-α Ti alloy. *J. Mater. Res. Technol.* **2023**, *23*, 2275–2287.

crystals

MDPI

Article

On Modeling X-Ray Diffraction Intensity Using Heavy-Tailed Probability Distributions: A Comparative Study

Farouq Mohammad A. Alam

Department of Statistics, Faculty of Science, King Abdulaziz University, Jeddah 21589, Saudi Arabia; fmalam@kau.edu.sa

Abstract

Crystallography, a cornerstone of materials science, provides critical insights into material structures through techniques such as X-ray diffraction (XRD). Among the metrics derived from XRD, intensity serves as a key parameter, reflecting the electron density distribution and offering information about atomic arrangements and sample quality. Due to its inherent variability and susceptibility to extreme values, intensity is best modeled using heavy-tailed, location-scale probability distributions. This paper investigates the model parameter estimation problem for three such distributions—log-Cauchy, half-Cauchy, and Cauchy Birnbaum–Saunders—using several methods, including maximum likelihood and the maximum product of spacings estimation methods. Monte Carlo simulations are conducted to assess the performance of these methods across various scenarios. Additionally, two real XRD intensity datasets are analyzed to compare the applicability and effectiveness of the proposed models. The results demonstrate the potential of heavy-tailed distributions for modeling XRD intensity data, providing a robust framework for future research and practical applications in material characterization.

Keywords: Cauchy Birnbaum–Saunders distribution; half-Cauchy distribution; log-Cauchy distribution; maximum likelihood estimation; maximum product of spacings; data analysis

MSC: 62F10; 62E10; 62P30; 62F40

1. Introduction

Materials science is at the forefront of technological and industrial innovation, as it seeks to understand and optimize the properties of materials for various applications. In this field, crystallography plays a critical role by providing structural insights that link materials' atomic and molecular arrangements to their macroscopic properties. Techniques such as X-ray diffraction (XRD) are widely used to study crystalline solids, offering valuable data that help researchers predict and enhance material performance across industries. Around eleven decades ago, the discovery of XRD in crystals was pioneered by Nobel Prize-winning physicists Max von Laue, William Henry Bragg, and William Lawrence Bragg [1]. Crystallography provides key metrics that inform material characterization, including alloys, which are critical for industrial applications. For instance, the full width at half maximum (FWHM), calculated using the Scherrer equation, indicates crystallite size and strain [2]. Another crucial metric is intensity, which measures the strength of diffracted X-rays at specific angles. Intensity is directly proportional to the electron density distribution in the crystal, offering insights into the abundance of crystallographic planes, atomic arrangements, and overall sample quality [3]. Recent research has advanced the understanding and application of XRD metrics, particularly focusing on the full width at half maximum (FWHM) and intensity measurements. For instance, the reference intensity ratio method has been explored for phase quantification in nanocrystals, emphasizing the importance of instrumental calibration to ensure accurate FWHM and intensity assessments [4]. Additionally, advancements in micro-structural analysis of clay minerals have demonstrated how variations in FWHM correlate with crystallite size, offering valuable insights into nanoscale structures [5]. Recent research utilizing the Scherrer method and FWHM includes but is not limited to [6], which focused on the synthesis and classification of pure and metal-doped hydroxyapatite (HAp), conducting an in-depth crystallographic analysis; [7], which investigated nanoscale HAp crystals synthesized from calcium carbonate and phosphoric acid using a wet chemical process with fish and cow bones as natural sources, employing various diffraction-based techniques to analyze crystallite size, stress, and strain; and [8], which examined gallium oxide nanostructures synthesized via hydrogen-reducing chemical vapor deposition at different substrate angles, using X-ray diffraction to confirm their polycrystalline monoclinic structure and identifying the size–strain plot method as the most suitable for estimating crystallite size and lattice strain. A study by [9] performed molecular dynamics simulations using the Born–Huggins–Mayer–Fumi–Tosi rigid ion potential, accurately predicting the structure factors and X-ray diffraction intensities of molten alkali halides, with the computed partial structure factors for six chemical compounds closely matching experimental neutron and X-ray diffraction data, while also identifying model limitations for molten fluorides and suggesting extensions for other alkali halides. Further studies have addressed challenges in intensity measurements during ultrafast diffraction experiments. Optimizing hard X-ray photon energies has been shown to enhance the coherently scattered signal strength while minimizing electronic radiation damage, which is critical for applications in silicon crystals and other materials [10]. Moreover, comprehensive reviews on XRD applications highlight the broader relevance of FWHM and intensity metrics in crystallographic evaluations, underscoring their role in characterizing diverse materials, including alloys [11]. Regarding other directions of XRD research, [12] recently proposed a novel method to significantly enhance the angular resolution in XRD experiments. By addressing the limitations associated with detector pixel size and sample dimensions, the researchers proposed a technique that achieves a nearly tenfold improvement in resolution compared to traditional single-pixel methods. This advancement sharpens peak shapes in diffraction data and substantially reduces noise, improving the signal-to-noise ratio and relevant measurement accuracy. In terms of machine learning, [13] addressed the challenge of analyzing vast amounts of XRD data by proposing a deep learning model for the automated classification of crystal systems and space groups, while [14] enhanced automated phase identification in XRD analysis by employing neural networks—one trained on XRD patterns and another on pair distribution functions (PDFs)—to improve accuracy by leveraging both reciprocal and real-space representations. A recent study by [15] proposed a multi-objective evolutionary search to solve complex powder X-ray diffraction (PXRD) patterns by optimizing enthalpy and pattern similarity. The proposed method overcomes computational and experimental limitations, improving accuracy and efficiency in structure prediction for various inorganic systems and significantly reducing computation time.

Intensity can be viewed as a phenomenon that can be modeled using a probabilistic framework, such as a probability distribution, from a statistical perspective. Statistical modeling of intensity data aims to capture its variability and underlying patterns, often influenced by factors such as experimental noise, crystallographic imperfections, and material heterogeneity. By employing appropriate probabilistic models, researchers can gain deeper insights into the structural characteristics of materials and evaluate the reliability and robustness of XRD measurements. Since the distribution of XRD measurements may exhibit heavy-tailed behavior, selecting an appropriate probability distribution is essential for accurately modeling the phenomenon. A heavy-tailed probability distribution, sometimes called a pathological distribution, is characterized by tails that decay more slowly than those of standard distributions. A well-known example of a pathological distribution is the Cauchy distribution. Mathematicians sometimes use the term *pathological* to describe mathematical functions designed to contradict or violate certain universally accepted assumptions, properties, or theorems; see, for example, [16]. This property increases the probability of extreme values, making heavy-tailed distributions particularly effective for capturing phenomena marked by significant outliers or high variability. Such distributions are widely applicable across diverse fields, including finance, physics, and materials science, where rare but impactful events or measurements are crucial. For general information about probability distributions, see [17,18]. For contributions about heavy-tailed probability distributions, see, for example, [19–22]. Recently, [23] examined the challenges that heavy-tailed distributions pose to traditional statistical inference methods. The researchers discussed the implications of heavy-tailed data in various fields. They emphasized the need for robust statistical methods that account for such distributions to ensure accurate and reliable inference. Furthermore, a heavy-tail model that may be appropriate for modeling phenomena and processes with extreme values called the Cauchy-Logistic Unit Distribution was proposed by [24]. They transformed the Cauchy distribution into a unit interval using generalized logistic mapping. Accordingly, they discussed distributional properties (e.g., moments and entropy) and parameter estimation. A recent study by [25] indicated that grain refinement could broaden diffraction peaks close to the Cauchy distribution in their XRD analysis.

In the context of XRD, heavy-tailed distributions are particularly valuable for modeling intensity data, as they effectively capture high-intensity diffraction peaks and the variability caused by structural irregularities or experimental conditions. This study addresses the problem of estimating the parameters of three heavy-tailed, location-scale probability distributions, which are well suited for modeling intensity metrics derived from XRD data. A location-scale family of probability distributions consists of distributions that can be expressed in terms of a standardized form through location and scale parameters. Notable examples of these distributions include but are not limited to the normal and exponential distributions. These distributions are useful because they allow flexibility in modeling data by shifting and rescaling a base distribution. By employing these distributions, this study aims to improve the understanding and reliability of intensity modeling in materials science. To achieve this, two estimation methods—maximum likelihood and maximum product of spacings—are utilized. These methods are selected for their distinct approaches to parameter estimation and their applicability to heavy-tailed distributions. Monte Carlo simulations are conducted to evaluate the performance of each estimation method across the three distributions. Additionally, two intensity datasets of lead-tin alloy are analyzed to compare the effectiveness of the considered distributions in practical scenarios. The datasets, originally obtained from [26], explore the structural and mechanical properties of lead-tin alloys using XRD methods and Vickers microhardness testing. Figures 1 and 2 show the histograms of the considered datasets.

The remainder of this paper is structured as follows. Section 2 overviews the heavy-tailed, location-scale probability distributions. Section 3 discusses the estimation methods of interest. Section 4 reports the analysis of the considered real datasets, while Section 5 reports the outcomes of the Monte Carlo simulation study conducted to examine estimation efficiency numerically. Section 6 concludes this paper.

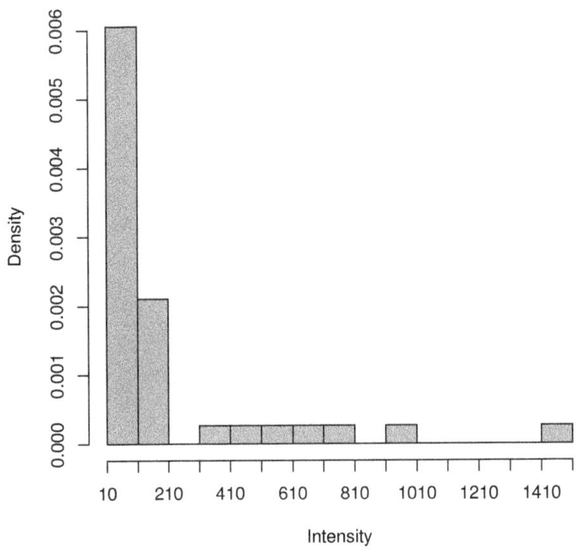

Figure 1. Histogram of intensity data from [26].

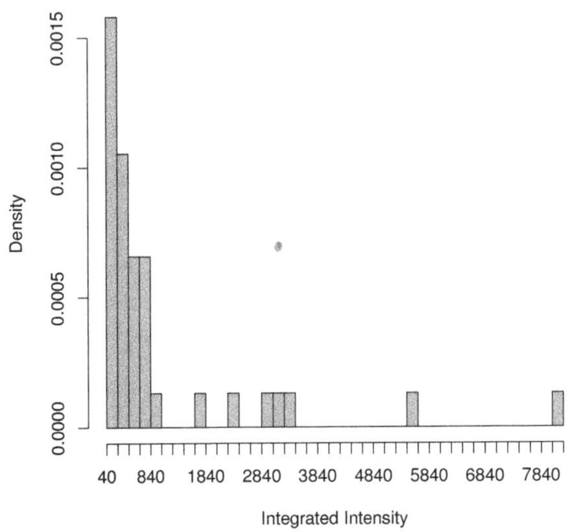

Figure 2. Histogram of integrated intensity data from [26].

2. The Candidate Probabilistic Models for Intensity

This section provides an overview of the three heavy-tailed, location-scale probability distributions considered in this study. For each distribution, the probability density function (PDF), cumulative distribution function (CDF), and quantile function (QF) are presented. The latter function is essential for simulating data from these models. In contrast, the remaining functions are vital for estimating the parameters using the maximum likelihood and maximum product of spacings methods. These distributions are related to the Cauchy distribution, an elliptical probability distribution notable for its counterintuitive properties, such as the absence of a well-defined mean.

2.1. The Half-Cauchy Distribution

The half-Cauchy (HCA) distribution is a continuous probability distribution derived from the Cauchy distribution by restricting its domain to non-negative values, i.e., truncating it at the location parameter. It is commonly used as a prior distribution in Bayesian statistics, particularly for scale parameters, due to its heavy-tailed nature, allowing for greater modeling robustness [27]. A recent application of the half-Cauchy distribution as a hyperprior in Bayesian models and its practical use in disease mapping is discussed in [28]. A random variable X is said to follow the half-Cauchy distribution with a location parameter $\mu \in \mathbb{R}$ and a scale parameter $\sigma \in \mathbb{R}_+$ if its PDF, CDF, and QF are given by

$$f(x;\mu,\sigma) = \frac{2}{\pi\sigma\left[1+\left(\frac{x-\mu}{\sigma}\right)^2\right]}, \quad x > \mu, \tag{1}$$

$$F(x;\mu,\sigma) = \frac{2}{\pi}\arctan\left(\frac{x-\mu}{\sigma}\right), \quad x \geq \mu, \tag{2}$$

and

$$Q(p;\mu,\sigma) = \mu + \sigma\tan\left(\frac{\pi p}{2}\right), \quad 0 < p < 1, \tag{3}$$

respectively. Although μ can be negative in this model, it is reasonable to impose $\mu > 0$ since the data considered in this study are positive, and the half-Cauchy distribution can be interpreted as a lifetime distribution, which must have positive support.

2.2. The log-Cauchy Distribution

The log-Cauchy (LCA) distribution is a continuous probability distribution derived from the Cauchy distribution by taking the exponential of a random variable that follows a Cauchy distribution. It is instrumental in modeling positive-valued data with heavy tails, in which extreme values play a significant role. The log-Cauchy distribution has applications in various fields (see, for example, [29,30]). A random variable X is said to follow the log-Cauchy distribution with a location parameter $\mu \in \mathbb{R}$ and a scale parameter $\sigma \in \mathbb{R}_+$ if its PDF, CDF, and QF are given by

$$f(x;\mu,\sigma) = \frac{1}{\pi(x-\mu)\left\{1+\left[\log\left(\frac{x-\mu}{\sigma}\right)\right]^2\right\}}, \quad x > \mu, \tag{4}$$

$$F(x;\mu,\sigma) = \frac{1}{\pi}\arctan\left(\log\left[\frac{x-\mu}{\sigma}\right]\right) + \frac{1}{2}, \quad x > \mu, \tag{5}$$

and

$$Q(p;\mu,\sigma) = \mu + \sigma\exp\left(\tan\left(\pi\left[p-\frac{1}{2}\right]\right)\right), \quad 0 < p < 1, \tag{6}$$

respectively.

2.3. The Cauchy Birnbaum–Saunders Distribution

In materials science, understanding the fatigue of material under periodic stress, such as pressure and cyclic loading, is crucial for assessing reliability. The Birnbaum–Saunders (BS) distribution, introduced in [31,32], has been extensively used to model fatigue data. It is a skewed probability distribution with a non-monotonic hazard rate function. Subsequent work [33] provided a biological justification for the BS distribution, broadening its applicability. The BS distribution shares properties with the log-normal distribution due to its relationship with the normal kernel. This connection has led to significant attention in the statistical literature, with hundreds of publications and a dedicated monograph [34] exploring its properties and applications. Generalizations of the BS distribution have also been proposed by replacing the normal kernel with other symmetric kernels, such as that of the Cauchy distribution [35]. The Cauchy Birnbaum–Saunders (CBS) distribution is obtained by replacing the normal kernel with the Cauchy kernel. Recently, [36] mathematically discussed the shape of the CBS distribution's density function and failure rate function. They also studied a quasi-maximum likelihood estimator for the shape parameter, assuming that the point estimate of the scale parameter of the CBS distribution is the sample median. Furthermore, they acquired regression and logarithmic moment estimations for the model parameters. In contrast, [37] considered several frequentist estimators different from those considered by [36]. A random variable X is said to follow the CBS distribution with a location parameter $\mu \in \mathbb{R}$ and a scale parameter $\sigma \in \mathbb{R}_+$ if its PDF, CDF, and QF are given by

$$f(x;\mu,\sigma) = \frac{B(x;\mu,\sigma)}{2\pi(x-\mu)\left(1+[A(x;\mu,\sigma)]^2\right)}, \quad x > \mu, \tag{7}$$

$$F(x;\mu,\sigma) = \frac{1}{\pi}\arctan\left(A(x;\mu,\sigma)\right) + \frac{1}{2}, \quad x > \mu, \tag{8}$$

and

$$Q(p,\mu,\sigma) = \mu + \frac{\sigma}{4}\left\{q(p) + \sqrt{[q(p)]^2+4}\right\}^2, \quad 0 < p < 1, \tag{9}$$

respectively, where

$$A(x;\mu,\sigma) = \sqrt{\frac{x-\mu}{\sigma}} - \sqrt{\frac{\sigma}{x-\mu}}, \tag{10}$$

$$B(x;\mu,\beta) = \sqrt{\frac{x-\mu}{\sigma}} + \sqrt{\frac{\sigma}{x-\mu}}, \tag{11}$$

and $q(p) = \tan(\pi[p-0.5])$.

Before concluding this section, it is essential to mention that these probability distributions have undefined statistical properties, such as the mean and variance. Figure 3 shows the PDFs of these three models, assuming $\mu = 0$ and $\sigma = 1$ (i.e., standardized forms).

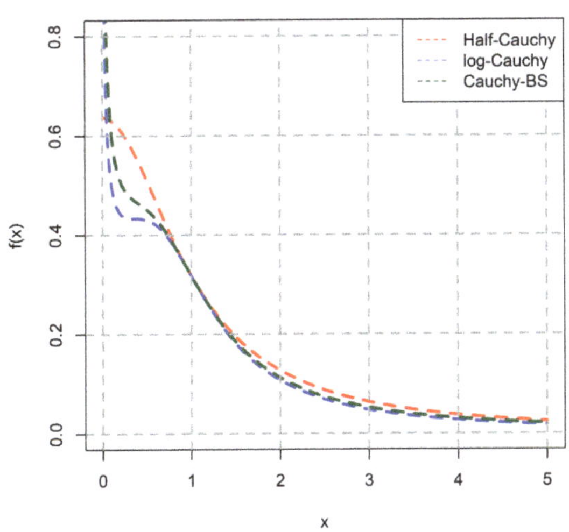

Figure 3. The PDFs of the standard probabilistic models for intensity analysis ($\mu = 0, \sigma = 1$).

3. Parameter Estimation Methodologies

This section discusses estimation methods for the probability distributions mentioned in the previous section.

3.1. Maximum Likelihood Estimation

Maximum likelihood estimation (MLE) is a widely used statistical method for estimating the parameters of probability distributions by maximizing the likelihood function, which represents the probability of observed data given specific parameter values. The versatility and efficiency of MLE have made it a fundamental tool in various scientific disciplines, including physics, where precise modeling and analysis of experimental data are crucial. In the context of X-ray Powder Diffraction (XRD) analysis, MLE has been employed to improve the accuracy of structural parameter estimation. For example, [38] demonstrated the application of MLE to analyze XRD data for cubic and tetragonal systems, leading to more reliable evaluations of lattice parameters and their standard deviations. This approach was successfully applied to materials such as yttrium-doped $BaTiO_3$, showcasing MLE's superiority over traditional estimation methods. Another notable application of MLE in XRD analysis is its use in refining observed diffraction intensity data. An analytical method based on MLE to optimize Rietveld refinement parameters and better quantify errors in statistical models was considered in [39]. This method has been shown to improve the accuracy of structural refinements for compounds such as fluorapatite ($Ca_5(PO_4)_3F$), anglesite ($PbSO_4$), and barite ($BaSO_4$). Although various estimation methods exist in the statistical literature, MLE remains superior, as some alternatives rely on distributional properties such as moments. Methods that cannot be applied to the considered models include the Method of Moments and its generalization [40], as well as the L-Moments method [41].

Suppose that $\mathbf{x} = (x_1, \ldots, x_n)$ represents a random sample of size n. If the model parameters are denoted by $\boldsymbol{\theta} = (\mu, \sigma) \in \boldsymbol{\Theta} \subset \mathbb{R}^2$, consider the following maximization problem:

$$\begin{aligned} \underset{\boldsymbol{\theta}\in\boldsymbol{\Theta}}{\text{maximize}} \; & \ell(\boldsymbol{\theta}|\mathbf{x}) = \sum_{i=1}^{n} \log f(x_i; \mu, \sigma) \\ \text{s.t. } & \mu \in \mathbb{R}, \sigma \in \mathbb{R}_+, \end{aligned} \tag{12}$$

where $f(\cdot; \mu, \sigma)$ is given by (1), (4), or (7). By solving the above problem, the maximum likelihood estimators (MLEs) of $\boldsymbol{\theta}$, say, $\hat{\boldsymbol{\theta}} = (\hat{\mu}, \hat{\sigma})$, are obtained as follows:

$$\hat{\boldsymbol{\theta}} = \underset{\mu\in\mathbb{R}, \sigma\in\mathbb{R}_+}{\arg\max} \; \ell(\boldsymbol{\theta}|\mathbf{x}).$$

3.2. Maximum Product of Spacings

The maximum product of spacings (MPS) methodology was introduced by [42,43] and independently formalized by [44]. This method emerged as an alternative to maximum likelihood estimation (MLE), particularly to address certain limitations of MLE, such as inconsistencies in parameter estimation for heavy-tailed distributions or models with unknown shift parameters. MPS is a special case of a broader class of estimation methods using spacings [45]. The methodology maximizes the product of spacings between ordered data points, leading to robust parameter estimation in scenarios where likelihood-based approaches might fail, for instance, when the likelihood function becomes unbounded. MPS is especially advantageous in cases involving heavy-tailed distributions or models with unknown shift parameters, as the likelihood function can become unbounded, leading to inconsistencies in MLE. This issue is prevalent in fields such as reliability studies and hydrology, where accurate modeling of extreme events is crucial [46]. Furthermore, unbounded MLE occurs when the likelihood function (or its natural logarithm) becomes arbitrarily large without converging to a finite maximum, commonly observed in heavy-tailed probability distributions. The challenges of applying MLE to such distributions with limited data were discussed in [47], who suggested that the estimators might fail to exhibit the expected asymptotic properties, potentially leading to inaccuracies in parameter estimation.

Suppose that $\mathbf{x} = (x_{1:n}, \ldots, x_{n:n})$ are the ordinary order statistics of a random sample of size n, i.e., $x_{1:n} < \cdots < x_{n:n}$. Under the MPS framework, consider the following maximization problem:

$$\begin{aligned} \underset{\boldsymbol{\theta} \in \boldsymbol{\Theta}}{\text{maximize}}\ S(\boldsymbol{\theta}|\mathbf{x}) &\propto \sum_{i=1}^{n+1} \log \Delta_i \\ \text{s.t. } \mu \in \mathbb{R}&, \sigma \in \mathbb{R}_+, \end{aligned} \tag{13}$$

such that

$$\Delta_i = \begin{cases} F(x_{1:n}; \mu, \sigma) & \text{if } i = 1 \\ F(x_{i:n}; \mu, \sigma) - F(x_{i-1:n}; \mu, \sigma) & \text{if } 1 < i \leq n \\ 1 - F(x_{n:n}; \mu, \sigma) & \text{if } i = n+1 \end{cases}$$

where $F(\cdot; \mu, \sigma)$ is given by (2), (5), or (8). By solving the above problem, the maximum product of spacings estimators (MPSEs) of $\boldsymbol{\theta}$, say, $\tilde{\boldsymbol{\theta}} = (\tilde{\mu}, \tilde{\sigma})$, are obtained as follows:

$$\tilde{\boldsymbol{\theta}} = \underset{\mu \in \mathbb{R}, \sigma \in \mathbb{R}_+}{\arg\max}\ S(\boldsymbol{\theta}|\mathbf{x}).$$

4. Intensity Data Analysis

The data for this study were derived from an experimental analysis of a lead-tin alloy conducted using X-ray diffraction (XRD) methods. This study focuses on three samples: a standard lead-tin sample, a recycled ingot, and a sample prepared from powdered recycled ingot treated thermally. The XRD analysis captures critical diffraction properties, including intensity and integrated intensity, which are primary indicators of the material's structural and phase composition. The intensity measurements, recorded as peak heights in the diffraction spectra, indicate the relative abundance and orientation of specific crystalline phases. Peaks corresponding to the lead (Pb), tin (Sn), and tin oxide (SnO) phases were identified. The recycled and thermally treated samples revealed the emergence of a stable tin oxide phase, as evidenced by distinct diffraction peaks. The integrated intensity, representing the area under each diffraction peak, provides a more comprehensive measure of the contribution of each phase. It accounts for the peak height and width, offering insights into phase proportions and grain sizes. For instance, this study highlights how the integrated intensity varies across phases in the samples, reflecting changes in composition due to thermal and mechanical treatments. These intensity-based metrics were pivotal in determining the structural transformations and phase distributions across the samples. Their analysis facilitated a deeper understanding of the effects of recycling and thermal treatments on the alloy's properties, particularly the emergence of the SnO phase and its role in mechanical strengthening [26].

All numerical results in this section and the next section were obtained using `R`, a statistical computing environment [48]. The data analysis involved several key steps. Since the data were assumed to originate from a heavy-tailed population, summary statistics, such as the mean and standard deviation, are not meaningful. Instead, robust measures, including the minimum, first quartile (Q_1), median, third quartile (Q_3), maximum, and median absolute deviation (MAD), were computed for each dataset to characterize the distribution. To estimate the parameters of interest, the objective functions (12) and (13) were maximized using the `R` function `optim` with the `L-BFGS-B` algorithm [49–51]. The L-BFGS-B method is an efficient optimization technique for problems with box constraints, leveraging a limited-memory BFGS approximation of the Hessian matrix to handle optimization problems effectively. This approach directly incorporates parameter bounds without requiring additional transformations or barrier methods. For initialization, the MAD statistic was used as the starting value for the scale parameter σ. However, the location parameter μ lies within the domain of the random variable, which violates one of the regularity conditions for maximum likelihood estimators (MLEs). Several regularity conditions ensure the existence, consistency, asymptotic normality, and efficiency of MLEs. These include but are not limited to the log-likelihood function's differentiability and the Fisher information's finiteness. For further details, see [52,53]. To address this issue, a re-parameterization was introduced, defining $\mu = x_{1:n} - \delta$, where $0 < \delta < x_{1:n}$. This transformation ensured that the regularity conditions were satisfied. The gradients of the objective functions with respect to δ and σ, necessary for the BFGS algorithm, are provided in Appendix A. For each estimation method and dataset, the goodness of fit was assessed using the Kolmogorov–Smirnov (KS) [54,55], Anderson–Darling (AD) [56], and Cramér–von Mises (CvM) [57,58] statistics. Since the data contained ties and the model parameters were estimated, the p-values for these statistics were calculated using a parametric bootstrap approach with $B = 1000$ resamples as follows: First, the model parameters μ and σ were estimated from the observed data, yielding $\hat{\mu}$ and $\hat{\sigma}$. These estimates were then used to generate random samples from the specified distribution. For each bootstrap sample, the goodness-of-fit statistic was computed, and this process was repeated B times to produce a distribution of the test statistics. The p-value was calculated as the proportion of bootstrap statistics exceeding the observed value. This approach accounts for ties in the data and the estimation of model parameters. For foundational knowledge, readers should consider [59], which provides an excellent introduction, while Davison and Hinkley [60] offer a comprehensive treatment of bootstrap methods and their applications. For those seeking an accessible introduction, Chernick [61] explains the bootstrap with an emphasis on conceptual understanding.

Table 1 presents the previously mentioned statistics. These statistics reveal that both datasets are heavy-tailed since the distance between Q_3 and the maximum is drastically large. Table 2 shows the data analysis for the intensity datasets alongside the considered goodness-of-fit metrics, while Table 3 provides similar information but for the integrated intensity dataset. It is important to note that the presence of ties complicated the calculation of the MPSEs, as the standard method could not be applied directly. Instead, a generalized version of the MPS method was employed to obtain the required estimators. Additional details about this generalization can be found in [62]. The aforementioned tables indicate that the HCA distribution fits the data better than the LCA and CBS models. To strengthen this conclusion, Figures 4–7 visually support the claim that the HCA model is suitable for both datasets, while Table 4 presents the information criteria based on MLE, namely the log-likelihood (ℓ), Akaike Information Criterion (AIC) [63], and Bayesian Information Criterion (BIC) [64]. All information criteria emphasize that the HCA model should be considered for the two datasets.

Table 1. Summary statistics for intensity and integrated intensity datasets. Clearly, the gaps between the median, the third quartile, and the maximum are large, which indicates that the data might follow a heavy-tailed probability distribution.

Data	Minimum	Q_1	Median	Q_3	Maximum	MAD
Intensity	13	36.25	73.5	171.25	1482	70.4235
Integrated Intensity	42	193.8	364.5	681	8106	364.7196

Table 2. Intensity data analysis and goodness of fit. The HCA model yielded the lowest KS and CvM statistics and the highest AD statistic. All parametric bootstrap p-values exceeded commonly used significance levels, providing insufficient evidence to reject the assumption that these models adequately fit the data. Notably, the HCA model had the highest p-value.

Model	Method	$\hat{\mu}$	$\hat{\sigma}$	KS	p-Value	AD	p-Value	CvM	p-Value
HCA	MLE	13	65.316	0.078	0.823	0.731	0.846	0.035	0.817
	MPS	9.916	72.580	0.069	0.917	0.314	0.651	0.040	0.677
LCA	MLE	13	62.011	0.111	0.451	0.675	0.342	0.073	0.479
	MPS	12.753	65.154	0.110	0.573	0.717	0.391	0.073	0.594
CBS	MLE	13	61.033	0.091	0.677	0.654	0.511	0.053	0.626
	MPS	11.943	65.618	0.095	0.622	0.502	0.485	0.052	0.639

Table 3. Integrated intensity data analysis and goodness of fit. The HCA model recorded the smallest KS and CvM statistics while achieving the largest AD statistic. Since all parametric bootstrap p-values were above standard significance thresholds, there is no substantial evidence to reject the hypothesis that these models appropriately describe the data. Notably, the HCA model had the highest p-value.

Model	Method	$\hat{\mu}$	$\hat{\sigma}$	KS	p-Value	AD	p-Value	CvM	p-Value
HCA	MLE	42	344.446	0.074	0.875	0.710	0.939	0.030	0.891
	MPS	25.669	384.754	0.097	0.436	0.292	0.703	0.045	0.601
LCA	MLE	42	345.985	0.098	0.636	0.657	0.356	0.076	0.462
	MPS	41.417	372.070	0.102	0.680	0.744	0.374	0.088	0.516
CBS	MLE	42	340.019	0.080	0.839	0.617	0.609	0.050	0.646
	MPS	37.582	370.683	0.098	0.572	0.494	0.491	0.061	0.525

Table 4. Information criteria calculated using MLEs, showing that the HCA outperforms the other models.

Data	Model	ℓ	AIC	BIC
Intensity	HCA	−235.35	474.69	477.97
	LCA	−241.29	486.59	489.86
	CBS	−238.74	481.47	484.75
Integrated Intensity	HCA	−294.22	592.43	595.71
	LCA	−299.82	603.65	606.92
	CBS	−297.22	598.44	601.71

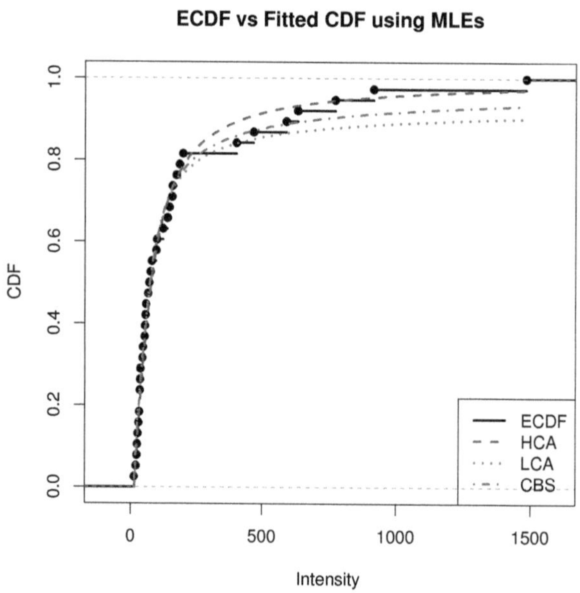

Figure 4. Fitting CDFs to intensity data and comparing them to the empirical CDF (ECDF), assuming MLEs.

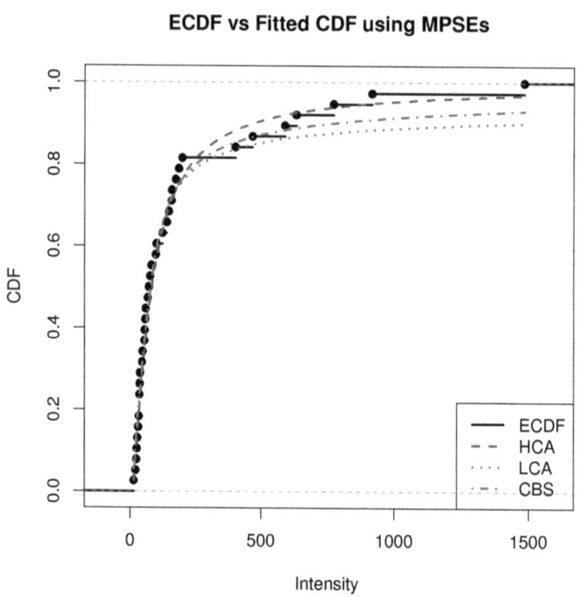

Figure 5. Fitting CDFs to intensity data and comparing them to the ECDF, assuming MPSEs rather than MLEs.

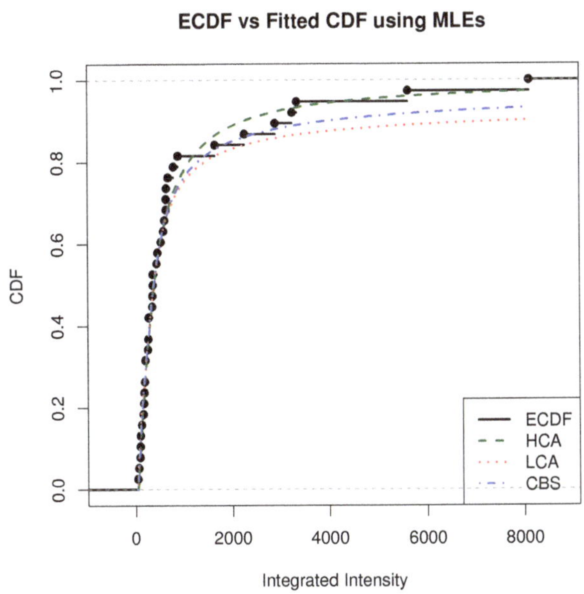

Figure 6. Fitted CDFs and the ECDF based on integrated intensity data under the MLE framework.

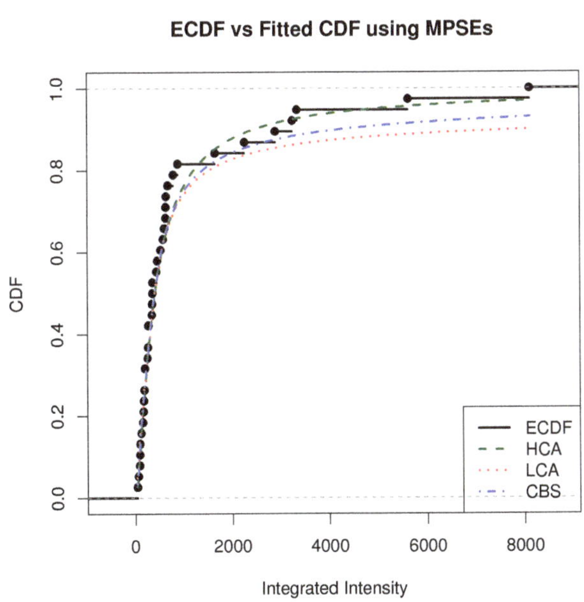

Figure 7. Fitted CDFs and the ECDF based on integrated intensity data under the MPS framework.

5. Monte Carlo Simulation Study

Simulation studies are crucial for assessing the performance of estimation methods under various scenarios, providing a robust framework for comparing their efficiency and accuracy. In practice, various estimators can be obtained for the parameters of a given model. To evaluate and compare the performance of these estimators, researchers often conduct Monte Carlo simulation studies, which offer numerical insights from both statistical and computational perspectives. Examples of such studies include [65–73], among others.

This section presents the outcomes of Monte Carlo simulation experiments designed to evaluate the performance of two estimation methods under varying sample sizes and parameter configurations. These simulations provide a basis for identifying the most reliable and efficient estimation method for the considered model. The simulation results are based on 1000 random samples from the considered distributions with different combinations of values for the model parameters and sample sizes. That is, the simulation study assumes $n = 10, 20, \ldots, 100$, $\mu = 0$, and $\sigma = 1$, without loss of generality. The root mean squared error (RMSE) is used as the evaluation metric to measure estimation efficiency, and it is defined as

$$\mathrm{RMSE}(\hat{\mu}) = \sqrt{\frac{1}{N}\sum_{i=1}^{N}(\hat{\mu}_i - \mu)^2} \text{ and } \mathrm{RMSE}(\hat{\sigma}) = \sqrt{\frac{1}{N}\sum_{i=1}^{N}(\hat{\sigma}_i - \sigma)^2},$$

where $N = 1000$ and $\hat{\mu}_i$ ($\hat{\sigma}_i$) is an estimator of the model parameter μ (σ) based on simulation repetition i. In addition to estimation efficiency, the goodness of fit of the estimated model parameters is evaluated using two metrics: the average absolute difference between the true and estimated CDFs (D_{abs}) and the maximum absolute difference between the true and estimated CDFs (D_{max}). These are determined as

$$D_{\mathrm{abs}} = \frac{1}{n \times N}\sum_{i=1}^{N}\sum_{j=1}^{n}\left|F(x_j; \mu, \sigma) - F(x_j; \hat{\mu}_i, \hat{\sigma}_i)\right|,$$

and

$$D_{\mathrm{max}} = \frac{1}{N}\sum_{i=1}^{N}\max_{j=1,\ldots,n}\left|F(x_j; \mu, \sigma) - F(x_j; \hat{\mu}_i, \hat{\sigma}_i)\right|,$$

respectively, where $F(x; \mu, \sigma)$ is given by (2), (5), or (8), and $\hat{\mu}_i$ ($\hat{\sigma}_i$) is an estimator of the model parameter μ (σ) based on simulation repetition i.

An estimator is considered computationally efficient when the simulated RMSE approaches zero as the sample size increases, regardless of the underlying probabilistic model. Figure 8 demonstrates that the MPSEs of μ are more efficient than the MLEs, whereas Figure 9 indicates that the MLEs of σ outperform their MPSE counterparts. In the goodness-of-fit analysis, a pair of estimators is deemed effective if the metrics D_{abs} and D_{max} approach zero as the sample size increases and remain robust to the heaviness of the distribution's tail. Such a pair minimizes the differences between the true and estimated CDFs. Figures 10 and 11 show that the MPSEs exhibit superior performance in terms of the goodness of fit compared to the MLEs. As the sample size increases, both methods have an indistinguishable behavior.

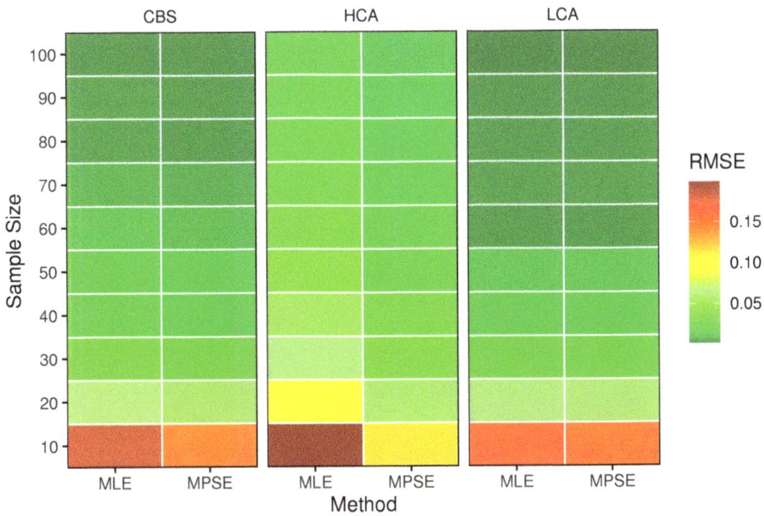

Figure 8. Simulated RMSEs for the estimators of μ. The MPSE outperforms the MLE for small samples. However, as the sample size increases, the behavior becomes similar, except for the HCA distribution, where the MPSE of μ is superior to its MLE counterpart even as the sample size increases.

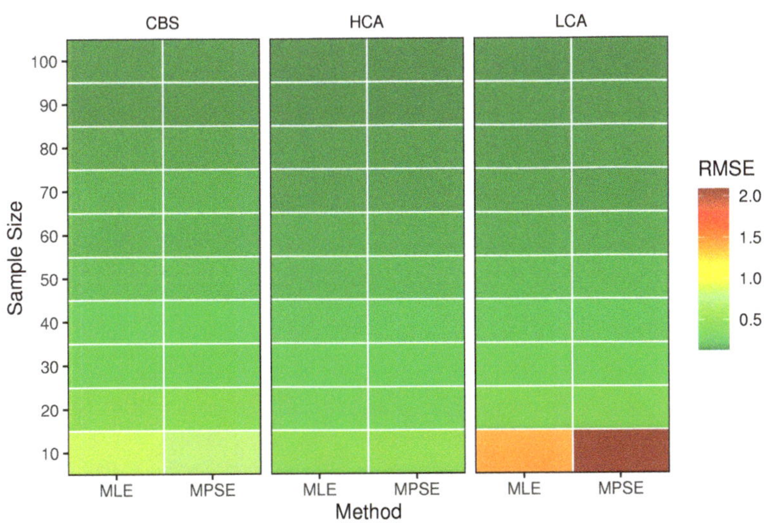

Figure 9. Simulated RMSEs for the estimators of σ. As the sample size increases, both methods behave similarly. However, the MLE performs better than the MPSE for small sample sizes.

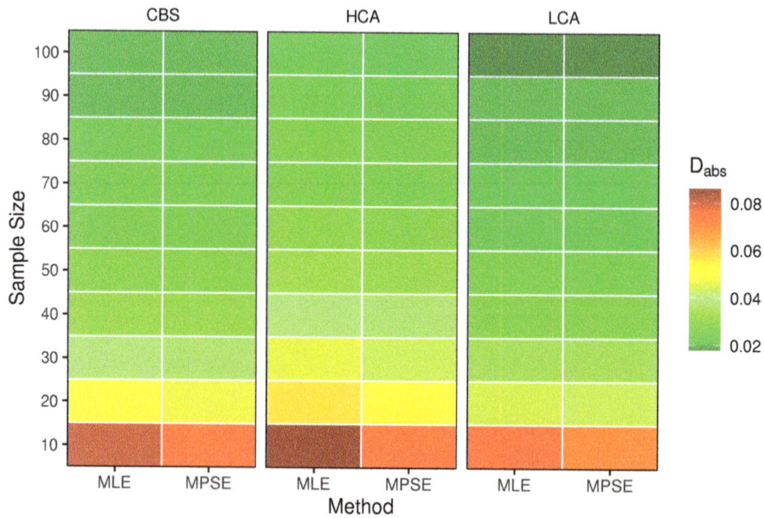

Figure 10. Simulated average absolute difference between the true and estimated CDFs. The MPSE framework provides fitted models that are better than those of its MLE counterpart. As the sample size increases, both methods behave similarly.

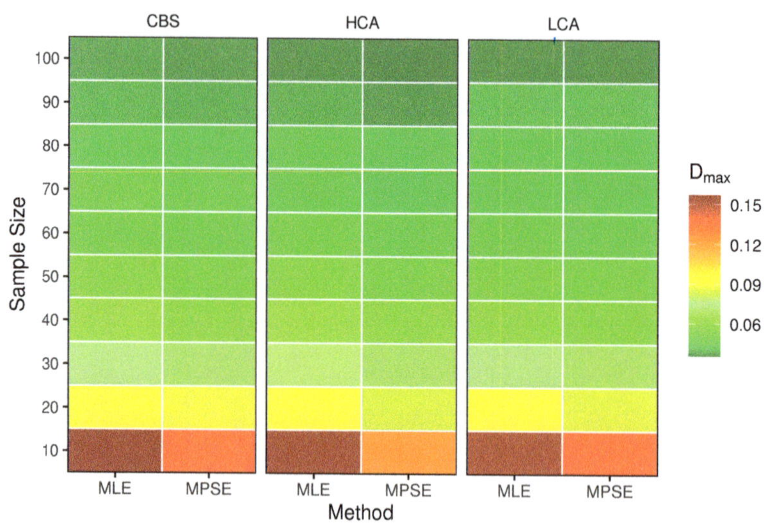

Figure 11. Simulated maximum absolute difference between the true and estimated CDFs. Once again, the MPSE framework outperforms its MLE counterpart in terms of the goodness-of-fit criterion. As the sample size increases, both methods behave similarly.

6. Conclusions

This study emphasizes the critical role of statistical modeling in enhancing the reliability and understanding of X-ray diffraction (XRD) intensity data, mainly through the utilization of heavy-tailed location-scale probability distributions. By addressing the challenges associated with high-intensity diffraction peaks and structural variability, this research demonstrates the applicability of heavy-tailed distributions in capturing the inherent variability of XRD intensity metrics. Three heavy-tail, location-scale probabilistic models are considered in this research. For each model, maximum likelihood and maximum product of spacings estimators are obtained for the model parameters. This study evaluated the performance of these estimators via Monte Carlo simulations and applied them to real-world datasets of lead-tin alloys. Data analysis revealed that the half-Cauchy distribution is potentially the best model to fit the intensity and integrated intensity data, followed by the Cauchy Birnbaum–Saunders distribution. Monte Carlo simulations showed that the MPS is the optimal framework for estimating μ, while the MLEs of σ are the preferred estimators for small sample sizes. In terms of the goodness of fit, the MPSEs are the optimal choice. Nevertheless, as the sample size increases, all methods perform similarly.

There remain several research directions to explore for these heavy-tail models. For example, could estimation under right censoring improve efficiency? An important avenue for future research is investigating estimation under conventional, progressive, and hybrid censoring schemes, assuming the considered probability distributions or their generalizations and extensions. Another promising direction is exploring alternative estimation methods, such as least-squares-based approaches. Lastly, both MLEs and MPSEs are frequentist estimation methods. A significant contribution to the statistical literature would be a future study examining Bayesian or other estimation frameworks and comparing their performance to MLEs and MPSEs under heavy-tail models.

Funding: This research received no external funding.

Institutional Review Board Statement: Not applicable.

Informed Consent Statement: Not applicable.

Data Availability Statement: No new data were created or analyzed in this study. Data sharing is not applicable to this article.

Acknowledgments: The author wishes to express sincere gratitude to editors and the two anonymous reviewers for their constructive and insightful comments and recommendations, which significantly improved the quality of this article.

Conflicts of Interest: The author declares no conflicts of interest.

Appendix A. Mathematical Expressions

This appendix presents the mathematical expressions for the objective functions and the elements of their gradients required by the L-BFGS-B algorithm to perform the maximization process. Note that $z_i = \sigma^{-1}(x_{i:n} - \mu)$, and it is straightforward to show that $\frac{\partial \ell}{\partial \delta} = -\frac{\partial \ell}{\partial \mu}$ since $\mu = x_{1:n} - \delta$. Moreover, the first-order derivatives of the objective function (13) under the MPS framework with respect to μ and σ are defined as

$$\frac{\partial S}{\partial \mu} = \sum_{i=1}^{n} \frac{D_\mu \Delta_i}{\Delta_i}, \text{ and } \frac{\partial S}{\partial \mu} = \sum_{i=1}^{n} \frac{D_\sigma \Delta_i}{\Delta_i}$$

respectively, where

$$D_\mu \Delta_i = \begin{cases} -f(x_{1:n};\mu,\sigma) & \text{if } i = 1 \\ -[f(x_{i:n};\mu,\sigma) - f(x_{i-1:n};\mu,\sigma)] & \text{if } 1 < i \leq n \\ f(x_{n:n};\mu,\sigma) & \text{if } i = n+1 \end{cases}$$

and

$$D_\sigma \Delta_i = \begin{cases} -z_i f(x_{1:n};\mu,\sigma) & \text{if } i = 1 \\ -z_i[f(x_{i:n};\mu,\sigma) - f(x_{i-1:n};\mu,\sigma)] & \text{if } 1 < i \leq n \\ z_i f(x_{n:n};\mu,\sigma) & \text{if } i = n+1 \end{cases},$$

where $f(\cdot;\mu,\sigma)$ is given by (1), (4), or (7).

Appendix A.1. Half-Cauchy Distribution

The first log-likelihood function and its first-order derivatives with respect to μ and σ are given by

$$\ell(\boldsymbol{\theta}|\mathbf{x}) = \text{Constant} - n\log\sigma - \sum_{i=1}^{n}\log\left[1+z_i^2\right],$$

$$\frac{\partial\ell}{\partial\mu} = \frac{2}{\sigma}\sum_{i=1}^{n}\frac{z_i}{(1+z_i^2)}, \text{ and } \frac{\partial\ell}{\partial\sigma} = \frac{2}{\sigma}\sum_{i=1}^{n}\frac{z_i^2}{(1+z_i^2)} - \frac{n}{\sigma}.$$

Appendix A.2. log-Cauchy Distribution

For this model, the log-likelihood function and its first-order derivatives with respect to μ and σ are given by

$$\ell(\boldsymbol{\theta}|\mathbf{x}) = \text{Constant} - \sum_{i=1}^{n}\log(x_{i:n}-\mu) - \sum_{i=1}^{n}\log\left\{1+[\log(z_i)]^2\right\},$$

$$\frac{\partial\ell}{\partial\mu} = \frac{2}{\sigma}\sum_{i=1}^{n}\frac{\log(z_i)}{z_i(1+[\log(z_i)]^2)} + \sigma^{-1}\sum_{i=1}^{n}\frac{1}{z_i}, \text{ and } \frac{\partial\ell}{\partial\sigma} = \frac{2}{\sigma}\sum_{i=1}^{n}\frac{\log(z_i)}{(1+[\log(z_i)]^2)}.$$

Appendix A.3. Cauchy Birnbaum–Saunders Distribution

The final log-likelihood function and its first-order derivatives with respect to μ and σ are given by

$$\ell(\boldsymbol{\theta}|\mathbf{x}) = \text{Constant} + \sum_{i=1}^{n}\log B(x_{i:n},\mu,\sigma) - \sum_{i=1}^{n}\log(x_{i:n}-\mu) - \sum_{i=1}^{n}\log\left\{1+[A(x_{i:n},\mu,\sigma)]^2\right\},$$

$$\frac{\partial\ell}{\partial\mu} = -(2\sigma)^{-1}\sum_{i=1}^{n}\frac{A(x_{i:n},\mu,\sigma)}{z_iB(x_{i:n},\mu,\sigma)} + \sigma^{-1}\sum_{i=1}^{n}\frac{A(x_{i:n},\mu,\sigma)B(x_{i:n},\mu,\sigma)}{z_i\left\{1+[A(x_{i:n},\mu,\sigma)]^2\right\}} + \sigma^{-1}\sum_{i=1}^{n}\frac{1}{z_i},$$

and

$$\frac{\partial\ell}{\partial\sigma} = -(2\sigma)^{-1}\sum_{i=1}^{n}\frac{A(x_{i:n},\mu,\sigma)}{B(x_{i:n},\mu,\sigma)} + \sigma^{-1}\sum_{i=1}^{n}\frac{A(x_{i:n},\mu,\sigma)B(x_{i:n},\mu,\sigma)}{\left\{1+[A(x_{i:n},\mu,\sigma)]^2\right\}}.$$

References

1. Eckert, M. Disputed discovery: The beginnings of X-ray diffraction in crystals in 1912 and its repercussions. *Z. Krist.* **2012**, *227*, 27–35. [CrossRef]
2. Scherrer, P. *Determination of Crystallite Size and Strain Using X-Ray Diffraction*; Springer: Berlin/Heidelberg, Germany, 1918.
3. Waseda, Y.; Matsubara, E.; Shinoda, K. *X-Ray Diffraction Crystallography: Introduction, Examples and Solved Problems*; Springer: Berlin/Heidelberg, Germany, 2011. [CrossRef]
4. Lee, A.; Brown, M. Quantitative Deviation of Nanocrystals Using the RIR Method in X-ray Diffraction. *Nanomaterials* **2022**, *12*, 2320. [CrossRef]
5. Green, S.; Patel, R. Applications of X-ray Powder Diffraction Microstructural Analysis to Clay Minerals. *Minerals* **2023**, *14*, 584. [CrossRef]
6. Disha, S.A.; Sahadat Hossain, M.; Habib, M.L.; Ahmed, S. Calculation of crystallite sizes of pure and metals doped hydroxyapatite engaging Scherrer method, Halder-Wagner method, Williamson-Hall model, and size-strain plot. *Results Mater.* **2024**, *21*, 100496. [CrossRef]
7. Alam, M.K.; Hossain, M.S.; Bahadur, N.M.; Ahmed, S. A comparative study in estimating of crystallite sizes of synthesized and natural hydroxyapatites using Scherrer Method, Williamson-Hall model, Size-Strain Plot and Halder-Wagner Method. *J. Mol. Struct.* **2024**, *1306*, 137820. [CrossRef]
8. Lim, J.S.; Yam, F.K. Structural parameters of CVD synthesized Ga_2O_3 nanostructures from X-ray diffraction analysis derived by Scherrer, Williamson-Hall, Size-Strain Plot and Halder-Wagner methods–A comparative study. *Phys. B Condens. Matter* **2025**, *699*, 416798. [CrossRef]

9. Abramo, M.C.; Costa, D.; Malescio, G.; Munaò, G.; Pellicane, G.; Prestipino, S.; Caccamo, C. Structure factors and X-ray diffraction intensities in molten alkali halides. *J. Phys. Commun.* **2020**, *4*, 075017. [CrossRef]
10. Chen, W.; Kumar, A. Advantages of Using Hard X-ray Photons for Ultrafast Diffraction Studies of Crystals. *Photonics* **2023**, *10*, 948. [CrossRef]
11. Ameh, E.S.; Martinez, C. A review of basic crystallography and X-ray diffraction applications. *Int. J. Adv. Manuf. Technol.* **2019**, *105*, 3289–3302. [CrossRef]
12. Zhang, X.M.; Zheng, X.; Li, X.L.; Meng, F.Q.; Yin, S.S. A method with ultra-high angular resolution for X-ray diffraction experiments. *J. Synchrotron Radiat.* **2024**, *31*, 35–41. [CrossRef] [PubMed]
13. Salgado, J.E.; Lerman, S.; Du, Z.; Xu, C.; Abdolrahim, N. Automated classification of big X-ray diffraction data using deep learning models. *npj Comput. Mater.* **2023**, *9*, 214. [CrossRef]
14. Szymanski, N.J.; Fu, S.; Persson, E.; Ceder, G. Integrated analysis of X-ray diffraction patterns and pair distribution functions for machine-learned phase identification. *npj Comput. Mater.* **2024**, *10*, 45. [CrossRef]
15. Racioppi, S.; Otero-de-la Roza, A.; Hajinazar, S.; Zurek, E. Powder X-ray diffraction assisted evolutionary algorithm for crystal structure prediction. *Digit. Discov.* **2025**, *4*, 73–83. [CrossRef]
16. Kleiner, I. Evolution of the Function Concept: A Brief Survey. *Coll. Math. J.* **1989**, *20*, 282–300. [CrossRef]
17. Johnson, N.L.; Kotz, S.; Balakrishnan, N. *Continuous Univariate Distributions*, 2nd ed.; Wiley: Hoboken, NJ, USA, 1994; Volume 1.
18. Johnson, N.L.; Kotz, S.; Balakrishnan, N. *Continuous Univariate Distributions*, 2nd ed.; Wiley: Hoboken, NJ, USA, 1994; Volume 2.
19. Foss, S.; Korshunov, D.; Zachary, S. *An Introduction to Heavy-Tailed and Subexponential Distributions*; Springer Series in Operations Research and Financial Engineering; Springer: Berlin/Heidelberg, Germany, 2013. [CrossRef]
20. Resnick, S.I. Heavy-Tailed Distributions and Their Properties. In *Extreme Values, Regular Variation, and Point Processes*; Springer: Berlin/Heidelberg, Germany, 2013; pp. 1–45. [CrossRef]
21. Suhov, Y.; Kelbert, M. Heavy-Tailed and Long-Tailed Distributions. In *An Introduction to Heavy-Tailed and Subexponential Distributions*; Springer: Berlin/Heidelberg, Germany, 2014; Chapter 2, pp. 25–50. [CrossRef]
22. Suhov, Y.; Kelbert, M. Heavy-Tailed Probability Distributions: Some Examples of Their Use. *Mathematics* **2023**, *11*, 3094. [CrossRef]
23. Vogel, R.M.; Papalexiou, S.M.; Lamontagne, J.R.; Dolan, F.C. When Heavy Tails Disrupt Statistical Inference. *Am. Stat.* **2024**. [CrossRef]
24. Stojanović, V.S.; Jovanović Spasojević, T.; Bojičić, R.; Pažun, B.; Langović, Z. Cauchy–Logistic Unit Distribution: Properties and Application in Modeling Data Extremes. *Mathematics* **2025**, *13*, 255. [CrossRef]
25. Liu, F.; Lu, C.; Zhao, Y.; Jiang, C.; Yuan, Z.; Shi, T.; Su, B. Severe Plastic Deformation of Bulk Polycrystalline Beryllium at Room Temperature. *SSRN* **2025**. [CrossRef]
26. Abdul Razzaq, A.H. Analytical Study Using X-Ray Diffraction Methods and Hardness Test for Lead-Tin Alloy. *Iraqi J. Sci.* **2012**, *53*, 322–329.
27. Polson, N. G.; Scott, J. G. On the Half-Cauchy Prior for a Global Scale Parameter. *Bayesian Anal.* **2012**, *7*, 887–902. [CrossRef]
28. Jaya, I.G.N.M.; Kristiani, F.; Andriyana, Y.; Chadidjah, A. Sensitivity Analysis on Hyperprior Distribution of the Variance Components in Bayesian Spatiotemporal Disease Mapping. *Mathematics* **2023**, *12*, 451. [CrossRef]
29. Lemmens, J.; Matthys, P.; Vandewalle, J. Modeling with the Log-Cauchy Distribution: Analyzing Heavy-Tailed Data. *Ann. Oper. Res.* **2004**, *135*, 111–123. [CrossRef]
30. Yin, Z.Y.; Peng, S.L.; Ren, H.; Guo, Q.; Chen, Z.H. LogCauchy, log-sech and lognormal distributions of species abundances in forest communities. *Ecol. Model.* **2005**, *184*, 329–340. [CrossRef]
31. Birnbaum, Z.; Saunders, S. A new family of life distributions. *J. Appl. Probab.* **1969**, *6*, 319–327. [CrossRef]
32. Birnbaum, Z.W.; Saunders, S.C. Estimation for a Family of Life Distributions with Applications to Fatigue. *J. Appl. Probab.* **1969**, *6*, 328–347. [CrossRef]
33. Desmond, A. Stochastic models of failure in random environments. *Can. J. Stat.* **1985**, *13*, 171–183. [CrossRef]
34. Balakrishnan, N.; Kundu, D. Birnbaum-Saunders distribution: A review of models, analysis, and applications. *Appl. Stoch. Model. Bus. Ind.* **2019**, *35*, 4–49. [CrossRef]
35. Diáz-Garciá, J.A.; Leiva-Sánchez, V. A new family of life distributions based on the elliptically contoured distributions. *J. Statist. Plann. Inference* **2005**, *128*, 445–457; Erratum in *J. Stat. Plan. Inference* **2007**, *137*, 1512–1513. [CrossRef]
36. Wang, R.; Xu, X.; Gu, B. Statistical Analysis of two-Parameter Generalized Birnbaum-Saunders Cauchy Distribution. *J. Appl. Anal. Comput.* **2020**, *10*, 1614–1635. [CrossRef]
37. Alam, F.M.A. On Comparing Different Methods of Estimation for the Parameters of a Pathological Distribution with Application to Climate Data. *Austrian J. Stat.* **2022**, *51*, 120–147. [CrossRef]
38. Yoshioka, S.; Murakami, T.; Kuroiwa, Y.; Kimura, S. Analysis of X-ray powder diffraction data using the maximum likelihood method for cubic and tetragonal systems. *J. Am. Ceram. Soc.* **2000**, *83*, 1777–1781. [CrossRef]

39. Toraya, H. An analytical method for observed powder diffraction intensity data based on maximum likelihood estimation. *Adv. X-Ray Anal.* **1994**, *37*, 189–199.
40. Hansen, L.P. Large Sample Properties of Generalized Method of Moments Estimators. *Econometrica* **1982**, *50*, 1029. [CrossRef]
41. Hosking, J.R.M. L-Moments: Analysis and Estimation of Distributions Using Linear Combinations of Order Statistics. *J. R. Stat. Soc. Ser. B Stat. Methodol.* **1990**, *52*, 105–124. [CrossRef]
42. Cheng, R.C.H.; Amin, N.A.K. Maximum product-of-spacings estimation with applications to the lognormal distribution. In *Technical Report 1*; University of Wales IST, Department of Mathematics: Cardiff, Wales, 1979.
43. Cheng, R.C.H.; Amin, N.A.K. Estimating Parameters in Continuous Univariate Distributions with a Shifted Origin. *J. R. Stat. Soc. Ser. B (Methodol.)* **1983**, *45*, 394–403. [CrossRef]
44. Ranneby, B. The Maximum Spacing Method. An Estimation Method Related to the Maximum Likelihood Method. *Scand. J. Stat.* **1984**, *11*, 93–112.
45. Ghosh, K.; Jammalamadaka, S. A general estimation method using spacings. *J. Stat. Plan. Inference* **2001**, *93*, 71–82. [CrossRef]
46. Kateri, M.; Nikolov, N.I. Maximum Product of Spacings Estimator Under Type-I Censoring with an Application to a Step-Stress Model with Weibull Lifetimes. *IEEE Trans. Reliab.* **2024**, *54*, 375–383. [CrossRef]
47. Larsen, P. Operational risk models and asymptotic normality of maximum likelihood estimation. *J. Oper. Risk* **2016**, *11*, 55–78. [CrossRef]
48. R Core Team. *R: A Language and Environment for Statistical Computing;* R Foundation for Statistical Computing: Vienna, Austria, 2024.
49. Nocedal, J.; Wright, S.J. *Numerical Optimization*; Springer: New York, NY, USA, 2006. [CrossRef]
50. Byrd, R.H.; Lu, P.; Nocedal, J.; Zhu, C. A Limited Memory Algorithm for Bound Constrained Optimization. *SIAM J. Sci. Comput.* **1995**, *16*, 1190–1208. [CrossRef]
51. Zhu, C.; Byrd, R.H.; Lu, P.; Nocedal, J. Algorithm 778: L-BFGS-B: Fortran Subroutines for Large-Scale Bound-Constrained Optimization. *ACM Trans. Math. Softw.* **1997**, *23*, 550–560. [CrossRef]
52. Self, S.G.; Liang, K.Y. Asymptotic Properties of Maximum Likelihood Estimators and Likelihood Ratio Tests under Nonstandard Conditions. *J. Am. Stat. Assoc.* **1987**, *82*, 605–610. [CrossRef]
53. Casella, G.; Berger, R.L. *Statistical Inference*, 2nd ed.; Duxbury Press: Belmont, CA, USA, 2002.
54. Kolmogorov, A. Sulla determinazione empirica di una legge di distribuzione. *G. Dell'Istituto Ital. Degli Attuari* **1933**, *4*, 83–91.
55. Smirnov, N. Table for Estimating the Goodness of Fit of Empirical Distributions. *Ann. Math. Stat.* **1948**, *19*, 279–281. [CrossRef]
56. Anderson, T.W.; Darling, D.A. Asymptotic Theory of Certain "Goodness-of-Fit" Criteria Based on Stochastic Processes. *Ann. Math. Stat.* **1952**, *23*, 193–212. [CrossRef]
57. Cramér, H. On the Composition of Elementary Errors. *Scand. Actuar. J.* **1928**, *1928*, 13–74. [CrossRef]
58. von Mises, R. *Wahrscheinlichkeit, Statistik und Wahrheit;* Julius Springer: Vienna, Austria, 1928.
59. Efron, B.; Tibshirani, R.J. *An Introduction to the Bootstrap;* Chapman and Hall/CRC: Boca Raton, FL, USA, 1994.
60. Davison, A.C.; Hinkley, D.V. *Bootstrap Methods and Their Applications;* Cambridge University Press: Cambridge, UK, 1997.
61. Chernick, M.R. *Bootstrap Methods: A Guide for Practitioners and Researchers*, 2nd ed.; Wiley-Interscience: Hoboken, NJ, USA, 2008.
62. Murage, P.; Mungátu, J.; Odero, E. Optimal Threshold Determination for the Maximum Product of Spacing Methodology with Ties for Extreme Events. *Open J. Model. Simul.* **2019**, *07*, 149–168. [CrossRef]
63. Akaike, H. A new look at the statistical model identification. *IEEE Trans. Autom. Control* **1974**, *19*, 716–723. [CrossRef]
64. Schwarz, G.E. Estimating the Dimension of a Model. *Ann. Stat.* **1978**, *6*, 461–464. [CrossRef]
65. Gupta, R.D.; Kundu, D. Generalized exponential distribution: Different method of estimations. *J. Stat. Comput. Simul.* **2001**, *69*, 315–337. [CrossRef]
66. Alkasasbeh, M.R.; Raqab, M.Z. Estimation of the generalized logistic distribution parameters: Comparative study. *Stat. Methodol.* **2009**, *6*, 262–279. [CrossRef]
67. Mazucheli, J.; Louzada, F.; Ghitany, M. Comparison of estimation methods for the parameters of the weighted Lindley distribution. *Appl. Math. Comput.* **2013**, *220*, 463–471. [CrossRef]
68. do Espirito Santo, A.; Mazucheli, J. Comparison of estimation methods for the Marshall–Olkin extended Lindley distribution. *J. Stat. Comput. Simul.* **2014**, *85*, 3437–3450. [CrossRef]
69. Balakrishnan, N.; Alam, F.M.A. Maximum likelihood estimation of the parameters of Student's *t* Birnbaum-Saunders distribution: a comparative study. *Commun. Stat. Simul. Comput.* **2019**, *51*, 793–822. [CrossRef]
70. Nassar, M.; Okasha, H.; Albassam, M. E-Bayesian Estimation and Associated Properties of Simple Step–Stress Model for Exponential Distribution Based on Type-II Censoring. *Qual. Reliab. Eng. Int.* **2021**, *37*, 997–1016. [CrossRef]
71. Alam, F.M.A.; Nassar, M. On Modeling Concrete Compressive Strength Data Using Laplace Birnbaum-Saunders Distribution Assuming Contaminated Information. *Crystals* **2021**, *11*, 830. [CrossRef]

72. Moloy, D.J.; Ali, M.A.; Alam, F.M.A. Modeling Climate data using the Quartic Transmuted Weibull Distribution and Different Estimation Methods. *Pak. J. Stat. Oper. Res.* **2023**, *19*, 649–669. [CrossRef]
73. Alosaimi, B.S.; Alam, F.M.; Baaqeel, H.M. Modeling Air Pollution Data Using a Generalized Birnbaum-Saunders Distribution with Different Estimation Procedures. In *Mathematical Modeling in Physical Sciences*; Springer: Cham, Switzerland, 2024; pp. 587–618. [CrossRef]

crystals

MDPI

Article

Effects of Y Additions on the Microstructure and Mechanical Properties of $CoCr_{1.7}Ni$ Medium-Entropy Alloys

Shaoshuai Zhou [1], Xiaoyong Shu [1,*], Linli Hu [2], Xunyu Yuan [1], Panpan Qiu [1] and Xiwen Xu [1]

1 Jiangxi Provincial Engineering Research Center for Surface Technology of Aeronautical Materials, Nanchang Hangkong University, Nanchang 330063, China; 15797674461@163.com (X.Y.); panpanqiu@126.com (P.Q.); nchu2012224@163.com (X.X.)

2 School of Mechanical Engineering, Jiangxi Vocational College of Mechanical and Electrical Technology, Nanchang 330013, China; hulinli98012@126.com

* Correspondence: xiaoyong202@126.com

Abstract

In order to improve the room temperature yield strength of X and enhance its engineering applicability, a series of $CoCr_{1.7}NiY_x$ (x = 0, 0.01, 0.02, 0.03, 0.04, and 0.1 at.%) medium-entropy alloys were synthesized to investigate the effect of Y addition on the microstructures and mechanical properties of the $CoCr_{1.7}Ni$-based alloy. The X-ray diffraction results show that the alloys exhibit face-centered cubic (FCC) + body-centered cubic (BCC) + hexagonal close packing (HCP) triphasic structure when the Y is adopted, whereas the $CoCr_{1.7}Ni$-based alloy has a FCC+BCC biphasic structure. The volume fraction of BCC and HCP phase increased with increasing Y content, which led to alloy grain refinement. As a result, the microhardness and strength of alloys were both enhanced. The addition of Y resulted in dispersion strengthening and solid solution strengthening of $CoCr_{1.7}Ni$ alloy, the appearance of HCP, and an increase in BCC, which improved the room temperature yield strength and hardness of $CoCr_{1.7}Ni$ alloy. In particular, for $CoCr_{1.7}NiY_{0.1}$ alloy, its microhardness and yield strength, respectively, increased by 98.18% and 260.59% as compared with those of $CoCr_{1.7}Ni$ alloy.

Keywords: medium-entropy alloys; rare earth element; yttrium addition; microstructure; mechanical properties

1. Introduction

In recent years, medium-entropy alloys (MEAs) and high-entropy alloys (HEAs) have become a quite popular topic; they contain at least three principal metallic elements rather than only one, as in traditional alloys [1–3]. Due to the thermodynamic high-entropy effect, structural lattice distortion effect, kinetic hysteresis diffusion effect, and property "cocktail" effect [4], MEAs and HEAs have excellent properties, such as cryogenic strength and toughness, thermal phase stability, wear resistance, corrosion resistance, and so on [5–8]. CoCrNi MEAs with single face-centered cubic (FCC) structures have unique low-temperature mechanical properties and outperform most HEAs in terms of strength and plasticity [9–11]. Despite this, CoCrNi MEAs struggle to meet the demands of practical engineering applications because of low room temperature yield strength. Therefore, it is of scientific significance to improve the room temperature yield strength of CoCrNi MEAs.

To improve the mechanical properties of MEAs, most researchers generally use the method of changing element contents and adding elements. Our previous work investigated the effects of Cr content on the microstructure and mechanical properties of $CoCr_xNi$

MEAs [12]. The yield strength is enhanced with increasing Cr content due to the precipitated hard phase of Cr-rich BCC precipitated from the FCC phase matrix. In particular, for $CoCr_{1.7}Ni$ MEA, it has excellent yield strength, and meanwhile maintains relatively good plasticity. Hong et al. [13] add 1.0 at.% Y into CoCrNi MEAs to improve the strength of the alloy, and the yield strength increases to 782.6 $\pm$ 22.8 MPa. Lee et al. [14] added 7.5 at.% Al into CoCrNi MEAs to precipitate the NiAl-rich B2 phase from the FCC phase matrix, resulting in significant grain refinement and yield strength increases to 680 MPa. Increasing the Al, Ti, and Ta elements not only promotes precipitation, but also improves the thermal stability of the nanoscale γ-phase and enhances the high-temperature yield strength (700 °C) of $(CoCrNi)_{95}Al_2Ti_2Ta$ MEA up to 620 MPa [15]. Furthermore, elements such as Hf [16], Nb [17,18], Si [19], and Mo [20–22] were also used to enhance the mechanical properties of CoCrNi MEAs.

Although the positive effect of the above elements on the mechanical properties of HEAs/MEAs has been confirmed, they can only improve the yield strength to a small extent (most of the improvement rates are limited to 100%); CoCrNi alloys have to face the dilemma of substandard yield strength. The Y element, with a hexagonal close-packed (HCP) structure, is perfectly suited to solve this problem. With increasing Y content, two hexagonal structure phases ($CaCu_5$ type and Ni_3Y type) precipitated from the FCC phase matrix substantially boosted the yield strength of $CoCrFeNiY_x$ HEAs (from 202 MPa to 1440 MPa, with an improvement rate of 712.87%) [23]. The Y has more outstanding properties for improving the yield strength of HEAs/MEAs than other elements. Y induced the FCC phase precipitatation from the BCC phase matrix, leading to excellent ultimate tensile strength and plasticity of $CoCr_3Fe_5NiY_xAl_y$ HEAs [24]. Furthermore, the ability of Y to improve high-temperature oxidation resistance [25,26] and corrosion resistance [27] has been confirmed.

Therefore, combined with our previous work [12], to further improve the room temperature yield strength of $CoCr_{1.7}Ni$ MEA and make it more suitable for practical engineering applications, a small amount (~0.1 at.%) of Y was added to $CoCr_{1.7}Ni$ MEA. The effects of Y content on the crystal structure, microstructure, and room temperature mechanical properties of $CoCr_{1.7}NiY_x$ MEAs were studied, and the mechanism will be discussed.

2. Material and Methods

2.1. Preparation of $CoCr_{1.7}NiY_x$ MEAs

$CoCr_{1.7}NiY_x$ MEAs (x = 0, 0.01, 0.02, 0.03, 0.04 and 0.1) were prepared by a vacuum electric arc melting furnace (WK-II). The raw materials adopted Co, Cr, Ni, and Y metal particles with a purity higher than 99.9%. To ensure all metals were melted completely, the raw materials were placed in the copper crucible in descending order of melting point. The ingot was repeatedly turned over six times under the protection of high-purity argon gas to ensure the uniformity of the alloy composition.

2.2. Composition and Structure Characterization

The crystal structure of the alloys was analyzed using an X-ray diffractometer (XRD, D8ADVANCE-A25, Bruker, Billerica, MA, USA) with Cu-Kα ray radiation operated in the scan 2θ range of 20–100° at a scan rate of 5°/min. Microstructures were characterized by scanning electron microscopy (SEM, FEI Nova Nano SEM450, IOCB Prague, Praha, Czech Republic) equipped with electron backscatter diffraction (EBSD) and an energy-dispersive X-ray spectrometer (EDS) detector, and the volume fraction of the precipitated phase was calculated. All samples for XRD and SEM were sliced to 10 $\times$ 10 $\times$ 3 mm by wire electrical emancipation machining (WEDM), sanded to 2000 # SiC sandpaper, and polished with diamond polishing paste.

A transmission electron microscope (TEM, Talos F200X, Thermo Fisher Scientific, Waltham, MA, USA) with an EDS detector was used to further observe the microstructure and confirm the elemental distribution. Specimens of 10 × 10 × 0.3 mm were sliced by an EDM wire cutter and polished to 100 μm by SiC sandpaper, after which they were punched into φ3 mm discs and pre-thinned, and finally thinned by a pit thinner and an ion thinner. Finally, the TEM results were analyzed and processed using Digital Micrograph software (Gatan Digital Micrograph basic).

2.3. Mechanical Property Measurements

To evaluate the mechanical properties of the alloys at room temperature, microhardness tests and compression tests were carried out at room temperature. The microhardness tests were performed with a micro Vickers hardness tester (HVS-1000A, Jinan Hensgrand Instrument Co., Ltd., Jinan, China) under a load of 0.5 kg for 15 s. For each sample, 20 points were measured and averaged to ensure the accuracy of the measurement. The samples for microhardness tests were sliced into 10 × 10 × 0.3 mm by WEDM. Compression tests were performed using a universal electronic compression tester with a strain rate of $1 \times 10^{-3}\ s^{-1}$. The samples for compression tests were sliced into φ 4 × 6 mm by WEDM.

3. Results and Discussion

3.1. Microstructure and Phase Identification

Figure 1a shows the XRD patterns of $CoCr_{1.7}NiY_x$ MEAs. To facilitate the observation of the HCP diffraction peaks, the XRD pattern of Y-0.1 alloy was enlarged, as shown in Figure 1b. It can be found that Y-0 alloy without the addition of Y has a biphasic FCC + BCC structure. No new phase precipitation was seen in the XRD pattern after the addition of a small amount of Y; the Y-0, Y-0.01, Y-0.02, Y-0.03 and Y-0.04 MEAs were still FCC + BCC structures. The crystal structure of the alloys transformed from FCC + BCC biphasic structure to FCC + BCC + HCP triphasic structure at $x = 0.1$; a new third phase was recognized as a hexagonal structure (YNi_5, $CaCu_5$ type), and the space group of the HCP phase was identified as P6/mm. The peak intensity of the HCP precipitates rose together with the Y content, indicating an increasing volume fraction of precipitates. This indicated that the addition of Y induced the formation of the HCP phase in $CoCr_{1.7}NiY_x$ MEAs.

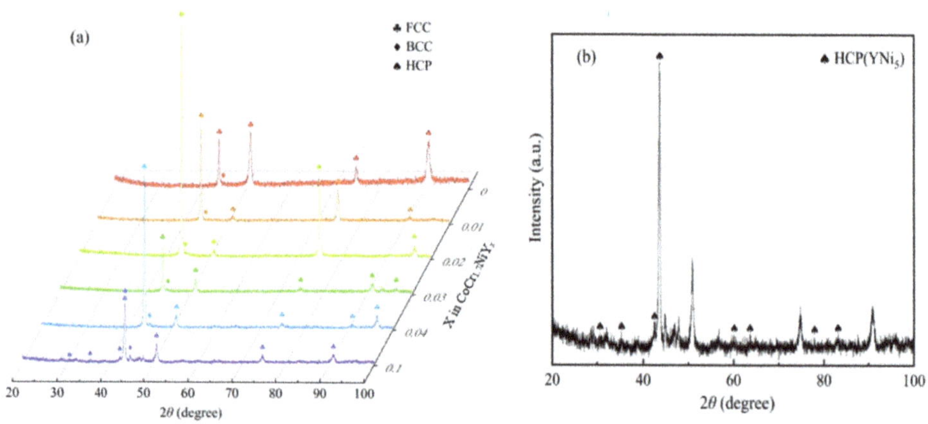

Figure 1. (**a**) The XRD patterns of $CoCr_{1.7}NiY_x$ MEAs with a scan rate of 5°/min. (**b**) The enlarged XRD pattern of Y-0.1 MEA.

The SEM images in Figure 2a–f show the microstructures of $CoCr_{1.7}NiY_x$ MEAs. In the FCC phase matrix of Y-0 MEA, a small number of BCC phases were scattered (Figure 2a). After adding Y, precipitates appeared in Y-0.01 (Figure 2b) with random distributions at the grain boundaries and inside the grains. Although the atomic radius of the Y element (r Y = 1.80 Å) is larger than those of Co, Cr, and Ni (r Co = 1.25 Å, r Cr = 1.28 Å, r Ni = 1.24 Å), the dissolution of Y in the CoCrNi FCC matrix would cause lattice distortion. Combined with the XRD phase analysis of Y-0.1 MEA, it is inferred that the non-spot precipitates were YNi_5 with an HCP phase structure. This diffraction peak was not reflected in the XRD patterns of Y-0.01, Y-0.02, Y-0.03, and Y-0.04 MEAs, probably due to the small volume fraction of the HCP phase and the limited detection accuracy of the XRD equipment. The volume proportion of both precipitates rose with the rise in Y content, according to SEM analysis.

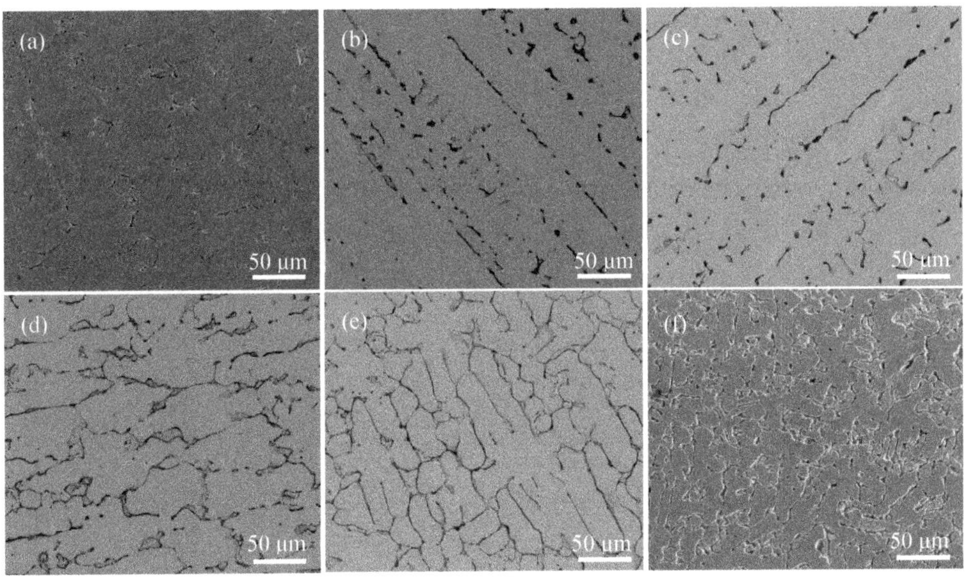

Figure 2. SEM images of $CoCr_{1.7}NiY_x$ MEAs for (**a**) Y-0, (**b**) Y-0.01, (**c**) Y-0.02, (**d**) Y-0.03, (**e**) Y-0.04, and (**f**) Y-0.1.

Figure 3 displays a succession of EBSD maps for Y-0.01, Y-0.02, Y-0.03, Y-0.04, and Y-0.1. The alloy possessed equiaxed grains, but the grain sizes were not consistent. Additionally, the photos showed deformation twinning. Twins typically arise when the energy of the stacking fault is low [28]. The emergence of twin boundaries restricts the dislocations' motion, which lowers SFE. As a result, the twins in the metal would promote the high yield strength of the alloys. The fraction of twin borders only marginally rose with the rise in Y-doping content, according to the results.

Since no HCP phase diffraction peaks were found in the XRD patterns of Y-0.01, Y-0.02, Y-0.03, and Y-0.04 MEAs, the elemental distribution of Y-0.03 MEA was qualitatively characterized by EBSD and TEM-EDS to further confirm the presence of HCP precipitated phase, and the analytical results are shown in Figures 4a–e and 5a–e, respectively. The variation of precipitation phase volume fraction with Y content is shown in Figure 4f, which increases with the increase in Y content, and the precipitation phase volume fraction gradually increases.

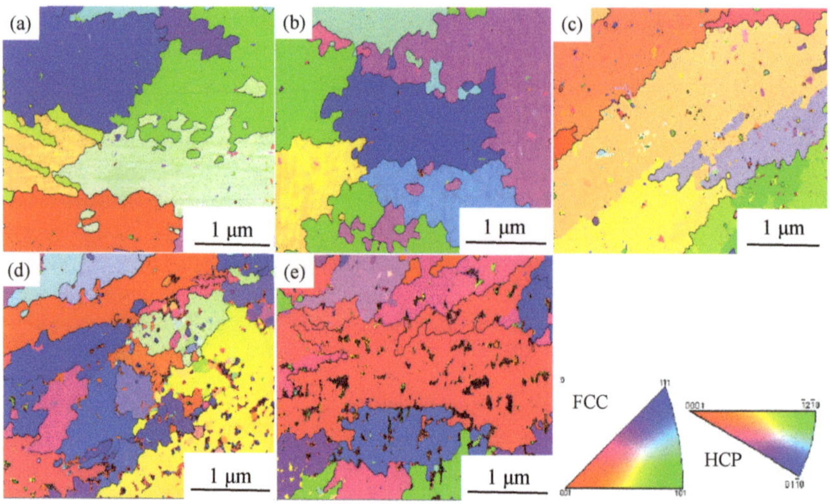

Figure 3. EBSD images of (**a**) Y-0.01, (**b**) Y-0.02, (**c**) Y-0.03, (**d**) Y-0.04 and (**e**) Y-0.1 showing twins in alloys.

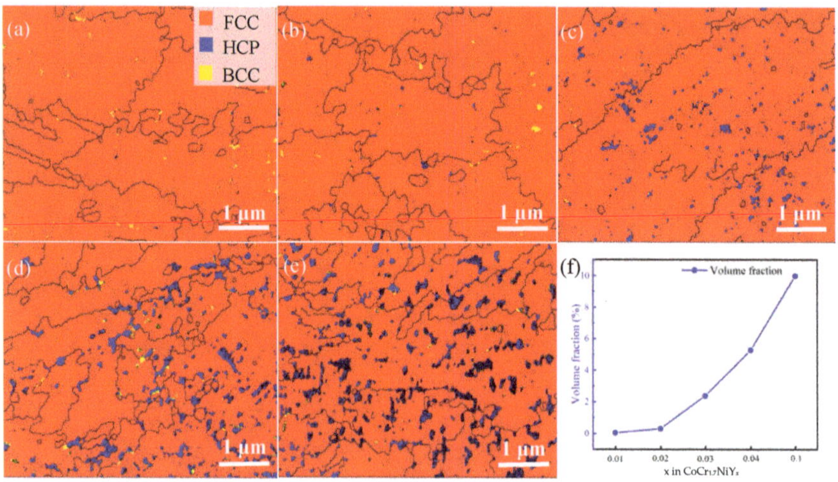

Figure 4. EBSD images of (**a**) Y-0.01, (**b**) Y-0.02, (**c**) Y-0.03, (**d**) Y-0.04 and (**e**) Y-0.1 showing grain boundaries (black line), precipitates (blue) and Cr-rich organization (yellow) in alloys. (**f**) The volume fraction of precipitates and matrix grain size as functions of Y-content in the alloy.

The EBSD images of Y-0.01, Y-0.02, Y-0.03, Y-0.04 and Y-0.1 are presented in Figure 4a–e, with the grain boundaries and precipitates highlighted with black lines and blue, respectively, for a clear view of the grain boundaries and precipitates. It was evident from the illustration that the precipitates (in blue) were dispersed at random along the grain borders and within the grains. Figure 4f depicts the volume fraction of precipitates as a function curve of Y content and it rose as Y content increased, peaking at about 9.97% for Y-0.1. Additionally, a small amount of Cr-rich structure (in yellow) was visible both inside and at the grain borders. The grain size reduces as the Y content increases, as illustrated in Figure 2a–f. This occurrence was brought about by the inhibition of grain formation, which was brought about by the Y element in the solid solution matrix as well as the presence of precipitates [14].

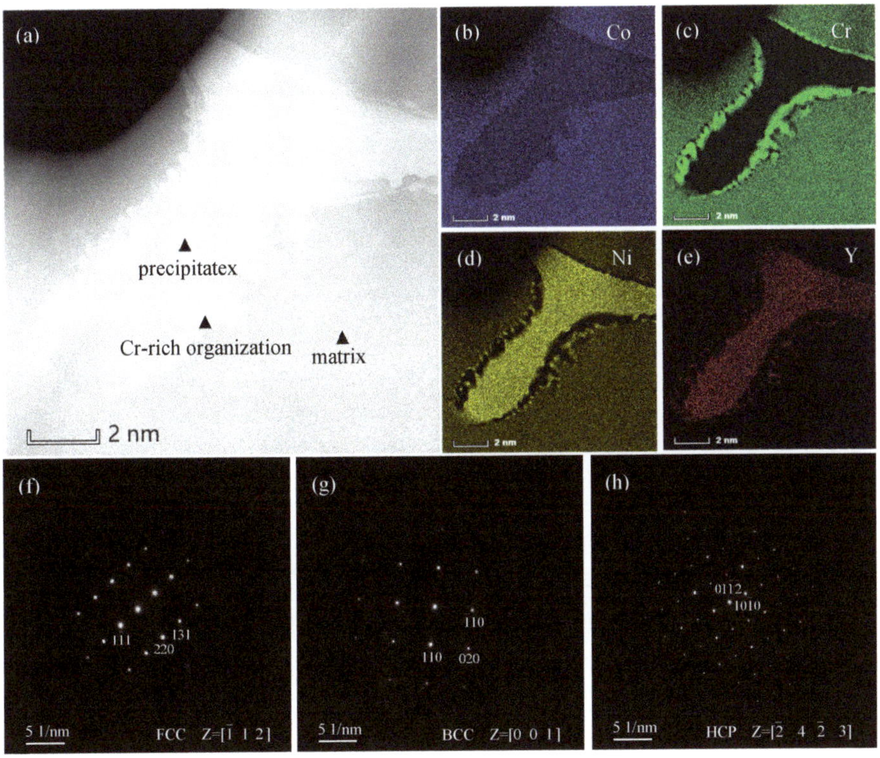

Figure 5. (**a**) TEM bright-field image, TEM-EDS patterns of (**b**) Co, (**c**) Cr, (**d**) Ni, and (**e**) Y, and SAED pattern of (**f**) matrix, (**g**) precipitates with spots, and (**h**) precipitates without spots of Y-0.03 MEA.

TEM pattern observations were carried out to further investigate the precipitates. A micro-scaled precipitate was seen in the TEM bright-field image of Y-0.03, shown in Figure 5a. The corresponding selected area electron diffraction (SAED) images for the matrix, the precipitate and the Cr-rich organization are shown in Figure 5f–h, respectively. The main diffraction spots along the [$\overline{1}$ 1 2] crystalline band axis (Figure 5f) confirmed the FCC phase structure of the matrix. The main diffraction spots along the [0 0 1] zone axis SAED of the matrix (Figure 5g) confirmed the BCC phase for Cr-rich organization, while the SAED pattern of the precipitate taken along the [$\overline{2}$ 4 $\overline{2}$ 3] zone axis (Figure 5h) confirmed the HCP phase for the precipitate. From the TEM-EDS images in Figure 5b–e, it could be found that the precipitated HCP phase consisted of a large amount of Ni and Y elements, and a small amount of Co element. The Cr-rich organization of the BCC phase structure was distributed at the edge of the HCP phase. In contrast, the Co, Cr, and Ni elements were more uniformly distributed in the FCC phase matrix. Moreover, the lattice constants of the three different phase structures were consistent with the PDF cards, and the results of the TEM analysis of Y-0.03 MEA were also compatible with the XRD and BSE analysis results, which further confirmed that $CoCr_{1.7}NiY_x$ MEAs are in FCC + BCC + HCP triphasic structure, and the precipitates of two kinds are Cr-rich organization in BCC phase structure and YNi_5 in HCP phase structure, respectively.

3.2. Mechanical Properties at Room Temperature

Figure 6a shows the experimental result of the microhardness of $CoCr_{1.7}NiY_x$ MEAs at room temperature. Table 1 shows the increase in (compared to Y-0) and the averages of microhardness of $CoCr_{1.7}NiY_x$ MEAs at room temperature. Compared with Y-0 MEA, the microhardness of Y-0.01 and Y-0.02 MEAs increased slightly (less than 10.00%), the microhardness of Y-0.03 and Y-0.04 MEAs was improved significantly (25–35%), while the microhardness of Y-0.1 MEA increased to 342.358 HV (an increase up to 98.17%). As shown in Figure 6a, the maximum, minimum and range values of the hardness data measured for each alloy also maintained an increasing trend. Since the test sites of microhardness were randomly selected, the hardness measured was relatively low when the test site was on the FCC matrix and relatively high when the test site was on the HCP precipitate. The maximum value of each set of data increased with the increase in Y content, indicating that the increase in Y content raised the volume fraction of the HCP precipitated phases. In this process, the range value of each dataset also increased with the increase in Y content, indicating that the increase in hardness of the HCP precipitated phase is larger than that of the FCC matrix hardness. The above results show that the increase in Y content would increase the microhardness of $CoCr_{1.7}NiY_x$ significantly through solid solution strengthening and precipitation strengthening, and the contribution of precipitation strengthening to hardness is higher than that of solid solution strengthening in this system. Therefore, we believed that the improvement of the mechanical properties of $CoCr_{1.7}Ni$ HEA by adding Y was mainly attributed to precipitation strengthening.

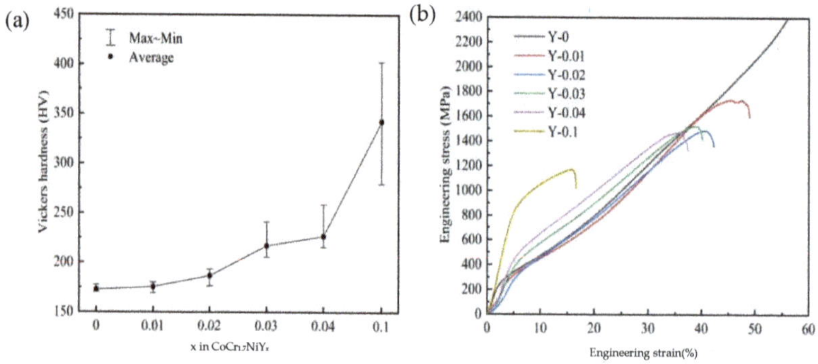

Figure 6. (**a**) Microhardness obtained by Vickers hardness test, and (**b**) the compressive stress–strain curves of CoCr1.7NiY$_x$ MEAs.

Table 1. The increase in (compared to Y-0) and the averages of microhardness of $CoCr_{1.7}NiY_x$ MEAs at room temperature.

	Y-0	Y-0.01	Y-0.02	Y-0.03	Y-0.04	Y-0.1
Average microhardness (HV)	172.75	175.47	186.69	216.76	226.66	342.35
Rate of increase (%)	-	1.57	8.07	25.47	31.20	98.17

Figure 6b shows the compressive stress–strain curves of $CoCr_{1.7}NiY_x$ MEAs at room temperature. The measured yield strength, ultimate compressive strength and compression are listed in Table 2. The compressive properties of $CoCr_{1.7}NiY_x$ MEAs at room temperature were closely related to the amount of Y added, and their yield strength would increase significantly with the increase in Y content. The yield strength of Y-0.01 MEA was increased to 266 MPa, which was 12.71% higher than Y-0 MEA, while the plasticity of the alloys deteriorated. The ultimate compressive strength was 1732 MPa, and the compression decreased to 47.5%. The yield strengths of Y-0.02, Y-0.03, Y-0.04, and Y-0.1 MEAs were further improved with the increase in Y content. In particular, the yield strength of Y-0.1 MEA was up to 851 MPa, which was 260.59% higher than Y-0 MEA. However, the ultimate compressive strength decreased to 1170 MPa, and the plasticity of this alloy decreased much further, with the compression rate falling to 15.7%. From this, it can be seen that with the increase in the amount of Y added, the yield strength increases, but the ultimate yield strength decreases, especially for Y-0.1 MEAs. Moreover, with the increase in the amount of Y added, the deformation of the sample after failure and cracking decreases continuously after reaching the ultimate yield strength, indicating that the plasticity of the alloy decreases with the increase in the amount of Y added.

Table 2. The measured yield strength, ultimate compressive strength, and compression of $CoCr_{1.7}NiY_x$ MEAs.

	Yield Strength (MPa)	UCS (MPa)	Compression (%)
Y-0	236	-	-
Y-0.01	266	1732	47.5
Y-0.02	307	1485	40.6
Y-0.03	398	1522	39.0
Y-0.04	455	1473	35.9
Y-0.1	851	1170	15.7

3.3. Strengthening Mechanism

Since the phase structure has an essential influence on the alloys, it is an important characteristic that affects the microstructure and mechanical properties. Therefore, in this section, the effects of the Y content variation on the mechanical properties of $CoCr_{1.7}Ni$ MEAs and the corresponding strengthening mechanism are illustrated by analyzing the mechanism of phase structure formation of $CoCr_{1.7}NiY_x$ MEAs.

Based on the measurement of the hardness of the matrix grains (Figure 6a), both solid solution strengthening and grain boundary strengthening contributed sporadically to the increase in hardness that resulted from the addition of Y to the alloy. It is clear from Figure 6a that the presence of HCP precipitates, whose hardness was significantly higher than that of the FCC matrix, contributed significantly to the hardening. A minor strength improvement of Y-0.01 was seen in the compression test (Figure 6b) compared to the $CoCr_{1.7}Ni$ MEA, but the alloy's strength rose dramatically in the presence of precipitates. Therefore, it was evident from both the hardness measurements and the compression tests that precipitation had a significant role in the strengthening and hardening of our system. However, an increase in the volume percent of the HCP phase would lead to a decrease in ductility since the FCC matrix includes more slip systems than HCP precipitates. Therefore, the increased Y content resulted in a significant increase in the strength of $CoCr_{1.7}NiY_x$ MEAs, but simultaneously, it also led to a deterioration in the plasticity of the alloys. The presence of deformation twins in Figure 3 does help to strengthen, but the inclusion of Y has essentially little impact on the boundary fraction of deformation twins, it should be highlighted. Therefore, we think that precipitation strengthening was mostly responsible for the increase in the mechanical characteristics of CoCr1.7Ni HEA caused by the addition of Y.

The relationship between the microhardness and yield strength of $CoCr_{1.7}NiY_x$ MEAs and the integral number of HCP precipitates are shown in Figure 7, and a good linear fit is obtained. Compared to the study by Hong et al. [13], we added less Y content but obtained better mechanical properties. Under the same Y content, the microhardness and yield strength were almost twice that of them. Most importantly, we did not conduct subsequent rolling and heat treatment.

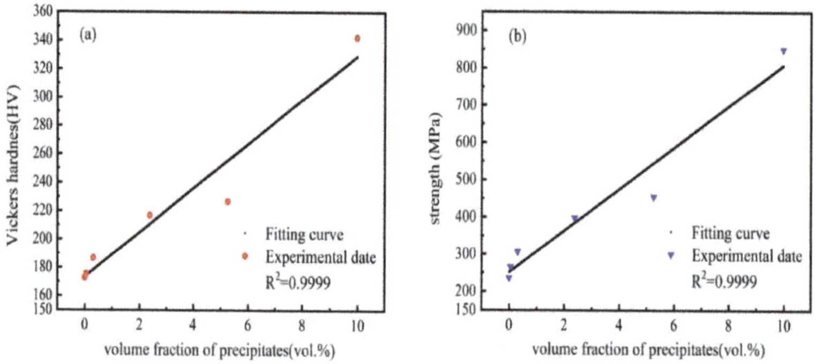

Figure 7. Microhardness (**a**) and yield strength (**b**) as functions of volume fraction of precipitates.

The fitting equation for the relationship between microhardness and volume fraction of HCP precipitates is as follows:

$$\mathrm{HV} = 172.7 + 64.8V_{HCP} - 36V_{HCP}^{1.5} + 3V_{HCP}^{2.5} - 0.27V_{HCP}^{3} \tag{1}$$

The fitting equation for the relationship between yield strength and HCP precipitation phase volume fraction is as follows:

$$\mathrm{F} = 236.2 + 137.5V_{HCP}^{0.5} - 16.7V_{HCP} - 3.4V_{HCP}^{2.5} + 1.4V_{HCP}^{3} \tag{2}$$

where HV and F are microhardness and yield strength, respectively; V_{HCP} is the volume fraction of HCP precipitates.

From the above two equations, it can be seen that the relationship between microhardness, yield strength, and volume fraction of HCP precipitates is positively correlated. The addition of Y increases the content of volume fraction of HCP precipitates, thus increasing microhardness and yield strength.

From the crystal structure and microstructure analysis in Section 3.1, it was clear that the volume fraction of the BCC and HCP phase structures increased gradually with the increase in Y content. For multi-component alloys, the Hume–Rothery law is usually used to analyze the mechanism of phase structure formation of the alloys [29–33]. And the main influencing factors include the valence electron concentration (VEC), mixing enthalpy (ΔH_{mix}), mixing entropy (ΔS_{mix}), melting point (T_m), relative coefficient of enthalpy–entropy (Ω), atomic size difference (δ), and Gibbs free energy (ΔG_{mix}). The specific calculation method is shown in Equations (3)–(9).

$$VEC = \sum_{i=1}^{n} c_i (VEC)_i \tag{3}$$

$$\Delta H_{mix} = \sum_{i=1}^{n-1} \sum_{j=i+1}^{n} 4\Delta H_{mix}^{ij} c_i c_j \tag{4}$$

$$\Delta S_{mix} = -R\sum_{i=1}^{n} c_i \ln c_i \tag{5}$$

$$T_m = \sum_{i=1}^{n} c_i (T_m)_i \tag{6}$$

$$\Omega = \frac{T_m \Delta S_{mix}}{|\Delta H_{mix}|} \tag{7}$$

$$\delta = \sqrt{\sum_{i=1}^{n} c_i \left(1 - \frac{r_i}{r}\right)^2} \quad (8)$$

$$\Delta G_{mix} = \Delta H_{mix} - T_m \Delta S_{mix} \quad (9)$$

In the equation, c_i represents the molar percentage of the ith element, $(VEC)_i$ represents the number of valence electrons of the ith element, ΔH_{mix}^{ij} represents the mixing enthalpy of the ij two-element alloy system, c_i and c_j represent the molar percentages of the ith and jth elements, R represents the gas constant (8.314 J/mol·K), $(T_m)_i$ represents the melting point of the ith element, r represents the average atomic radius of the alloys, and r_i represents the atomic radius of the ith element. The relevant parameters of $CoCr_{1.7}NiY_x$ MEAs (Table 3) were brought into Equations (3)–(8) to obtain the VEC, ΔH_{mix}, ΔS_{mix}, T_m, Ω, δ, and ΔG_{mix} of $CoCr_{1.7}NiY_x$ MEAs in Table 4.

Table 3. Mixing enthalpies, ΔH (kJ/mol), for different binary components [34].

Element (Atomic Radius (Å), Melting Point (°C), Valence Electrons)	Co	Cr	Ni	Y
Co (1.26, 1495, 9)	-	−4	0	−22
Cr (1.27, 1907, 6)	-	-	−7	11
Ni (1.24, 1453, 10)	-	-	-	−32
Y (1.80, 1522, 3)	-	-	-	-

Table 4. The valence electron concentration (VEC), mixing enthalpy (ΔH_{mix}), mixing entropy (ΔS_{mix}), melting point (T_m), relative coefficient of enthalpy–entropy (Ω), atomic size difference (δ), and Gibbs free energy (ΔG_{mix}) of CoCr1.7NiYx MEAs.

	ΔH_{mix} (kJ/mol)	ΔS_{mix} (J/mol·k)	Ω	δ/%	VEC	T_m(K)	ΔG_{mix}
Y-0	−5.46	8.85	2709	0.98	7.89	1672.94	−14,812
Y-0.01	−5.54	8.98	2713	2.43	7.88	1672.53	−15,028
Y-0.02	−5.61	9.08	2707	3.28	7.87	1672.13	−15,190
Y-0.03	−5.68	9.17	2698	3.95	7.85	1671.73	−15,333
Y-0.04	−5.75	9.25	2687	4.50	7.84	1671.33	−15,462
Y-0.1	−6.16	9.63	2609	6.87	7.76	1668.97	−16,077

According to the method proposed by Guo et al. [35] to forecast FCC and BCC phase formation in HEAs by valence electron concentration (VEC), when VEC $\geq$ 8, only the single FCC solid solution structure is formed in the alloys. When VEC $\leq$ 6.87, only the single BCC solid solution structure is formed in the alloys, and when 6.87 < VEC < 8, both FCC solid solution structure and BCC solid solution structure will be present in the alloys. From Table 4, it can be found that the VEC values of $CoCr_{1.7}NiY_x$ MEAs were all between 6.87 and 8, which was consistent with the coexistence of FCC and BCC phase structures. Combined with the analysis results of SEM and TEM, it was further confirmed that both FCC and BCC phase structures existed in $CoCr_{1.7}NiY_x$ MEAs. And with the addition of Y element, a three-phase structure of FCC, BCC, and HCP was formed in $CoCr_{1.7}NiY_x$ MEAs, further improving the mechanical properties, especially the microhardness and yield strength of $CoCr_{1.7}NiY_x$ MEAs.

As shown in Table 4, with the increase in Y content, the mixing enthalpy of the alloys kept decreasing, while the mixing entropy kept increasing. The enthalpy–entropy ratio of the alloys kept decreasing, while the atomic size difference kept increasing. In conjunction with [36], a higher enthalpy–entropy ratio promotes the formation of solid solution, while a lower enthalpy–entropy ratio facilitates the formation of intermetallic compounds, such as YNi_5. The mixing enthalpy between the elements (Table 3) can also reflect the binding force between them to some extent [36]. In the Co-Cr-Ni-Y system, the mixing enthalpy between Y and Ni (−32 kJ/mol) is the smallest. Therefore, it can be concluded that Y and Ni are more likely to aggregate and form a Y-Ni enriched phase than other atoms. And the lattice distortion in the alloys is more severe with the atomic size difference increasing. Table 4 shows the Gibbs free energy change trends of $CoCr_{1.7}NiY_x$ MEAs. It decreases with the increase in Y content, which indicates that the system becomes more and more stable due to the addition of Y.

The above analysis shows that the increase in Y content would promote the enrichment of Y and Ni, which led to the increasing volume fraction of YNi_5 HCP precipitates. According to the TEM-EDS patterns of Y-0.3 (Figure 5b–e), the HCP phase contains only a small amount of Co in addition to Y and Ni. This indicated that YNi_5 replaces the original Cr position and thus promotes the enrichment of Cr to precipitate the BCC phase. Therefore, the volume fraction of the BCC phase also increased with the increase in Y content.

Y has a certain degree of solid solution in CoCrNi MEAs, which means that Y elements with a larger atomic radius (1.80 Å) would displace Co, Cr, and Ni elements, resulting in changes of lattice constants, which in turn causes lattice distortion and solid solution strengthening. With the increase in Y content, the atomic size difference of the alloys was increased, which led to the increasing lattice distortion effect of the matrix of the FCC phase structure. Therefore, the alloys' matrix increased in hardness and strength under the influence of solid solution strengthening and dispersion strengthening. With the increase in Y content, the effects of solid solution strengthening and dispersion strengthening are more significant; the minimum values of hardness measured by Y-0.03, Y-0.04, and Y-0.1 MEAs showed a significant increase compared to Y-0 MEA (Figure 6a), which also indicated that the solid solution strengthening and dispersion strengthening increases the hardness of $CoCr_{1.7}NiY_x$ MEAs matrix. However, the solid solution strengthening and dispersion strengthening effects were smaller for small amounts of Y (Y-0.01 and Y-0.02 MEAS), which is due to the fact that solid solution strengthening is not only related to lattice mismatch but also parameters such as shear modulus mismatch [37,38]. And when the Y content is low, the sediment content is also called high, and the dispersion strengthening effect is not significant.

It can be found from Figure 2 that the number of precipitated object integrals in the HCP and BCC phase structures increased with increasing Y content, which led to the significant refinement of the grains of $CoCr_{1.7}NiY_x$ MEAs. When plastic deformation occurred in fine grains by external forces, it could be dispersed in more grains, and the plastic deformation was more uniform than coarse grains, and the stress concentration was more minor. Fine grains have more curved grain boundaries and larger surfaces, a feature that is not conducive to continuous crack expansion. As a result, the strength of the alloys is continuously increased by the effect of fine grain strengthening. In Figure 2, the Y-0.03 MEA shows a significant grain refinement compared to Y-0.02 MEA, combined with the increase in yield strength of Y-0.03 MEA compared to Y-0.02 MEA of up to 29.64% in Table 2. And the ultimate compressive strength increased instead of decreasing, which indicates that the fine grain strengthening contributed to the strength increase of $CoCr_{1.7}NiY_x$ MEAs.

The addition of Y promoted the increase in the volume fraction of both precipitates; the interaction between precipitates and dislocations blocked the dislocation motion and improved the deformation resistance of the alloys, so the strength and hardness of the alloys increased significantly with the increase in the volume fraction of precipitates. However, the plasticity of the alloys deteriorated continuously with the increase in the volume fraction of precipitates because the slip system of the HCP phase (one slip plane × three slip directions) is less than that of the FCC phase (four slip planes × three slip directions). In Y-0.1 MEA, the precipitates were uniformly distributed in the matrix as fine diffuse particles (Figure 2f). Compared with Y-0 MEA, its hardness increased up to 98.18%, and the yield strength increased up to 260.59%, indicating that the precipitation strengthening contributed significantly to the hardness and strength improvement of $CoCr_{1.7}NiY_x$ MEAs. The range value of hardness measured in Figure 7 increased with the increase in Y content, indicating that the hardness enhancement rate of the precipitated phase was much higher than that of the matrix, which also confirmed that the contribution of precipitation strengthening to the hardness of $CoCr_{1.7}NiY_x$ MEAs was significantly higher than that of solid solution strengthening.

4. Conclusions

In this study, the effects of Y addition on microstructures and room temperature mechanical properties of $CoCr_{1.7}NiY_x$ MEAs were investigated, and the following conclusions were summarized.

1. Without the addition of Y, the Y-0 alloy matrix has an FCC phase structure. The volume fraction of Cr-rich BCC precipitates was lower. The YNi_5 HCP precipitate formed in Y-0.01 MEA. The volume fraction of BCC and HCP phase precipitates increased with Y content, which led to the precipitation and dispersion strengthening of $CoCr_{1.7}NiY_x$ MEAs.
2. Most of the added Y elements were enriched with Ni to form HCP phase YNi_5 precipitates, and the remaining Y elements were solidly soluble in the FCC phase matrix. As the relatively large atomic radius of Y element replaces the position of Co, Cr and Ni elements in the matrix, lattice distortion is caused. At the same time, the degree of lattice distortion increased with the increase in Y addition, resulting in solid solution strengthening effect on $CoCr_{1.7}NiY_x$ MEAs.
3. The addition of element Y significantly improved the strength and hardness of the $CoCr_{1.7}NiY_x$ MEAs, and the degree of improvement increased with the increase in Y content, and the highest level of improvement is achieved when the Y content is 0.1. The microhardness and yield strength of Y-0.1 MEA reached 342 HV and 851 MPa, respectively; compared with Y-0 MEA, the increase reached 98.18% and 260.59%, respectively. However, the plasticity of $CoCr_{1.7}NiY_x$ MEAs deteriorated due to the less slip system of the HCP phase.

Author Contributions: Conceptualization, X.S. and P.Q.; methodology, P.Q.; software, L.H.; validation, X.X., S.Z. and X.S.; formal analysis, P.Q.; investigation, X.Y.; resources, X.X.; data curation, X.S.; writing—original draft preparation, P.Q.; writing—review and editing, S.Z.; visualization, L.H.; supervision, L.H.; project administration, X.Y.; funding acquisition, X.S. All authors have read and agreed to the published version of the manuscript.

Funding: The work was supported by the Graduate Innovation Special Fund Project of Nanchang Hangkong University (project Grant No. YC2022-008) and Aeronautical science foundation (project Grant No. 2023Z056056001).

Data Availability Statement: The data that support the findings of this study are available from the corresponding author upon reasonable request.

Acknowledgments: Thank you to Nanchang Hangkong University for providing funding and experimental support for this project through the Graduate Innovation Special Fund Project of Nanchang Hangkong University.

Conflicts of Interest: The authors declare no conflicts of interest.

References

1. Yeh, J.W.; Chen, S.K.; Lin, S.J.; Gan, J.Y.; Chin, T.S.; Shun, T.T.; Tsau, C.H.; Chang, S.Y. Nanostructured high-entropy alloys with multiple principal elements: Novel alloy design concepts and outcomes. *Adv. Eng. Mater.* **2004**, *6*, 299–303. [CrossRef]
2. Yu-sheng, T.; Wen-zhe, Z.; Qing-biao, T.; Ming-xu, W.; Shen, Q.; Guo-liang, Z.; Da, S.; Bao-de, S. A review of refractory high-entropy alloys. *Trans. Nonferrous Met. Soc. China* **2022**, *32*, 3487–3515.
3. Zhuo, L.; Xie, Y.; Chen, B. A review on recent progress of refractory high entropy alloys: From fundamental research to engineering applications. *J. Mater. Res. Technol.* **2024**, *33*, 1097–1129. [CrossRef]
4. Yeh, J.W. Recent progress in high-entropy alloys. *Ann. Chim. Sci. Des Mater.* **2006**, *31*, 633–648. [CrossRef]
5. Gou, X.; Cao, R.; Zhou, W.; Shen, Z.; Li, Y. Microstructures, mechanical properties, and strengthening mechanisms of the $(NbMoTa)_{100-x}C_x$ refractory medium-entropy alloys. *J. Mater. Sci. Technol.* **2024**, *214*, 105–119. [CrossRef]
6. Shi, C.; Tan, X.; Xiao, S.; Su, C.; Guo, N.; Guo, S. $Al_xHfTaTi$ ($0 \leq x \leq 0.5$) refractory medium entropy alloys with excellent room-temperature tensile properties. *Mater. Sci. Eng. A* **2024**, *909*, 146849. [CrossRef]
7. Zhou, Z.; Wang, M.; Li, W.; Zhao, S.; Lu, Y. A novel Cu-containing TiZrNb-based medium entropy alloy with excellent mechanical properties, antibacterial activity and biocompatibility. *Intermetallics* **2024**, *171*, 108347. [CrossRef]
8. Laplanche, G.; Gadaud, P.; Horst, O.; Otto, F.; Eggeler, G.; George, E.P. Temperature dependencies of the elastic moduli and thermal expansion coefficient of an equiatomic. single-phase CoCrFeMnNi high-entropy alloy. *J. Alloys Compd.* **2015**, *623*, 348–353. [CrossRef]
9. Haiping, Z.; Jingxiang, M.; Hui, J.; Hongbin, Z.; Wenqing, W.; Shengxue, Q.; Lv, T.; Jian, X. Effect of rare-earth element Y addition on microstructure and mechanical properties of $CrFeNi_2$ medium entropy alloy. *Intermetallics* **2023**, *163*, 108079.
10. Laplanche, G.; Kostka, A.; Reinhart, C.; Hunfeld, J.; Eggeler, G.; George, E.P. Reasons for the superior mechanical properties of medium-entropy CrCoNi compared to high-entropy CrMnFeCoNi. *Acta Mater.* **2017**, *128*, 292–303. [CrossRef]
11. Wu, Z.; Bei, H.; Pharr, G.M.; George, E.P. Temperature dependence of the mechanical properties of equiatomic solid solution alloys with face-centered cubic crystal structures. *Acta Mater.* **2014**, *81*, 428–441. [CrossRef]
12. Shu, X.Y.; Qiu, P.P.; Hu, L.L. Research on Microstructure and Mechanical Properties of $CoCr_xNi$ Medium-Entropy Alloy. *Rare Met. Mater. Eng.* **2023**, *52*, 511–558.
13. Hong, X.W.; Hsueh, C.H. Effects of yttrium addition on microstructures and mechanical properties of CoCrNi medium entropy alloy. *Intermetallics* **2022**, *140*, 107405. [CrossRef]
14. Lee, D.; Agustianingrum, M.P.; Park, N.; Tsuji, N. Synergistic effect by Al addition in improving mechanical performance of CoCrNi medium-entropy alloy. *J. Alloys Compd.* **2019**, *800*, 372–378. [CrossRef]
15. An, N.; Sun, Y.N.; Wu, Y.D.; Tian, J.J.; Li, Z.R.; Li, Q.; Chen, J.Y.; Hui, X.D. High temperature strengthening via nanoscale precipitation in wrought CoCrNi-based medium-entropy alloys. *Mater. Sci. Eng. A* **2020**, *798*, 140213. [CrossRef]
16. Du, Y.; Pei, X.H.; Tang, Z.W.; Zhang, F.; Zhou, Q.; Wang, H.F.; Liu, W.M. Mechanical and tribological performance of CoCrNiHfx eutectic medium-entropy alloys. *J. Mater. Sci. Technol.* **2021**, *90*, 194–204. [CrossRef]
17. Liu, Y.L.; Zhang, F.; Huang, Z.Y.; Zhou, Q.; Ren, Y.; Du, Y.; Wang, H.F. Mechanical and dry sliding tribological properties of $CoCrNiNb_x$ medium-entropy alloys at room temperature. *Tribol. Int.* **2021**, *163*, 107160. [CrossRef]
18. Lu, W.J.; Luo, X.; Yang, Y.Q.; Huang, B. Effects of Nb additions on structure and mechanical properties evolution of CoCrNi medium-entropy alloy. *Mater. Express* **2019**, *9*, 291–298. [CrossRef]
19. Liu, S.F.; Lin, W.T.; Zhao, Y.L.; Chen, D.; Yeli, G.M.; He, F.; Zhao, S.J.; Kai, J.J. Effect of silicon addition on the microstructures. mechanical properties and helium irradiation resistance of NiCoCr-based medium-entropy alloys. *J. Alloys Compd.* **2020**, *844*, 156162. [CrossRef]
20. Wang, J.Y.; Yang, H.L.; Huang, H.; Ruan, J.M.; Ji, S.X. In-situ Mo nanoparticles strengthened CoCrNi medium entropy alloy. *J. Alloys Compd.* **2019**, *798*, 576–586. [CrossRef]
21. Chang, R.B.; Fang, W.; Yan, J.H.; Yu, H.Y.; Bai, X.; Li, J.; Wang, S.Y.; Zheng, S.J.; Yin, F.X. Microstructure and mechanical properties of CoCrNi-Mo medium entropy alloys: Experiments and first-principle calculations. *J. Mater. Sci. Technol.* **2021**, *62*, 25–33. [CrossRef]
22. Li, N.; Gu, J.; Gan, B.; Qiao, Q.; Ni, S.; Song, M. Effects of Mo-doping on the microstructure and mechanical properties of CoCrNi medium entropy alloy. *J. Mater. Res.* **2020**, *35*, 2726–2736. [CrossRef]
23. Zhang, L.J.; Zhang, M.D.; Zhou, Z.; Fan, J.T.; Cui, P.; Yu, P.F.; Jing, Q.; Ma, M.Z.; Liaw, P.K.; Li, G.; et al. Effects of rare-earth element, Y, additions on the microstructure and mechanical properties of CoCrFeNi high entropy alloy. *Mater. Sci. Eng. A* **2018**, *275*, 437–446. [CrossRef]
24. Zhang, J.F.; Zhu, H.G.; Xie, Z.H. Effects of Y and Al additions on the microstructure and tensile properties of $CoCr_3Fe_5Ni$ high entropy alloys. *Mater. Lett.* **2021**, *299*, 130110. [CrossRef]
25. Du, Y.H.; Ding, D.Y.; Lai, L.M.; Xiao, S.M.; Guo, N.; Song, B.; Guo, S.F. Effect of Y on the high-temperature oxidation behavior of CrMoTaTi refractory high entropy alloy. *Int. J. Refract. Met. Hard Mater.* **2022**, *103*, 105755. [CrossRef]

26. Lu, J.; Li, L.; Chen, Y.; Liu, X.Z.; Zhao, X.F.; Guo, F.W.; Xiao, P. Y-Hf co-doped AlCoCrFeNi high-entropy alloy coating with superior oxidation and spallation resistance at 1100 °C. *Corros. Sci.* **2021**, *182*, 109267. [CrossRef]
27. Gu, Z.; Mao, P.; Gou, Y.F.; Chao, Y.; Xi, S.Q. Microstructure and properties of $MgMoNbFeTi_2Yx$ high entropy alloy coatings by laser cladding. *Surf. Coat. Technol.* **2020**, *402*, 126303. [CrossRef]
28. Caliari, F.R.; Candioto, K.C.G.; Reis, D.A.P. Effect of aging treatment on inconel 718 superalloy: Application in elevated temperatures. *Mater. Sci. Forum* **2015**, *825*, 199–203. [CrossRef]
29. Hume-Rothery, W. The Engel-Brewer theories of metals and alloys. *Prog. Mater. Sci.* **1968**, *130*, 229–265. [CrossRef]
30. Tong, Y.; Zhao, S.J.; Bei, H.B.; Egami, T.; Zhang, Y.W.; Zhang, F.X. Severe local lattice distortion in Zr-and/or Hf-containing refractory multi-principal element alloys. *Acta Mater.* **2020**, *183*, 172–181. [CrossRef]
31. Liang, D.S.; Zhao, C.C.; Zhu, W.W.; Wei, P.B.; Jiang, F.L.; Zhang, Y.W.; Sun, Q.P.; Ren, F.Z. Overcoming the strength-ductility trade-off via the formation of nanoscale Cr-rich precipitates in an ultrafine-grained FCC CrFeNi medium entropy alloy matrix. *Mater. Sci. Eng. A* **2019**, *762*, 138107. [CrossRef]
32. Liang, D.S.; Wei, C.X.; Ren, F.Z. Introducing Laves phase strengthening into an ultrafine-grained equiatomic CrFeNi alloy by niobium addition. *Mater. Sci. Eng. A* **2021**, *806*, 140611. [CrossRef]
33. Zhang, Y.; Zhou, Y.J.; Lin, J.P.; Chen, G.L.; Liaw, P. Solid-Solution Phase Formation Rules for Multi-component Alloys. *Adv. Eng. Mater.* **2008**, *10*, 534–538. [CrossRef]
34. Yang, X.; Zhang, Y. Prediction of high-entropy stabilized solid-solution in multi-component alloys. *Mater. Chem. Phys.* **2012**, *132*, 233–238. [CrossRef]
35. Guo, S.; Ng, C.; Lu, J.; Liu, C.T. Effect of valence electron concentration on stability of fcc or bcc phase in high entropy alloys. *J. Appl. Phys.* **2011**, *109*, 213. [CrossRef]
36. Takeuchi, A.; Inoue, A. Classification of Bulk Metallic Glasses by Atomic Size Difference, Heat of Mixing and Period of Constituent Elements and Its Application to Characterization of the Main Alloying Element. *Mater. Trans.* **2005**, *46*, 2817–2829. [CrossRef]
37. Wu, Z.G.; Guo, W.; Jin, K.; Poplawsky, J.D.; Gao, Y.F.; Bei, H.B. Enhanced strength and ductility of a tungsten-doped CoCrNi medium-entropy alloy. *J. Mater. Res.* **2018**, *33*, 3301–3309. [CrossRef]
38. Wu, Z.G.; Gao, Y.F.; Bei, H.B. Thermal activation mechanisms and Labusch-type strengthening analysis for a family of high-entropy and equiatomic solid-solution alloys. *Acta Mater.* **2016**, *120*, 108–119. [CrossRef]

crystals

Article

Research on the Microstructure and Properties of Arc-Sprayed Austenitic Stainless Steel and Nickel-Based Alloy Composite Coatings with Different Spraying Distances

Jingang Yan [1,2], Zhenming Yang [2], Limin Zhang [2] and Jianxin Wang [2,3,*]

1 Robotics and Intelligent Equipment Research Institute, Tianjin Vocational and Technical Normal University, Tianjin 300222, China; yjg0709@126.com
2 School of Mechanical Engineering, Tangshan Polytechnic University, Tangshan 063299, China; kr0522@126.com (Z.Y.); zhanglimindota1@163.com (L.Z.)
3 Tangshan Talent Development Group Co., Ltd., Tangshan 063000, China
* Correspondence: wangjianxin0402@163.com

Abstract

1Cr18Ni9Ti and Monel composite metal coatings with five different spraying distances were prepared by arc spraying technology. The density, hardness, friction, and wear properties and acid corrosion rate of the coatings with different spraying distances were studied by X-ray diffraction, scanning electron microscopy, Rockwell hardness test, and friction and wear test. Research shows that the spraying distance has a significant effect on the density, hardness, porosity, friction, and wear properties and corrosion rate of the coating. When the spraying distance is 250 mm, the coating has the maximum density and hardness, the minimum porosity and corrosion rate, and the minimum friction coefficient and wear volume. Cu3.8ni and cr0.19fe0.7ni0.11 compounds in the coating have significant effects on the friction, wear, and hardness of the coating. The results show that too-high or too-low spraying distance will lead to pores and large particle agglomeration in the coating, which will affect the surface physical properties of the coating.

Keywords: arc spraying; coating; friction and wear; corrosion behavior

1. Introduction

The surface of steel structures is prone to corrosion during contact with acidic, alkaline, and other liquid environments, leading to a decrease in material properties and significant economic losses. Arc spraying technology has been widely recognized in fields such as aerospace and petrochemicals [1–3] due to its advantages in corrosion resistance, long service life, and low cost, and it has been applied to large-scale steel structure protection projects.

1Cr18Ni9Ti is an austenitic stainless steel with superior corrosion resistance, good mechanical properties, and welding performance, especially in high-temperature and corrosive environments [4,5]. However, under specific erosion and wear conditions, the surface quality of 1Cr18Ni9Ti deteriorates, which can significantly reduce its service life [6]. It is particularly prone to galvanic corrosion in salt fog environments [7]. Monel alloy, a nickel-based alloy containing copper, iron, and manganese, features a unique single-phase austenitic structure that offers excellent resistance to high temperatures and highly saline seawater environments [8–10]. Monel exhibits superior corrosion resistance in alkaline solutions and dilute hydrochloric, sulfuric, and phosphoric acids [11]. However, its high nickel content makes it relatively expensive [12]. The alloy's exceptional corrosion

 https://doi.org/10.3390/cryst15020142

resistance stems from its strong passivation ability, which can be compromised by improper heat treatment. Therefore, precise control of heat treatment parameters is essential to optimize performance [13]. Previous studies have shown that elements such as Cu, Cr, and Fe play critical roles in influencing the microstructure, phase evolution, surface hardness, friction, and wear properties of nickel-based alloy coatings [14–22]. Most current research focuses on improving individual materials' properties and processing techniques, such as 1Cr18Ni9Ti and Monel alloy. Common approaches include heat treatment and surface coating to enhance corrosion and wear resistance. However, there is little research on developing composite coatings for these two materials using arc spraying technology.

In this study, the excellent physical properties of 1Cr18Ni9Ti are combined with the corrosion resistance of Monel alloy. Arc spraying technology prepares a composite metal coating on a 45# steel substrate. The performance of the composite coating is evaluated in detail through density and hardness measurements, scanning electron microscopy (SEM), friction wear testing, and acidic corrosion testing. This research aims to determine the optimal spraying distance for heterogeneous dual-wire composite spraying of 1Cr18Ni9Ti and Monel. The findings have significant potential to enhance steel structures' service life and reliability in complex environments.

2. Experimental Materials and Methods

2.1. Coating Preparation

The substrate material used in this experiment was 45# steel (Table 1), with dimensions of 400 mm × 100 mm × 8 mm. The substrate material was cleaned with acetone to remove surface oil stains and rust. Then, brown corundum particles were used for sandblasting to ensure that the surface roughness of the substrate material reached Ra2.5 or above. This step improves the bonding strength between the coating and the substrate material. After sandblasting the substrate, arc spraying was performed. The equipment used was a QD8-400 arc spraying machine produced by Shanghai Eurasia Spray Machinery Co., Ltd. and a Yaskawa robot. As shown in Figure 1, the arc spraying machine allows the adjustment of parameters such as wire feeding speed and spraying current. The wires used in the experiment were 1Cr18Ni9Ti and Monel alloys, and their chemical compositions are shown in Table 2. The diameter of both wires was 2.0 mm, the test filaments were sprayed according to the ratio of 1:1, and the compressed air pressure for spraying was 0.65 MPa. The average thickness of the sprayed coating is 1.5 mm, and the average thickness distribution of the five sprayed distances is shown in Figure 2. Samples were prepared using the process parameters in Table 3, with experiments conducted at five different spraying distances. The distance between the nozzle of the arc spraying machine and the substrate surface was set at 100 mm, 150 mm, 200 mm, 250 mm, and 300 mm. They were represented as H100, H150, H200, H250, and H300, respectively, in the following figures. These distances were used to explore the optimal spraying effect.

Table 1. Main chemical composition of 45# steel (wt.%).

	Fe	C	Si	Mn	P	S	Cr	Ni	Others
45#	≥97	0.42–0.5	≤0.35	0.5–0.8	≤0.04	≤0.05	≤0.25	≤0.25	<0.5

Table 2. Main Chemical Composition of the Two-Wire Materials (wt.%).

1Cr18Ni9Ti	Cr	Ni	Mn	Si	P	C	S	Ti	Fe	Others
	18	9	2	1	0.035	0.12	0.03	0.6	68.415	0.8
Monel	Ni	Cu	Fe	Mn	C	Al	Si	S	Others	
	65	28	2.5	2	0.3	0.5	0.5	0.024	1.176	

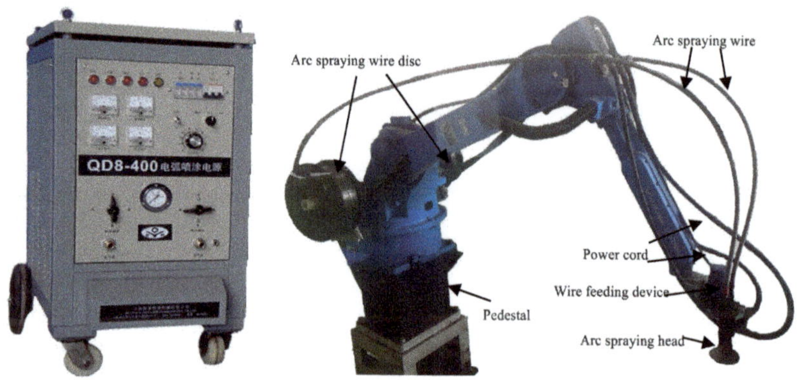

Figure 1. Arc Spraying Equipment.

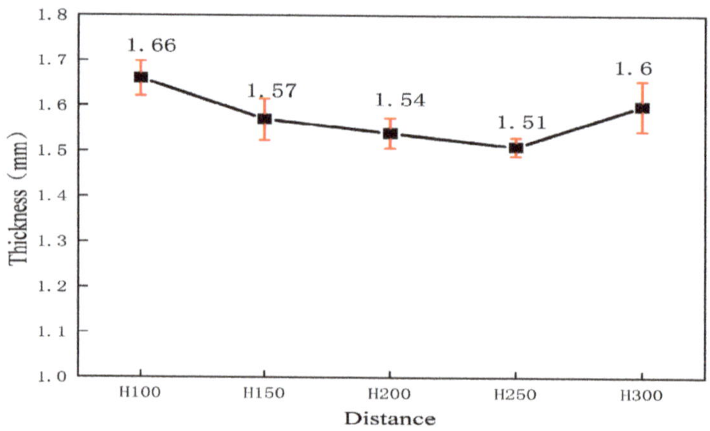

Figure 2. Distribution diagram of average coating thickness for five spraying distances.

Table 3. Spraying Process Parameters.

Types of Silk Materials	Wire Feeding Voltage/V	Wire Feeding Current/A	Spray Current/A	Spray Voltage/V
1Cr18Ni9Ti, Monel	16	25	200	36

2.2. Coating Characterization Methods and Performance Testing

(1) Use the HandySCAN BLACK (CREAFORM INC., Lévis, QC, Canada) three-dimensional scanner produced by the Quebec Provincial CREAFORM INC.to measure the volume of the five samples, The three-dimensional data were processed using Geomagic Wrap software (version 2021.0.0.3008) to calculate the volume of the samples, the scanned sample was weighed with a balance, and then the density of the sample was calculated. The hardness of the five samples was measured using a Huayin HR-150A (Laizhou, China) Rockwell hardness tester, with a load of 150 kg and a loading time of 15 s. Each sample was tested at six points, and the average value of the six test data was obtained. The surface morphology of the samples was observed using a Hitachi S-4800 (Tokyo, Japan) scanning electron microscope (SEM), and surface phase analysis was also performed.

(2) Porosity calculation. After the process of inlaying, grinding, and polishing of the spray-coated specimen section, the pictures were processed using Image-Pro Plus 6.0

software. Five SEM pictures were randomly selected for analysis and converted into images with large contrasting colours to ensure that the pore portion was displayed in red, and then the percentage of the pore area to the whole field of view area was calculated, and the average value was taken to obtain the porosity of different coatings.

(3) After polishing the sprayed coating samples, a PCM5100W metallurgical microscope (Jinan, China) was used to inspect them. A UMT-5 multifunctional friction (Billerica, MA, USA) wear tester, employing a reciprocating friction wear mode, was used to conduct wear-resistance tests on the coatings. The friction counterpart was a GCr15 steel ball with a diameter of 10 mm and a hardness of 62 HRC. The experimental load was 30 N, the friction speed was 20 mm/s, and the testing time was 900 s. After the test, the friction and wear properties of the coating were characterized and analyzed by friction coefficient, three-dimensional morphology of wear marks, and wear rate. The wear rate is calculated from (1):

$$W = \frac{V}{DL} \tag{1}$$

where W is the wear rate ($mm^3 \cdot n^{-1} \cdot m^{-1}$), V is the wear volume (mm^3), D is the sliding stroke (m), and l is the load (N).

(4) Corrosion rate calculation. The corrosion rate was determined by calculating the weight change of the specimen by immersing the coated specimen in a 5% sulfuric acid solution for a test period of 10 h. After the static corrosion test, clean and dry the sample and weigh it. The difference between the two masses is the mass loss caused by corrosion. The corrosion rate is calculated as $v = (W_1 - W_2)/St$, where W_1 is the mass of the sample before the test (mg); W_2 is the mass of the sample after the test (mg); S is the area of the sample (cm^2); and t is the test time (h).

(5) Acid invasion test. This test selected concentrated hydrochloric acid and concentrated nitric acid for the 3:1 ratio of acid solution, the use of cotton swabs dipped in the solution to wipe the surface of the specimen, the corrosion time of 6 s. After acid etching, the samples were washed with clean water, wiped with anhydrous ethanol, and dried with cold air. Finally, the changes of surface morphology were observed by Hitachi s-4800 scanning electron microscope (SEM).

3. Results and Discussion

3.1. Sample Density

The density of the arc-sprayed coating samples varies compared to the density of the original wire material. Coating density can reflect the compactness of the coating to a certain extent. The higher the density, the more compact the coating and the lower the porosity [23]. The variation in sample density with spraying distance is shown in Figure 3. When the spraying distance is 100 mm, the sample density reaches the minimum value of 6.1257 g/cm^3. As the spraying distance increases, the sample density gradually increases, reaching a maximum value of 6.4227 g/cm^3 at a spraying distance of 250 mm. After that, the density decreases gradually. When the distance between the spray nozzle and the substrate surface is too close, the metal droplets are not sufficiently atomized due to the effect of the high-speed airflow. Large metal droplets attach to the substrate surface, leading to particle accumulation. Conversely, the metal droplets are sufficiently atomized when the nozzle is too far from the substrate. However, they experience a longer exposure time in the air, causing their temperature to drop and their adhesion to weaken. This results in the formation of pores between the metal droplets, reducing the density of the sprayed coating.

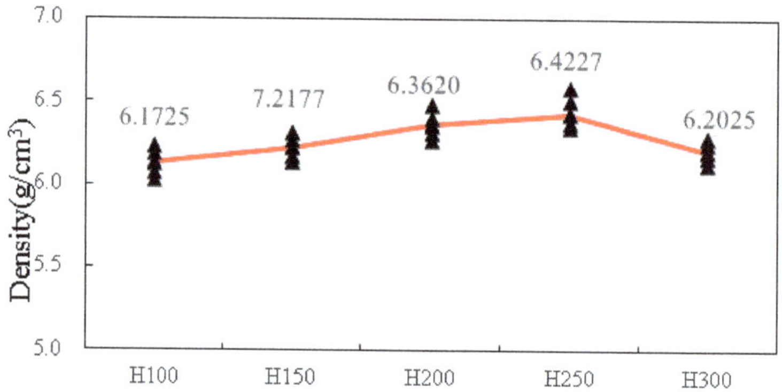

Figure 3. Density Distribution Curve of Sprayed Coatings at Different Distances.

3.2. Coating Hardness

The surface hardness of the 1Cr18Ni9Ti and Monel alloy coating samples was tested, and it was found that the average Rockwell hardness of the five sprayed samples ranged from 26.13 to 36.45 HRC. Figure 4 shows the distribution of hardness values for the five samples. The hardness distribution shows that the samples' average hardness values are related to the spraying distance. When the spraying distance is 100 mm, the average hardness is the lowest, at 26.13HRC. As the spraying distance increases, the hardness of the sprayed coating gradually increases, reaching a maximum value of 36.45 HRC at a spraying distance of 250 mm, after which the hardness gradually decreases. By comprehensively analyzing the scatter distribution of hardness, the variation curve of average hardness values, and the variation trend of density, it can be concluded that the spraying distance significantly affects the formation of the sprayed coating. Whether the metal droplets are fully atomized directly affects the formation of large particle agglomeration on the substrate surface, creating uneven porosity within the sprayed coating and indirectly affecting the coating's hardness.

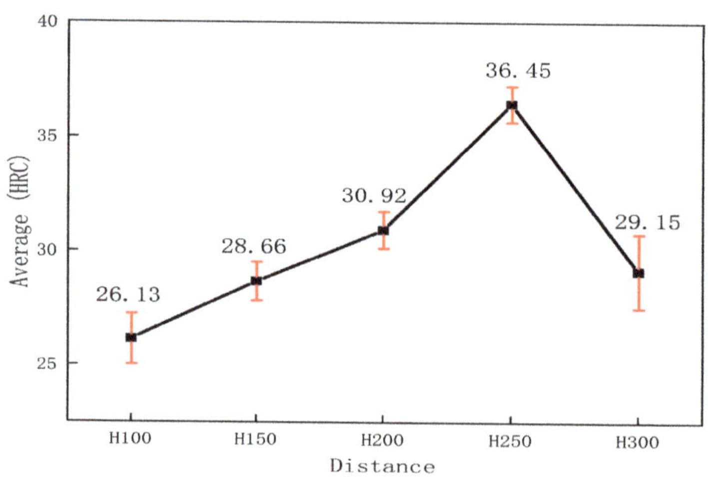

Figure 4. Scatter Distribution of Hardness Values for Sprayed Coatings at Different Distances.

3.3. Characterization of Coating Surface Morphology and Calculation of Porosity

The microstructure of the arc-sprayed coating samples was observed, as shown in Figure 5a–e. The surface of the alloy coating is rough with obvious ups and downs. The roughness of the five coating surfaces is different, and the roughness of the surface in Figure 5d is more uniform than that of other coating surfaces. During arc spraying, the wire melts due to short-circuit heating. The molten droplets are propelled onto the substrate surface under the impact of high-speed airflow. The droplets rapidly spread out and gradually deposit, forming flattened particles. These flattened particles overlap layer by layer, forming a tight lamellar structure and particle accumulation. The flat particles are stacked layer by layer and finally form a compact sheet structure and particle accumulation structure, which are closely attached to the substrate surface and play a good shielding role, as shown in Figure 5f.

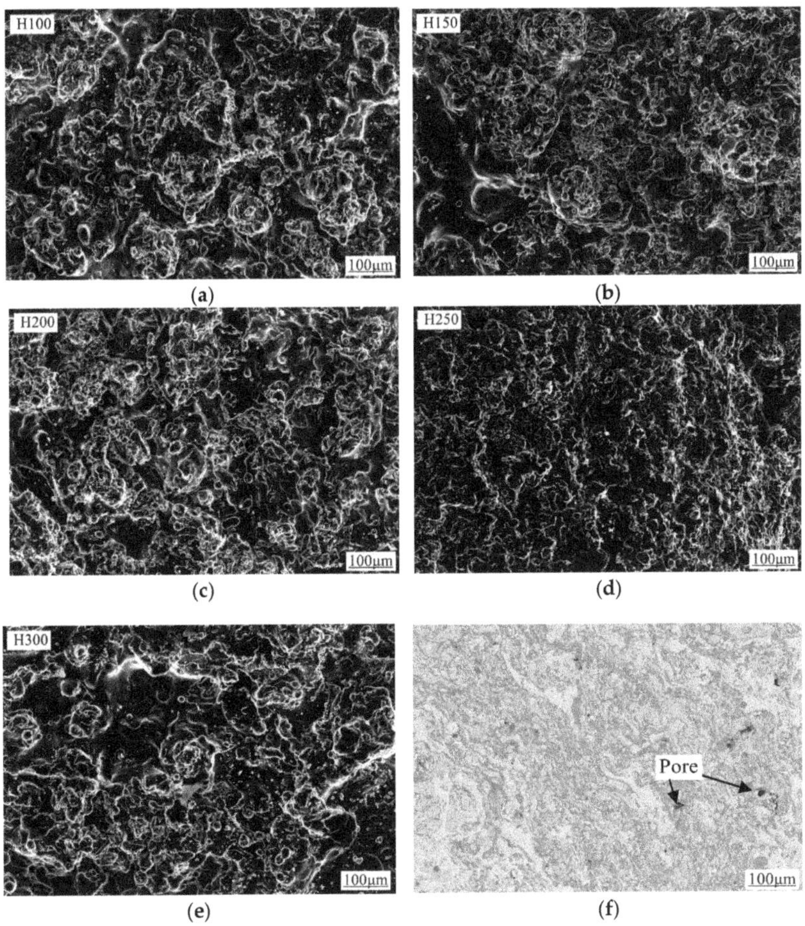

Figure 5. Microstructure of Sprayed Coatings at Different Spraying Distances. (**a**–**e**) micromorphology of the test surfaces at five spraying distances; (**f**) scanning photographs of the H150 cross-section.

Figure 6 is a scanning electron microscope photograph of the coating section when the spraying distance is 100 mm and 250 mm. It can be seen that the coating presents a typical compact layered structure, which is composed of flattened particles, oxides, and pores. The reason for the layered structure is that the huge energy generated when the wires at both ends are short-circuited makes the wires melt and form droplets. The droplets are atomized by high-pressure gas, and the small particles formed after instant atomization quickly hit the substrate surface, spreading and solidifying to form a layered structure. The whole section is mainly composed of oxides, large particle accumulation, etc. The existence of oxide phase is caused by the inevitable contact reaction between part of the droplet surface and oxygen in the atmosphere during the spraying process, and the oxide phase formed by these reactions will affect the spreading of the droplet to a certain extent, and then affect the continuity of the coating, forming pores or holes [24].

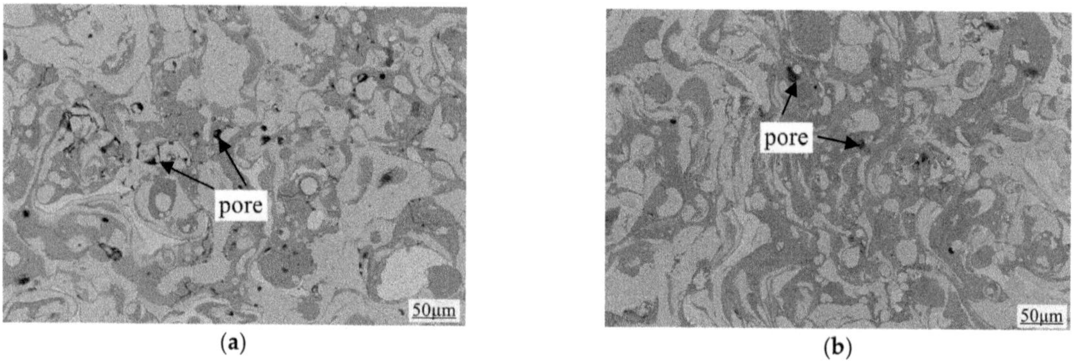

Figure 6. Scanning Electron Micrographs of Coating Cross-Sections. (**a**) Scanning photograph of H100 cross-section; (**b**) Scanning photograph of H250 cross-section.

Different spraying distances form different pores. Spraying distance affects the molten droplet particles in the air flight time and the molten droplet metal particles sprayed to the substrate material in the process. The size of the spraying speed, temperature, and kinetic energy, will affect the distribution of unmelted particles inside the coating, which in turn will affect the porosity inside the coating. When the spraying distance is 100 mm, there is a large particle agglomeration phenomenon and obvious porosity on the surface of the sprayed coating. The reason for this is that the spraying distance is relatively close. The molten metal droplets are not fully atomized under the action of high-speed airflow, and the coating sprayed onto the surface of the substrate does not solidify in time, leading to the phenomenon of hanging coating, which produces large particle agglomeration and porosity. When the spraying distance is 250 mm, the surface of the sprayed coating shows a uniform lamellar structure, with uniform particle size, analyzing the reason why the molten metal droplets can be completely atomized under the action of high-speed airflow and spread completely on the surface of the substrate to form uniform metal particles, reducing the generation of pores. With the further increase of the spraying distance, the molten metal droplets are exposed in the air for a longer period of time, the adhesion on the surface of the substrate decreases, the spatial mesh structure forms, and the internal porosity of the coating is significantly increased.

By scanning the sections of sprayed coatings with different heights, five SEM images were taken for each section. The coating sections were analyzed by Image Pro Plus 6.0 software, and the average value was calculated. The porosity distribution of coatings with different spraying distances was obtained, as shown in Figure 7.

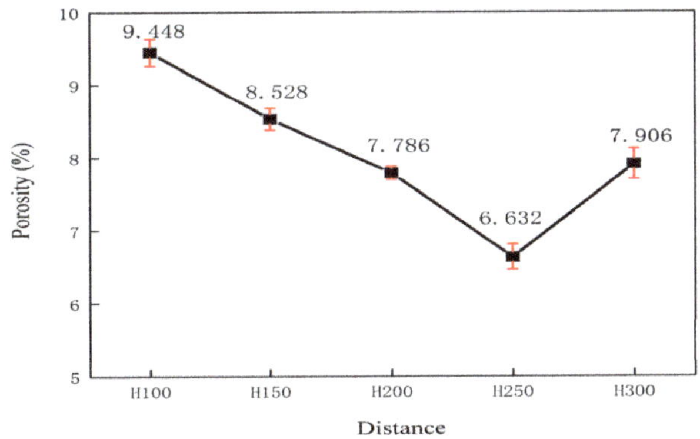

Figure 7. Porosity Distribution.

3.4. Analysis of Coating Friction and Wear Behavior

3.4.1. Coating Friction Factor Analysis

UMT-5 Multifunctional Friction and Wear Tester calculates the coefficient of friction by measuring the force generated during friction. The test is performed by sensor detection when a specific load of 30N is applied to the test and the reciprocating friction speed is 20 mm/s. During the friction test, the tester measures the friction force occurring between the contacting surfaces, which allows the coefficient of friction to be measured directly by the built-in sensor. The coefficient of friction (μ) is the ratio between the friction force (F_f) and the perpendicular load (F_n), where Ff is the measured friction force and F_n is the normal force acting on the specimen surface.

As shown in Figure 8, a and b are the curves of the friction coefficient of the coating of the arc-sprayed sample as a function of time. The polished sample surface is relatively smooth in the initial friction stage, resulting in a low friction coefficient. As dry friction continues, the friction coefficient rapidly increases due to the formation of wear particles and other substances during the friction process. These particles create dents on the coating surface, increasing the contact area between the wear ball and the coating. Over time, the wear particles gradually fill in the dents, further increasing the surface friction coefficient. The friction coefficient is closely related to the material's physical properties. The lower the friction coefficient, the better the material's wear resistance. After 200 s, friction enters the steady wear stage, and the friction coefficient remains relatively stable. Figure 6a shows the variation of the friction coefficient during the testing process, while Figure 6b shows the variation in friction coefficients for different spraying distances. When the spraying distance is 100 mm, the friction coefficient reaches a maximum value of 0.7752. As the spraying distance increases, the friction coefficient gradually decreases. It reaches a minimum of 0.6214 at 250 mm. After that, it rises again.

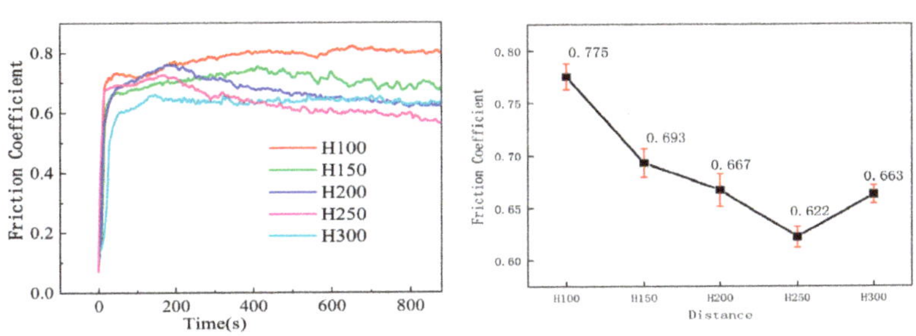

Figure 8. Variation of Friction Coefficient with Time. (**a**) Five coating friction coefficient curve change diagrams; (**b**) Average friction coefficient of five coatings.

By comparing the sample density, average hardness, and friction coefficient values, it was found that the trends in density and average hardness were similar. This phenomenon is attributed to the influence of pore structure within the samples' density and hardness. However, the variation trend in the friction coefficient is opposite to that of density and average hardness, indicating that the denser the coating, the higher its hardness, and the lower its friction coefficient. Phase analysis results showed that when the spraying distance was 250 mm, the friction coefficient of the sprayed coating was the lowest. The precipitation of elemental Cu and Cu3.8Ni in the coating was vital in reducing the friction coefficient.

3.4.2. Analysis of Coating Wear Performance

Figure 9 shows the three-dimensional morphology of the abrasion marks of five spray distance coatings after friction and wear tests. It can be seen that the shape of the abrasion marks of the five coatings is close to one another, showing a large abrasion mark width and abrasion mark depth. The wear interiors of the five coatings are not smooth, with many obvious bumps and pits. Among them, A coating has the largest abrasion mark width and depth, and exhibits the largest abrasion volume, the worst abrasion resistance, and the worst structural integrity under friction load. D coating is analyzed in comparison with the other coatings. It has a shallower depth of abrasion mark, the smallest width of abrasion mark, the smallest abrasion volume, and the best structural integrity under friction load, which reflects the good friction and abrasion resistance performance. The spraying distance of 250 mm showed the best wear performance of the coating.

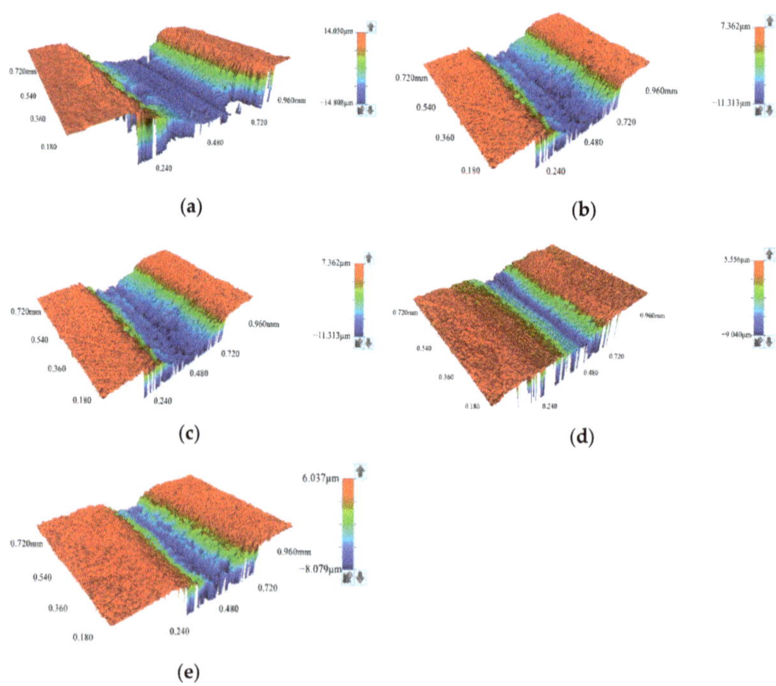

Figure 9. 3D Morphology of Abrasion Marks of Coatings. (**a**) Spray distance 100 mm; (**b**) Spray distance 150 mm; (c) Spray distance 200 mm; (**d**) Spray distance 250 mm; (**e**) Spray distance 300 mm.

Figure 10 shows the wear rate of the five coatings. The total volume loss of the wear marks of the coatings can be calculated by vision software. The average wear rates of the five coatings are 1.73×10^{-3}, 1.31×10^{-3}, 1.03×10^{-3}, 0.79×10^{-3}, and 1.11×10^{-3} $mm^3 \cdot N^{-1} \cdot m^{-1}$, respectively. Coating H100 has the highest wear rate and the worst wear resistance, while coating H250 has the lowest wear rate and shows good wear resistance. According to the previous comprehensive analysis of density, hardness, and porosity, the wear rate of the coating is affected by density, hardness, and porosity.

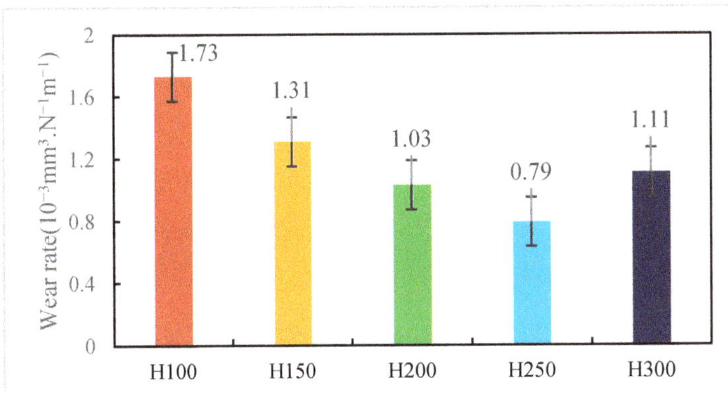

Figure 10. Wear Rate of Coating.

3.5. Coating Phase Analysis

The superposition of the scattering intensity of each atom forms the diffraction intensity in XRD. For elements present in relatively low concentrations, only a small number of atoms within each unit cell participate in forming the crystal structure, leading to weaker diffraction peaks [25]. The composition of the 1Cr18Ni9Ti and Monel wires is relatively complex, resulting in various compounds forming in the coating. In the phase analysis of the coating, some phases with weak diffraction intensity were excluded. Qualitative analysis of the phases in the sprayed coating was conducted using XRD. The five groups of sprayed samples were prepared with the same wire material and identical spraying parameters, with only the spraying distance being adjusted. XRD analysis showed that the surface composition of the sprayed samples was essentially the same, as shown in Figure 11.

The two spray wires contain higher concentrations of Fe and Ni. XRD spectrum analysis revealed that Fe forms multiple compounds, including Fe_3O_4, $MnFeO_4$, and Cr0.19Fe0.7Ni0.11. Ni is present in compounds such as Cu3.8Ni and Ni-Ti. Fe is chemically active during spraying in the high-temperature molten state and effortlessly combines with oxygen to form oxides. Ti has multiple crystal structures, including body-centered cubic, hexagonal close-packed, and face-centered cubic structures. Ti strongly bonds with Ni, Cr, and Fe, forming stable intermetallic compounds. These compounds have high melting points and hardness, improving the physical properties of the alloy [26]. At around 882 °C, Ti's body-centered cubic and hexagonal close-packed structures are relatively stable. However, in the molten state at high temperatures, the stability of Ti's crystal structure decreases significantly, and it easily forms intermetallic compounds with various elements. Ni-Ti compounds were observed in all coatings at diffraction angles of 50°, 62°, and 78°. Although Ti content is relatively low in the spray wire materials, its strong bonding ability with Ni in the molten state leads to the formation of Ni-Ti compounds, which exhibit high diffraction peak intensity in the phase analysis.

When the diffraction angle was 43°, all five sprayed coatings showed double diffraction peaks. Especially at a spraying distance of 250 mm, clear double diffraction peaks were formed. In the state of high-temperature melting, the metal grains undergo the cyclic transformation process of remelting and recrystallization, which promotes the redistribution and structural reorganization of elements. In this process, Cu reacts with Ni to form Cu3.8Ni. The formation of Cu3.8Ni is closely related to the almost unlimited dissolution characteristics of Cu in the grain. Cu can dissolve almost unlimitedly in the Ni lattice, usually forming a face-centered cubic solid solution with Ni, indicating that Cu3.8Ni has a face-centered cubic structure [27]. This stable alloy phase significantly improves the friction performance of the material. During friction, copper tends to be evenly distributed and slightly enriched in the grain boundary regions [28]. It helps to form a natural lubricating film, effectively reducing the friction coefficient at the interface. Thus, Cu3.8Ni not only plays a crucial role in improving the microstructure of the coating but also effectively reduces the friction coefficient and wear rate. Cu3.8Ni enhances the wear resistance of the coating.

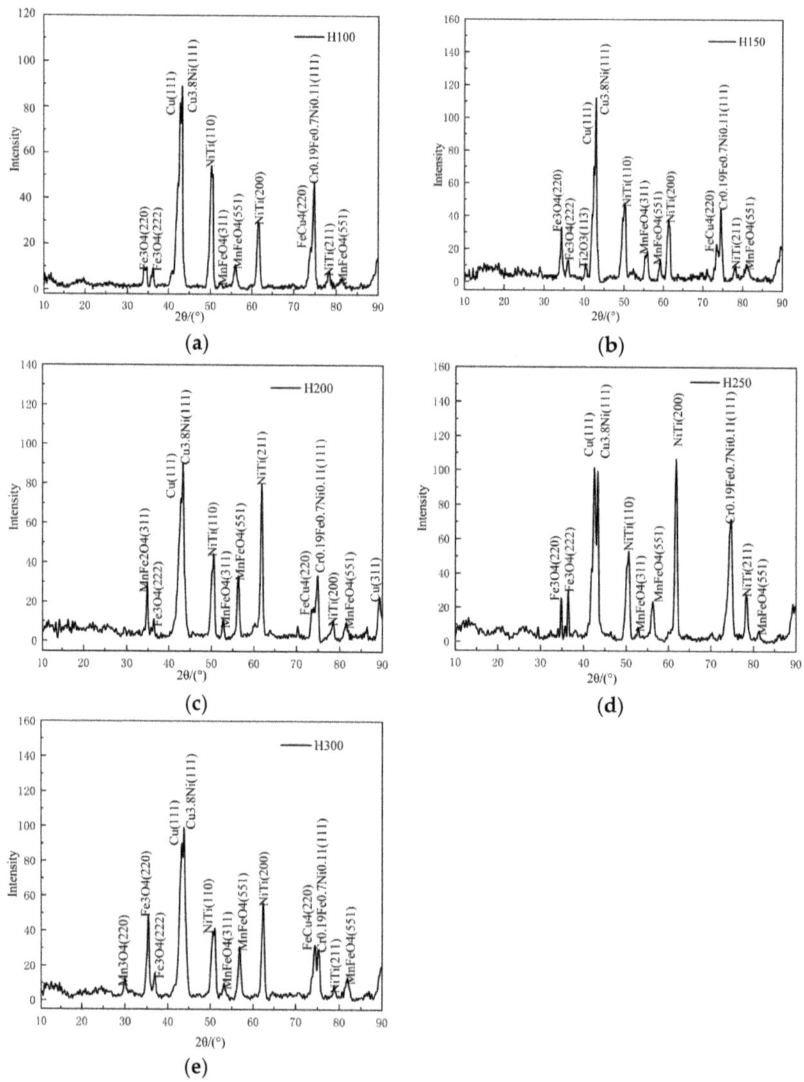

Figure 11. XRD Pattern of Sprayed Coatings. (**a**) Spray distance 100 mm; (**b**) Spray distance 150 mm; (**c**) Spray distance 200 mm; (**d**) Spray distance 250 mm; (**e**) Spray distance 300 mm.

At a diffraction angle of 74° the spraying distance was 250 mm and no double diffraction peaks were observed, but at other spraying distances, double diffraction peaks appeared. Due to the similar atomic radii and electronegativity of Fe, Ni, and Cr [29], Fe and Cr have relatively high solubility in Ni. These three elements' physical properties and interactions are crucial for forming specific solid solutions and compounds. Cr significantly enhances the alloy's corrosion resistance and thermal stability, Fe contributes to toughness, and Ni provides good wettability and corrosion resistance. The interactions and equilibrium among these elements in the molten state determine the final phase structure and distribution. During the spraying process, the short-circuiting of the metal wire generates intense thermal energy, causing the metal droplets to melt rapidly and bond closely with the substrate surface. The high Cr, Fe, and Ni concentrations in the spray wires diffuse, dissolve, and recombine to form new alloy phases. The high-speed airflow facilitates rapid cooling of the sprayed coating, promoting the formation of refined grains and reducing phase segregation [30]. This promotes the generation of complex, wear-resistant phases. A certain amount of Cr, Fe, and Ni compounds were detected in the XRD spectrum, forming cr0.19fe0.7ni0.11 phase. [31]. The supersaturation and precipitation of Cr, Fe, and Ni elements form this phase. Cr, Fe, and Ni atoms rearrange and crystallize in specific proportions, forming a stable Cr0.19Fe0.7Ni0.11 solid solution structure. The presence of Cr0.19Fe0.7Ni0.11 significantly increases the microhardness of the coating, which helps the coating resist wear and effectively reduces wear volume.

3.6. Coating Acid Corrosion Resistance Test

Figure 12 shows the corrosion rate of arc-sprayed coatings at five different spraying distances. The coatings were immersed in a 5% sulfuric acid solution. The main components of these sprayed samples include Fe, Cu, and Ni alloys. According to the corrosion performance tests of Ni-Cu alloys in air reported in the literature [32], nickel-copper alloys exhibit the fastest corrosion rate in a 5% sulfuric acid solution. The surface of the coatings primarily consists of Fe, Cu, and Ni compounds, and when reacting with 5% sulfuric acid, the oxide phases on the surface react first. The sample with a spraying distance of 100 mm had the fastest corrosion rate of 4.3787 $mg{\cdot}cm^{-2}{\cdot}h^{-1}$, while the sample with a spraying distance of 250 mm had the slowest corrosion rate of 3.5229 $mg{\cdot}cm^{-2}{\cdot}h^{-1}$.

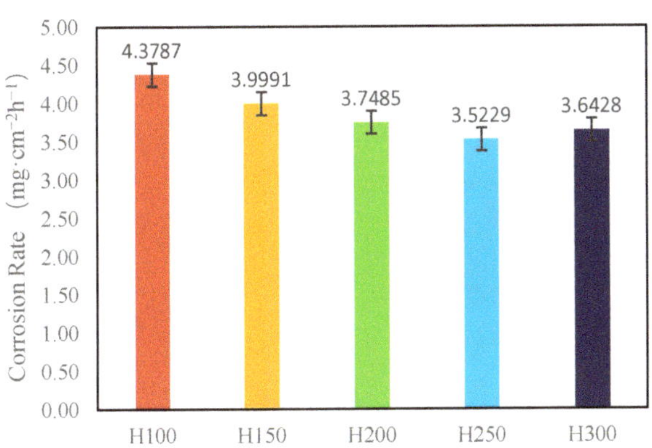

Figure 12. Corrosion Rate of Sprayed Coatings.

The arc spraying process involves molten particles stacking through high-speed airflow to form the coating. During the stacking process, pores are on and within the coating, reducing the coating's wear and corrosion resistance [33]. SEM observations revealed the presence of significant pores within the arc-sprayed coatings. These pores significantly reduce the density and hardness of the alloy. Furthermore, the presence of pores increases the contact area between the sprayed surface and the acid solution, thereby accelerating the corrosion rate of the coating.

Figure 13 shows the SEM analysis of the cross-sectional surface morphology of the coating after acid etching. Significant pores, compounds, and large particle agglomeration are visible in the images. At a spraying distance of 100 mm, the images show more internal pores, and the accumulation of compounds and large particles is more pronounced and scattered. As the spraying distance increases to 250 mm, the internal pores are significantly reduced, the diameter of the large particles becomes smaller, and the distribution is more uniform. The distribution of the compounds also appears more consistent.

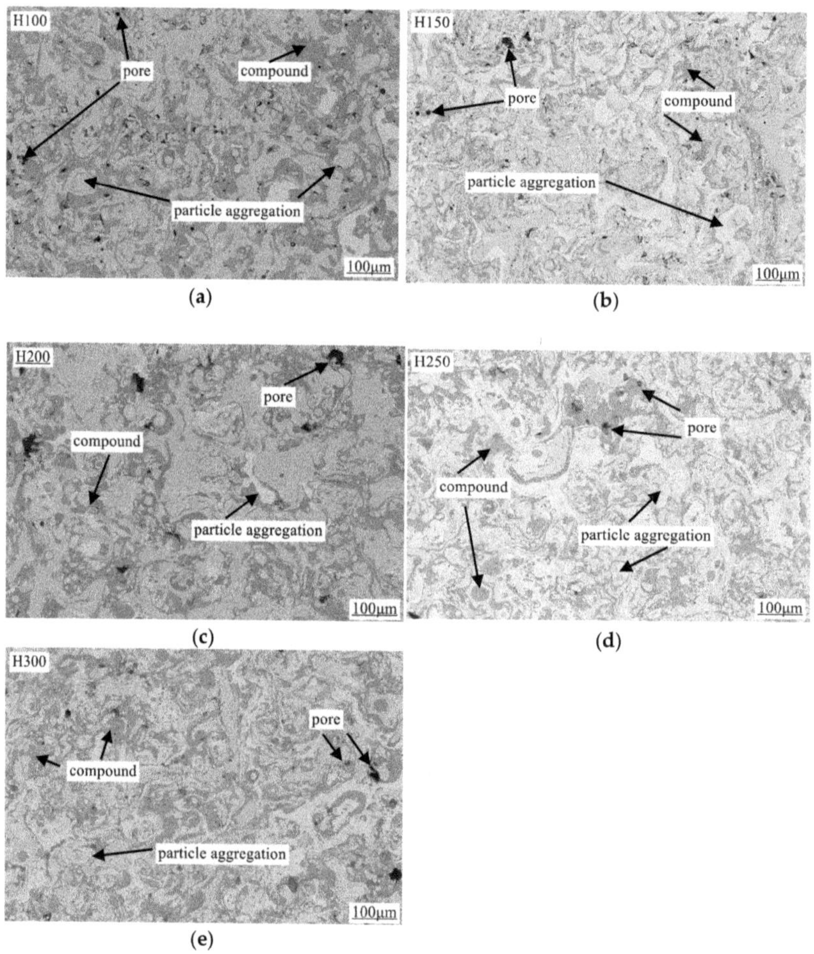

Figure 13. Surface Morphology of Sprayed Coating After Acid Etching. (**a**) Spray distance 100 mm; (**b**) Spray distance 150 mm; (**c**) Spray distance 200 mm; (**d**) Spray distance 250 mm; (**e**) Spray distance 300 mm.

4. Conclusions

This study used arc spraying technology to prepare 1Cr18Ni9Ti and Monel-sprayed coatings on the surface of 45# steel. The density, hardness, surface microstructure, friction, wear behavior, and corrosion resistance of the coatings sprayed at five different distances were investigated. The main conclusions are as follows:

1. The surface morphology of the sprayed coatings is relatively rough, with many pores between the particles. In the phase analysis of the sprayed coatings, at a spraying distance of 250 mm and a diffraction angle of 73°, a single Cr0.19Fe0.7Ni0.11 phase alloy precipitated, which contributed to the increase in the microhardness of the coating.
2. From the friction and wear behavior analysis, the friction coefficient of the sprayed coatings is closely related to the density and hardness of the coatings. When the spraying distance was 100 mm, the friction coefficient reached its maximum value of 0.7752. As the spraying distance increased to 250 mm, the friction coefficient decreased to a minimum of 0.6214. The variation curve of the friction coefficient is positively correlated with density and hardness. At a diffraction angle of 43°, Cu and Cu3.8Ni phases were observed, indicating that the composition of the coating plays an active role in reducing the friction coefficient and inhibiting wear.
3. In the acidic corrosion test, at a spraying distance of 250 mm, the coating exhibited the highest density of 6.4227 g/cm^3, the highest Rockwell hardness of 31.3 HRC, a porosity of 6.63%, a wear rate of 0.79×10^{-3} $mm^3 \cdot N^{-1} \cdot m^{-1}$, the lowest friction coefficient of 0.6214, and the lowest acidic corrosion rate of 3.5229 $mg \cdot cm^{-2} \cdot h^{-1}$. Coatings sprayed at other distances had more pores and lower compactness. This reduced compactness increased the contact area between the acid and the coating, accelerating the corrosion process.

Author Contributions: Conceptualization, J.Y.; Methodology, J.W. and J.Y.; Software, L.Z.; Writing—review and editing, Z.Y. All authors have read and agreed to the published version of the manuscript.

Funding: This research was funded by the Tangshan Science and Technology Bureau project "Microstructure Analysis and Performance Change Law of Nickel based Arc Spraying Anti-Corrosion Coating" (No. 131302118b).

Data Availability Statement: The data sets generated during and/or analyzed during the current study are available from the corresponding author on reasonable request.

Conflicts of Interest: Jianxin Wang is employed by the company Tangshan Talent Development Group Co., Ltd. The remaining authors declare that the research was conducted in the absence of any commercial or financial relationships that could be construed as a potential conflict of interest.

References

1. Qiang, W.; Kang, M.; Liu, J.; Ndumia, J.N. Microstructure and Wear Performance of High-Velocity Arc Sprayed FeMnCrNiBNbAl Coating. *Coatings* **2024**, *14*, 428. [CrossRef]
2. Li, B.; Yang, D.; Liu, Z.; He, J.; Bai, J.; Jiang, H.; Tian, Y.; Zhang, Z.; Liu, S. Fabrication, Microstructure and Corrosion Resistance of Zn/Al Composite Coating by Arc Spraying. *Coatings* **2023**, *13*, 1406. [CrossRef]
3. Rakhadilov, B.; Magazov, N.; Kakimzhanov, D.; Apsezhanova, A.; Molbossynov, Y.; Kengesbekov, A. Influence of Spraying Process Parameters on the Characteristics of Steel Coatings Produced by Arc Spraying Method. *Coatings* **2024**, *14*, 1145. [CrossRef]
4. Li, S.J.; Mao, D.Y.; He, Z.H.; Lu, B.; Liu, J.; Wang, Y.L.; Liu, H. Forming Performance of 1Cr18Ni9Ti Stainless Steel Laser Oscillating Welding Joints. *Hot Work. Technol.* **2025**, *51–55*, 64. [CrossRef]
5. Xu, X.J.; Lin, W.Q.; Lu, H. Defect Reason on Outer Surface of 1Cr18Ni9Ti Stainless Steel Oil Return Tube. *Corros. Prot.* **2020**, *41*, 70–74.
6. Wang, Y.F.; Zhao, Y.T.; Ma, H.H.; Ren, H.P. Effect of heat treatment on corrosion resistance of 1Cr18Ni9Ti and 2Cr13 steel plate welded joint. *Heat Treat. Met.* **2021**, *46*, 7–11. [CrossRef]
7. Yang, Y.; Jiang, L.H.; Luo, S.H.; Hu, H.B.; Tang, T.G.; Zhang, Q.M. Effect of strain on microstructure evolution of 1Cr18Ni9Ti stainless steel during adiabatic shearing. *J. Mater. Eng. Perform.* **2016**, *25*, 29–37. [CrossRef]
8. Gangloff, R.P.; Ha, H.M.; Burns, J.T.; Scully, J.R. Measurement and Modeling of Hydrogen Environment-Assisted Cracking in Monel K-500. *Metall. Mater. Trans. A* **2014**, *45*, 3814–3834. [CrossRef]
9. Yang, M.H.; Li, X.W.; Sun, C.H.; Ruan, Y. Microstructure and Properties of Monel K-500 Alloy in Synergetic Modulation of Directional Solidification and Thermal Processing. *Acta Metall. Sin.* **2024**, *61*, 1–13.

10. Jia, X.M.; Wang, J.; Liu, T.Y.; Zheng, H.B.; Wang, Q.Y.; Xi, Y.C.; Dong, L.J. Effect of aging temperature on microstructure and properties of Monel K500 alloy. *Iron Steel Vanadium Titan.* **2023**, *44*, 160–166. [CrossRef]
11. Zhu, Y.L.; Zheng, W.J.; Song, Z.G.; Feng, H. Development and Application of Nickel-copper Alloy in China. *Hot Work. Technol.* **2019**, *48*, 23–26. [CrossRef]
12. Luo, L.X.; Liu, R.Z.; Xiao, Q.; Sun, H.; Ma, Y.L.; Shi, Y.J.; Zhang, C.; Ma, X.L.; Zhang, Z. Microstructure and properties of Zn–Al–Ni–Cu composite coating prepared by arc spraying. *Electroplat. Finish.* **2023**, *42*, 27–36. [CrossRef]
13. Ge, Y.; Tan, P.; Li, Z.F.; Wang, J.Y.; Wang, Q.B.; Yang, B.J. Study on mechanical properties of Monel porous plate. *Powder Metall. Ind.* **2017**, *27*, 34–38. [CrossRef]
14. Siddiqui, A.A.; Dubey, A.K. Recent trends in laser cladding and surface alloying. *Opt. Laser Technol.* **2021**, *134*, 106619. [CrossRef]
15. Cheng, E.C.; Liu, J.; Zhang, J.; Wang, T.X.; Liu, L.H.; Li, Y.F. Microstructure and friction and wear properties of laser cladding Fe-Cr-Ni alloy layer. *Trans. Mater. Heat Treat.* **2022**, *43*, 143–152. [CrossRef]
16. Yang, X.T.; Zhou, J.; Wang, X.H.; Wei, H.L.; Zeng, R.; Yang, Q.B.; Li, W.S. Effect of Cu Content on Microstructure and Tribological Properties of Ni-Based Directional Structure Alloy Coatings. *Rare Met. Mater. Eng.* **2022**, *51*, 1420–1426.
17. Zhang, H.Y.; Wang, F.R.; Du, S.M.; Zeng, Z.X.; Cai, H.; Liu, E.Y. Effect of Ni Element on Microstructure and Tribological Properties of Plasma Sprayed Iron-Based Coatings. *Min. Metall. Eng.* **2021**, *41*, 150–155.
18. Elttayef, A.K.; Abass, L.N.; Abd Al-Latif, L.G. Improvement mechanical properties of Inconel and Monel alloys synthesis by laser coating. *Opt. Laser Technol.* **2019**, *109*, 49–54. [CrossRef]
19. Krelling, A.P.; Melo, F.S.; Almeida, E.A.S.; da Costa, C.E.; Milan, J.C.G. Microstructure and properties of borided Monel 400 alloy. *Mater. Res. Express* **2019**, *6*, 106410. [CrossRef]
20. Kostryzhev, A.G.; Marenych, O.O.; Pan, Z.; Li, H.; van Duin, S. Strengthening mechanisms in Monel K500 alloyed with Al and Ti. *J. Mater. Sci.* **2023**, *58*, 4150–4164. [CrossRef]
21. Wang, K.; Qi, H.; Ma, S.; Wang, L.; He, N.; Li, F. Effect of Nickel Addition on Solidification Microstructure and Tensile Properties of Cast 7075 Aluminum Alloy. *Crystals* **2023**, *13*, 1589. [CrossRef]
22. Karlina, A.I.; Karlina, Y.I.; Kondratiev, V.V.; Kononenko, R.V.; Breki, A.D. Study of Wear of an Alloyed Layer with Chromium Carbide Particles After Plasma Melting. *Crystals* **2023**, *13*, 1696. [CrossRef]
23. Wu, S.; Ning, Y.; Xie, H.; Tian, H.; Lv, J.; Liu, B. Study on the galvanic corrosion behavior of copper-nickel/titanium alloys under simulated seawater environment. *J. Solid. State Electrochem.* **2024**, *28*, 2315–2329. [CrossRef]
24. Wielage, B.; Pokhmurska, H.; Student, M.; Gvozdeckii, V.; Stupnyckyj, T.; Pokhmurskii, V. Iron-Based Coatings Arc-Sprayed with Cored Wires for Applications at Elevated Temperatures. *Surf. Coat. Technol.* **2013**, *220*, 27–35. [CrossRef]
25. Zeng, J.; Li, K.J.; LÜ, J.J.; Zhao, J.; Liu, Q.; He, Q.B. Effect of solution and two-stage aging on microstructure and hardness of Cu-containing nickel-based alloys. *Trans. Mater. Heat Treat.* **2024**, *45*, 113–122. [CrossRef]
26. Liu, J.; Li, J.; Yang, H.L.; Ruan, J.M. Microstructure of titanium-nickel alloy by mechanical alloying. *J. Cent. South Univ. (Sci. Technol.)* **2015**, *46*, 1202–1207.
27. Liu, X.Z.; Shen, Q.W.; Liu, X.Z.; Chen, J.; Zhu, L.W.; Qi, J. Effect of Heat Treatment Temperature on the Spectral Properties of Cu-Ni Coating. *Spectrosc. Spectr. Anal.* **2015**, *35*, 1094–1098.
28. Yang, X.; Wang, X.; Zhou, J.; Wei, H.; Zeng, R.; Li, W. Effect of Cu addition on the microstructure and tribological performance of Ni60 directional structure coating. *Int. J. Miner. Metall. Mater.* **2023**, *30*, 715–723. [CrossRef]
29. Yao, K.; Liu, Z.D.; Liu, Q.B. Effect of Laser Power on Microstructure and Properties of Ni-Based Alloy Coatings on 30CrMnSiA Steel. *J. Therm. Spray Technol.* **2022**, *31*, 2136–2146. [CrossRef]
30. Zhao, Y.; Chen, Y.; Zhang, T.; Yu, T. Laser fabricated nickel-based coating with different overlap modes. *Mater. Manuf. Process.* **2021**, *36*, 1618–1630. [CrossRef]
31. Zhao, Y.; Guan, C.; Chen, L.; Sun, J.; Yu, T. Effect of process parameters on the cladding track geometry fabricated by laser cladding. *Optik* **2020**, *223*, 165447. [CrossRef]
32. Lu, S.Y. *Super Stainless Steel and High Nickel Corrosion Resistant Alloys*; Chemical Industry Press: Beijing, China, 2012; p. 355.
33. Wang, H.Z.; Zhou, X.Y.; Liu, M.; Jia, L.; Huang, Y.F.; Wang, H.D. Research Progress of the Sealant Decreased Thermal Spray Coating Porosity. *Mater. Rep.* **2023**, *37*, 210–228.

crystals

MDPI

Article

Compositional Design, Microstructure, and Thermal Processing of Aluminum-Based Complex Concentrated Alloys

Spyridon Chaskis [1,2,*], Constantinos Tiktopoulos [1], Evangelos Gavalas [3], Marianthi Bouzouni [3], Fotis Tsiolis [3] and Spyros Papaefthymiou [1,*]

1 Laboratory of Physical Metallurgy, Division of Metallurgy and Materials, School of Mining and Metallurgical Engineering, National Technical University of Athens, 9, Heroon Polytechniou Street, 15780 Athens, Greece; mm17079@central.ntua.gr or ctiktopoulos@etemgestamp.com

2 Technology and R&D Department of HALCOR, Copper Tubes and Alloys Extrusion Division, ELVALHALCOR S.A., 62nd km Athens-Lamia National Road, 32011 Oinofyta, Greece

3 Department of Physical Metallurgy and Forming, Hellenic Research Centre for Metals (ELKEME S.A.), 61st km Athens-Lamia National Road, 32011 Oinofyta, Greece; egavalas@elkeme.gr (E.G.); mbouzouni@elkeme.gr (M.B.); ftsiolis@elkeme.gr (F.T.)

* Correspondence: schaskis@metal.ntua.gr or schaskis@halcor.com (S.C.); spapaef@metal.ntua.gr (S.P.); Tel.: +30-69-7958-7537 (S.C.)

Abstract

Three lightweight aluminum-based complex concentrated alloys with chemical compositions that have not been previously studied were manufactured and studied: $Al_{52}Mg_{9.6}Zn_{16}Cu_{15.5}Si_{6.9}$ w.t.% or $Al_{63}Mg_{13}Zn_{8}Cu_{8}Si_{8}$ a.t.% (alloy A), $Al_{44}Mg_{18}Zn_{19}Cu_{19}$ w.t.% or $Al_{55}Mg_{25}Zn_{10}Cu_{10}$ a.t.% (alloy B), and $Al_{47}Mg_{21.4}Zn_{12}Cu_{9.7}Si_{9.7}$ w.t.% or $Al_{52.7}Mg_{26.6}Zn_{5.6}Cu_{4.6}Si_{10.4}$ a.t.% (alloy AM), with low densities of 3.15 g/cm^3, 3.18 g/cm^3 and 2.73 g/cm^3, respectively. During alloy design, the CALPHAD method was used to calculate a variety of phase diagrams for the various chemical compositions and to predict possible phases that may form in the alloy. The CALPHAD methodology results showed good agreement with the experimental results. The potential of the designed alloys to be used in some industrial applications was examined by manufacturing them using standard industrial techniques, something that is a rarity in this field. The alloys were produced using an induction furnace and pour mold casting process, while industrial-grade raw materials were utilized. Heat treatments with different soaking times were performed in order to evaluate the possibility of improving the mechanical properties of the alloys. Alloys A and AM were characterized by a multiphase microstructure with a dendritic FCC-Al matrix phase and various secondary phases (Q-AlCuMgSi, Al_2Cu and Mg_2Si), while alloy B consisted of a parent phase T-$Mg_{32}(Al,Zn)_{49}$ and the secondary phases α-Al and Mg_2Si. The microstructure of the cast alloys did not appear to be affected by the heat treatments compared to the corresponding as-cast specimens. However, alterations were observed in terms of the elemental composition of the phases in alloy A. In order to investigate and evaluate the mechanical properties of the as-cast and heat-treated alloys, hardness testing along with electrical conductivity measurements were conducted at room temperature. Among the as-cast samples, alloy AM had the highest hardness (246 HV_4), while among the heat-treated ones, alloy A showed the highest value (256 HV_4). The electrical conductivity of all the alloys increased after the heat treatment, with the highest increase occurring during the first 4 h of the heat treatment.

Keywords: complex concentrated alloys; aluminum-based alloys; lightweight high-entropy alloys; CALPHAD; mechanical properties; heat treatment; application-based design

1. Introduction

In the automotive industry, reducing the vehicle's weight is the primary goal in order to lower fuel consumption and CO_2 emissions. Aluminum, with its low specific weight and excellent mechanical properties, plays a key role in achieving this objective [1]. An average car (1300–1400 kg) contains approximately 180 kg of aluminum, of which, 65% (117 kg) is cast aluminum—about 10% of the vehicle's total weight [2,3]. Notably, reducing a vehicle's weight by 1 kg can prevent 20 kg of CO_2 emissions over its lifecycle [4].

The key cast aluminum components include cylinder heads, engine blocks, pistons, brake calipers, rims, and crankcases. Pistons, subjected to high mechanical and thermal loads, require alloys with high strength, hardness, and thermal stability [5]. Cylinder heads and engine blocks prioritize strength, thermal conductivity, and recyclability, with recycled aluminum gaining importance over primary aluminum as a raw material [5]. Common alloys such as $AlCu_4MgSi$ (A206), $AlSi_{10}Mg$ (A360), and $AlSi_9Cu_3$ (A380) are widely used, though the adoption of cast aluminum in automotive applications is expected to decline [2]. Therefore, continuous research is being carried out to develop novel low-density alloys, which will have the required properties at a reasonable cost in order to replace Al cast components.

Aluminum alloys are essential in internal combustion engine (ICE) applications, where they must exhibit high strength, fatigue resistance, and thermal stability [6]. While traditional Al-Si alloys have dominated due to their casting properties and mechanical performance [6], sustainability goals have prompted exploration of complex concentrated alloys (CCAs). These CCAs can incorporate recycled aluminum scrap with a high alloying element content, transforming post-consumer waste into valuable raw materials. This approach conserves natural resources, while reducing energy consumption and the carbon footprint of aluminum production. The innovative use of "dirty" alloys, as highlighted by Raabe et al., enables the development of advanced aluminum alloys that meet stringent mechanical requirements while promoting a circular economy by leveraging sustainable raw materials [7].

In this direction, along with the CCAs, the Lightweight High-Entropy Alloys (LWHEAs) could offer an alternative. In general, High-Entropy Alloys (HEAs) constitute a relatively new field of research that is trying to subvert and revitalize the materials science sector's way of thinking about alloying design strategies. Lightweight High Entropy Alloys (LWHEAs) are defined as alloys with a density less than or equal to that of commercial titanium alloys (4.51 g/cm^3) [8]. In other publications, alloys with a density lower than the density of an alloy used for a particular application are considered LWHEAs [9].

By definition, HEAs or equiatomic multicomponent alloys are composed of at least five metallic elements in equimolar or near-equimolar ratios [10]. Yeh et al. proposed a classification of the alloys based on their configurational entropy (ΔSconf). The alloys were classified as HEAs if their ΔSconf in a random state is greater than 1.5 R (R being the gas constant); as Medium-Entropy Alloys (MEAs) when the value of their ΔSconf is in the range of 1R-1.5 R; and finally, traditional alloys are classified as Low-Entropy Alloys, with a ΔSconf less than 1R. The concept of HEAs was initially based on the theory that an increase in the number of alloying elements would result in an increase in the ΔSconf of such systems and it would promote solid-solution (SS) structures, such as FCC and BCC structures [10]. However, now that this field is more scientifically mature, this effect is not considered to be as great as was initially believed [11], since the addition of more components has been proven to lead to the formation of second phases [12]. Therefore, since the limitation of single-phase SS microstructures was surpassed, the term CCAs [8] was introduced, which enables a greater degree of freedom in the design and development of novel compositions with a more complex microstructure, which could provide broader potential comparing to HEAs. Multiphase HEAs, known as CCAs [13], are now mainly composed of two or more phases, sharing similar core effects with HEAs and are a unique subject of research [14]. High-Entropy Aluminum Alloys (HEAls), a promising derivative of HEAs, were recently introduced by Asadikiya et al. as a novel category of aluminum alloys [15]. Although, their entropy is lower than that of traditional HEAs, HEAls still take advantage of it to achieve enhanced properties [16]. The formation of intermetallics (ICs) in HEAls is inevitable due to the high negative enthalpy of mixing (ΔHmix) between Al and most of the commonly employed alloying elements [4]. As a result, HEAls generally feature multiphase microstructures with α-Al solid-solution phases reinforced by multicomponent ICs. Mundrha et al. proposed this novel approach as robust design direction to create high-performance Al alloys strengthened with multicomponent ICs [9].

HEAs, MEAs, and CCAs are usually manufactured using expensive high-purity raw materials and vacuum arc melting and vacuum induction melting techniques, in an Ar or N_2 protective atmosphere, and cast in water-cooled copper molds [15]. However, in order to scale up the production of these alloys, a more industrially applicable approach is necessary [9,16]. In the industry, instead of high-purity raw materials, industrial-grade raw materials are used, making the large-scale production of metals sustainable [14,17]. Thus, the concept of a circular economy is capitalized on and added value is created for lower-purity materials [18–20]. The aim of the current project was to design alloys using standard industrial techniques and economically viable raw materials.

Barnett et al. introduced a scrap-tolerant alloying concept based on HEAs, highlighting compositional flexibility as a pathway to create value from waste alloys [18]. HEAls are promising for high-wear, high-temperature applications, and as candidate materials for circular product design. Their multicomponent, multiphase compositions also render them suitable for manufacturing using recycled scrap materials.

There have been several studies of promising Lightweight HEAs (LWHEAs) using Al as the principal element. Lei Shao et al. developed a series of low-density alloys with Al and Mg as the main components, accounting for 90 at.% of the alloy, with the remaining components consisting of Cu, Zn, and Si [21]. These alloys were prepared using pure raw materials in an Ar protective atmosphere and they presented densities ranging from 2.64 to 2.75 g/cm^3 and compression strengths of up to 814 MPa. Their microstructure consisted of various large intermetallic phases such as Mg_2Si and $Mg_{32}(AlZn)_{49}$ within an aluminum matrix. Using the same components, Sanchez et al, through Thermocalc software (TCAL5 database), designed two systems, $Al_{65}Cu_5Mg_5Si_{15}Zn_5X_5$ and $Al_{70}Cr_5Cu_5Mg_5Si_{10}Zn_5X_5$, where the elements Fe, Ni, Cr, Mn, and Zr were introduced in place of X. The best results were from the $Al_{70}Cr_5Cu_5Mg_5Si_{10}Zn_5$ at.% alloy, which consisted of Al_2Cu, Mg_2Si, and $Al_{13}Cr_4Si_4$ intermetallic phases in an aluminum matrix with compressive strength of 490 MPa, plastic deformation of 6%, and microhardness of 200HV$_{0.1}$. The rest of the alloys demonstrated slightly higher strengths and hardnesses, but the majority exhibited embrittlement behaviors [13]. Following his original research, Sanchez designed a number of LWHEAs by testing various combinations of compositions with the primary elements Al, Mg, Cu, Zn, and Si and additional elements Ca, Ti, Sn, Mn, Ni, and Fe [8,22,23]. The results of his work revealed that with the HEA design strategy, hardness values ranging from 437 HV to 916 HV with densities in the range of 3.15–5.10 g/cm^3 could be achieved. In further continuation of this design strategy, Sanchez et al. designed two scrap-tolerant LWHEAs, $Al_{65}Cu_{10}Mg_{10}Si_{10}Zn_5$ and $Al_{80}Cu_5Mg_5Si_5Zn_5$ at.%, with high compressive strengths of 696 and 722 MPa, respectively, due to the formation of ICs [24]. In an attempt to produce alloys with even lower densities, Tun et al. used Li in the composition of their alloys. As a result, they successfully manufactured $Al_{70}Li_{10}Mg_{10}Cu_5Zn_5$ and

$Al_{75}Li_{10}Mg_5Cu_5Zn_5$ a.t% systems, with densities of 2.76 g/cm^3 and 2.7 g/cm^3, respectively. The alloys were prepared using disintegrated melt deposition and ultra-pure raw materials in a protective Ar atmosphere [25] [21]. In fact, the latter alloy, which presented the highest compressive strength (533 MPa), was also examined for its anti-corrosion behavior, with encouraging results [26]. However, Li is a scarce and expensive raw material. More generally, the density found in the LWHEA literature varies between 2.67 g/cm^3 and 5.20 g/cm^3 [12]. The largest high-entropy alloy reported in the literature is approximately 10 kg with a density of 3.08 g/cm^3 and it is the only Al-based CCA, which was prepared by electromagnetic stirring casting. Using this method, a microhardness of 177 HV and compressive strength of 554 MPa were achieved. The most important finding of this work, however, was that the electromagnetic frequency influenced the morphology and distribution of the Mg_2Si phase, whose average particle size decreased from 40.0 to 17.7 mm. The research on the Al-Mg-Zn-Cu-Si system is limited to few alloys [27]. Recently, Chaskis et al. designed an $Al_{58}Mg_{18}Zn_{12}Cu_5Si_7$ at.% ($Al_{47.4}Mg_{13}Zn_{23}Cu_9Si_6$ wt.%) alloy, which displayed a multiphase microstructure consisting of an Al matrix reinforced with various secondary phases, such as Mg_2Si and $MgZn_2$. In terms of mechanical properties, the developed alloy exhibited a high hardness value of 249 Vickers and compressive strength of 588 MPa. In this work, it was proven that, if the Mg_2Si phase is present in large amounts, it leads to embrittlement of the alloy [28,29]. Nonetheless, the importance of controlling Mg_2Si- and Mg-bearing eutectic phases was underlined, since plasticity can be compromised [30]. Therefore, a way to reduce the size and the percentage of this phase is deemed important.

However, most studies on these alloys have focused solely on hardness and compressive strength, with limited investigation into other mechanical properties. Additionally, alloy designs often lack application-specific targets, complicating the assessment of their potential to replace existing commercial materials. Moreover, the reliance on protective atmospheres and high-purity raw materials in HEA production significantly increases costs.

The motor contributes slightly to the overall vehicle weight, but sustainability demands innovative solutions to repurpose cast aluminum alloys from machine parts and utilize them as raw materials. These alloys, with their high alloying content, are unsuitable for malleable products and require alternatives for use in next-generation internal combustion engines (ICEs) or other cast aluminum components. For example, Tesla utilizes two die-cast aluminum parts for the chassis of its models, highlighting opportunities for aluminum-based CCAs in a circular product design approach.

To evaluate their industrial potential, the alloys in this study were manufactured in a standard induction furnace without a protective atmosphere, using industrial purity raw materials from the Greek metallurgical industry. Developed for cast aluminum automotive components, the alloys were compared with 3xx and 7xx series casting alloys since they combine elements from these groupings. This research aimed to develop lightweight aluminum-based alloys in the Al-Mg-Cu-Zn-Si and Al-Mg-Cu-Zn systems using an industry-relevant approach.

In this work, the term HEAs is used for single-phase alloys with a ΔSconf greater than 1.5 R. The term CCAs is used for multiphase alloys with a ΔSconf greater than 1.5 R. Finally, MEAs have a ΔSconf that lies between 1 R and 1.5 R. Since MEAs can also be multiphase alloys, in order to better describe the manufactured alloys under investigation, the term CCAs was adopted. The developed alloys could offer a lightweight alternative to some other aluminum alloys, while promoting a sustainable alloy design approach utilizing raw materials commonly used in industry practices. Moreover, the compositional design focuses on controlling the size and the percentage of specific phases found in earlier studies.

This work provides valuable information for researchers aiming to design and produce industry-like aluminum-based CCAs with an industrially relevant approach.

2. Materials and Methods

In this work, the Al-Mg-Cu-Zn-Si and Al-Mg-Cu-Zn systems were selected for study. Aluminum was chosen to be the main component of the alloy due to its low specific weight, combined with its excellent mechanical properties and recyclability [1]. The remaining components were selected based on certain criteria, such as

- Widely available components in the Greek metallurgical industry.
- Does not significantly increase the specific weight of the alloy.
- Provides positive properties to the alloy for its final application.

The components Mg, Zn, Cu, and Si satisfy most of the requirements and were therefore considered the most suitable. The design of the first system was based on the restriction of the Mg_2Si phase in a controlled quantity. Although this phase is known to enhance the hardness of commercial aluminum alloys [31], when present in large amounts, it has been proven to lead to embrittlement of the alloy [28,29] due to its high hardness in contrast to the neighboring phases. It is also known that Al-Cu alloys exhibit deteriorated casting properties, which significantly limits the production of these alloys. Additionally, they do not respond to aging heat treatments, preventing the achievement of higher levels of strengthening [30]. Thus, the chemical composition of the first alloy (alloy A—$Al_{63}Mg_{13}Zn_8Cu_8Si_8$ a.t.%) was chosen with the aim of having a very low percentage of the Mg_2Si phase according to the Thermocalc software's calculations (TCAL7 database). The design of the second system (alloy B—$Al_{55}Mg_{25}Zn_{10}Cu_{10}$ a.t.%) was based on the complete absence of the Mg_2Si phase, so that the properties of an Al-based HEA without its presence could be examined. Finally, the purpose of the third alloy (alloy AM—$Al_{52.7}Mg_{26.6}Zn_{5.6}Cu_{4.6}Si_{10.4}$ a.t.%) design was to test the limits of adding Mg to an HEA without ignition occurring in the melt pool.

For the purpose of designing HEAs and CCAs, some thermo-physical parameters, based on Hume–Rothery rules and thermodynamic parameters, were taken into account [32]. These criteria consist of parameters such as the enthalpy of mixing in the liquid phase (ΔHmix), atomic size difference (r), Pauling electronegativity difference ($\Delta\chi$), and Ω parameter. In the current study, the artificial neural network software pyMPEALab Toolkit (Upadesh Subedi et al.) was used for the thermodynamic calculations, as a means of calculating the thermo-physical parameters of the chosen alloys [33].

Thermocalc® software (Version 2022a, Thermo-Calc Software AB, Stockholm, Sweden) in conjunction with the TCAL7 [34] thermodynamic database was used to calculate a variety of phase diagrams for various chemical compositions with the selected components during the alloy development process. ThermoCalc® software utilizes the CALPHAD (CALculation of PHAse Diagrams) method which has been extensively used in the prediction of the thermodynamic equilibrium of multicomponent systems. In the CALPHAD method, the thermodynamic equilibrium is predicted by minimizing the total ΔG for a certain temperature, pressure, and chemical composition. The TCAL7 database is commonly used for commercial Al alloys and according to previous works, the use of this database for Al-based HEAs shows good agreement with experimental results [8,29]. This tool shows the equilibrium phases as a function of temperature in order to provide an approximate idea of the type of phases that would form in the selected alloy [35].

The experimental alloys were melted in an electrical resistance N 41/H Nabertherm furnace under atmospheric conditions and casted into conventional cast iron ingot molds. The raw materials used to produce the alloys consisted of master alloys of industrial purity (such as $Al_{65}Cu_{35}$, $Al_{50}Si_{50}$, and $Al_{95}Mg_5$, supporting the approach of scrap-based alloy design), sheets of pure zinc, and tablets of cast bars of pure magnesium. These master alloys are commercially available and their physical characteristics are listed in Table 1. To prevent the oxidation of the elements, casting flux was used to protect the melt pool surface and prior to casting the molten metal, it was skimmed (after each of the alloying additions, the melt was thoroughly stirred with a graphite rod to ensure a homogenous melt without disturbing the metal/flux interface). The ingot was manufactured using a typical furnace pour mold casting method similar to industry practices.

Table 1. Physical information about the utilized alloying elements and master alloys [29,36–45].

Alloying Element	Melting Temperature (°C)	Boiling Temperature (°C)	Density (g/cm^3)	Crystal Lattice Type/Lattice Constant (nm)	Atomic Radius (pm)	Electronegativity (Pauling Scale)
Al	660.3	2450	2.70	FCC/0.4	143	1.61
Mg	650.0	1005	1.74	HCP/0.32	141	1.31
Cu	1085.0	2570	8.96	FCC/0.36	128	1.90
Zn	419.5	907	7.13	HCP/0.1	122	1.65
Si	1410.0	3280	2.33	FCC/0.543	111	1.90

For alloys A and AM, master alloys Al65Cu35, Al50Si50 and Al95.5Mg4.5 were added to a steel mold, which was placed in the furnace at 1100 °C. After melting, a bath base at 800 °C was guaranteed. Subsequently, Mg and Zn were added to the molten pool and soaked for 1 h. The reason the direct addition of magnesium and zinc was avoided is because they have a boiling point lower than the initial bath temperature and they would volatilize (see Table 1). Finally, the melt was cooled down to 700 °C and manually poured into a steel mold (pouring temperature of melting: +50 °C superheat).

For alloy B, master alloys Al65Cu35 and Al95.5Mg4.5 were added to the mold together with the pieces of pure magnesium and Zn sheets and placed in the furnace at 700 °C. After melting, the melt was manually poured into a steel mold. For all 3 alloys, before casting, the melt was stirred for at least for 15 min in order to achieve a fully homogenous bath.

Ingots with dimensions of 35 mm × 35 mm × 40 mm were obtained for each alloy, as shown below (Figure 1). Following this, the microstructures of the as-cast and heat-treated alloys at room temperature were studied to evaluate the suitability and behavior of the developed CCAs. The alloys' density was determined experimentally using the Archimedes Principle [46].

The X-ray diffraction (XRD) equipment used to characterize the crystal structures of the alloys was a Phillips Xpert diffractometer with a long fine focus copper anode X-ray source and the diffraction diagrams were measured at a diffraction angle of 2θ, in the range from 10° to 110°, with a step size of 0.02°/sec.

Samples were sectioned from the center of the ingot, parallel to the casting direction. Metallographic preparation of the samples was conducted with the appropriate techniques and microscopic observation was performed on as-polished condition, without prior etching using a Nikon Epiphot 300 inverted metallographic microscope.

Specimens AH4, BH4, and AMH4 underwent a solid-solution heat treatment at 400 °C for 4 h, followed by water cooling, while AH24, BH24, and AMH24 were subjected to a heat treatment at 400 °C for 24 h, followed again by water cooling, with the purpose of examining the effect of soaking time. For the heat treatment of the samples, an N 60/85HA Nabertherm furnace was used.

The different regions, the average overall chemical composition of each sample and of the phases in it were investigated using a JEOL JSM-IT800 scanning electron microscope (SEM) equipped with an energy-dispersive X-ray spectrometer (EDS).

A Vickers hardness Duramin-40 M1 model (Struers, Copenhagen, Denmark) measurer was employed on the polished sample surface using a 4 kg load, which was applied for

10 s. At least 10 random individual measurements were made for each test. At least five specimens were recorded to ensure repeatability. To measure the electrical conductivity, a SIGMATEST D-2.068 (FOERSTER Instruments Inc., Pittsburgh, PA, USA) device was utilized and at least 5 random individual measurements were made for each specimen.

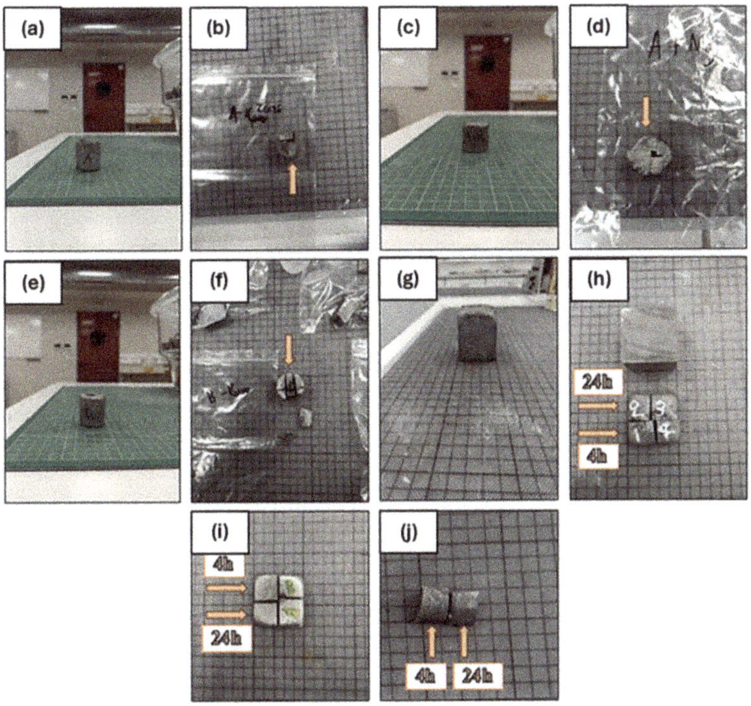

Figure 1. Specimens before and after sectioning (1 box is equal to 1cm): (**a**,**b**) as-cast sample A, (**c**,**d**) as-cast sample B, (**e**,**f**) as-cast sample AM, (**g**,**h**) heat-treated sample A, (**i**) heat-treated sample B, and (**j**) heat-treated sample AM.

3. Results

3.1. Thermo-Physical Parameters for Phase Formation in HEAs/CCAs

After the preliminary study of the thermo-physical parameters for phase formation in the HEAs and of various equilibrium phase diagrams for the different chemical compositions, the systems that were finally selected were the following: $Al_{52}Mg_{9.6}Zn_{16}Cu_{15.5}Si_{6.9}$ w.t.% or $Al_{63}Mg_{13}Zn_{8}Cu_{8}Si_{8}$ a.t.% (alloy A), $Al_{44}Mg_{18}Zn_{19}Cu_{19}$ w.t.% or $Al_{55}Mg_{25}Zn_{10}Cu_{10}$ a.t.% (alloy B), and $Al_{47}Mg_{21.4}Zn_{12}Cu_{9.7}Si_{9.7}$ w.t.% or $Al_{52.7}Mg_{26.6}Zn_{5.6}Cu_{4.6}Si_{10.4}$ a.t.% (alloy AM) (Table 2).

Table 2. Chemical compositions of alloys in wt.% and at.% in this work.

Alloy	Al (w.t.%)	Cu (w.t.%)	Mg (w.t.%)	Si (w.t.%)	Zn (w.t.%)	Al (a.t.%)	Cu (a.t.%)	Mg (a.t.%)	Si (a.t.%)	Zn (a.t.%)
A	52	15.5	9.6	6.9	16	63	8	13	8	8
B	44	19	18	-	19	55	10	25	-	10
AM	47	9.7	21.4	9.7	12	52.7	4.6	26.6	10.4	5.6

In Table 3, the calculated thermo-physical parameters of the defined systems are summarized. The ΔHmix of this system was in the range of −8 to −1 kJ/mole, which is outside the limits required for the formation of SS structures [23]. Additional criteria related to the formation of SS structures in alloys containing low-density elements are δ values below 4.5%, a $\Delta\chi$ equal or below 0.175, and an Ω higher than 10 [23,47]. It is expected that multiphase alloys were obtained regardless of the ΔSmix value, since the ΔHmix i–j between Al and other elements is highly negative [12]. Thus, based the alloy design approach of other studies [13,23,29,48–50], the parameters were kept as close as possible to the threshold parameters for SS formation, in order to restrain the formation of excessive amounts of ICs. Following the standard classification of HEAs, all three alloys were categorized as Medium-Entropy Alloys. Nonetheless, maximizing the ΔSmix was not necessary in the developed alloys. In contrast to the main goal of classic HEA design approaches, which is to promote the formation of SS phases, in this study, the formation of multiphase microstructures was promoted to enhance the mechanical properties of the aluminum alloys. To obtain a microstructure composed of a mix of SS structures and ICs, the ΔHmix was kept between −8 kJ/mol and −4 kJ/mol, avoiding excessively high negative values. A decrease in the Al content and absence of Si increased the value of ΔHmix, whereas increases in the Mg and Si contents further decreased the value of ΔHmix. The increased δ in all three alloys highlights the tendency to form intermetallics. On the other hand, in alloy A, the $\Delta\chi$ value indicates that SS structures can be formed, but it is well known that satisfying only one parameter does not promote the formation of SS structures [51,52]. The calculated Ω parameters (ratio between the ΔHmix and the ΔSmix in the liquid state of the alloy) were very far away from the literature's proposed value of $\Omega >$ ~10 for the formation of SS structures in low-density multiprincipal-component alloys [48]. A higher Ω is related to a lower tendency to form complex phases in Al-based multicomponent alloys [53]. In contrast, the Ω value in all three alloys indicated a tendency of promoting intermetallics according to the threshold value proposed for traditional HEAs [12]. Lastly, the $\Delta\chi$ was within the recommended values (under 0.175) for forming SS structures in HEAs for alloys A and B and above that limit for alloy AM [12]. Therefore, mixed microstructures consisting of SS structures and ICs were expected to form based on the known thermo-physical parameters. This was found to be consistent with the CALPHAD calculations for these alloys. The theoretical density of the alloys was calculated using the rule of mixtures.

Table 3. Thermo-physical parameters and the theoretical density of the developed alloys.

Alloy	ΔHmix (kJ/mol)	δ (%)	ΔSmix (J/K/mol)	Ω	$\Delta\chi$	VEC	Tm (K)	$\rho_{theoretical}$ (g/cm^3)
A	−5.63	7.64	9.66 (1.16R)	1.73	0.16	4.31	1006.5	3.15
B	−1.76	6.57	9.44 (1.14R)	5.09	0.17	4.45	948.8	3.18
AM	−7.04	9.03	10.21 (1.22R)	1.47	0.19	3.71	1014.4	2.73

In these systems, a microstructure consisting mainly of an Al-based face-centered cubic (FCC) matrix phase along with other intermetallic compounds was expected to be stable at room temperature.

3.2. CALPHAD Methodology and Equilibrium Phase Diagrams

The equilibrium phase diagrams of the developed alloys are shown in Figure 2. Alloy A was expected to be composed of a mixture of five different phases at equilibrium, namely Al_2Cu, α (Al-FCC), C14_Laves ($MgZn_2$), Q ($Al_4Cu_2Mg_8Si_7$), and V (AlCuMgZn), while alloy B was expected to be composed of just three phases, namely α (Al), C14_Laves ($MgZn_2$), and και T ($Al_2Mg_3Zn_3$ or $Mg_{32}(Al,Zn)_{49}$). As for alloy AM, it was expected to consist of Al_2Cu, α (Al-FCC), C14_Laves ($MgZn_2$), Mg_2Si, and T ($Mg_{32}(Al,Zn)_{49}$).

(a)

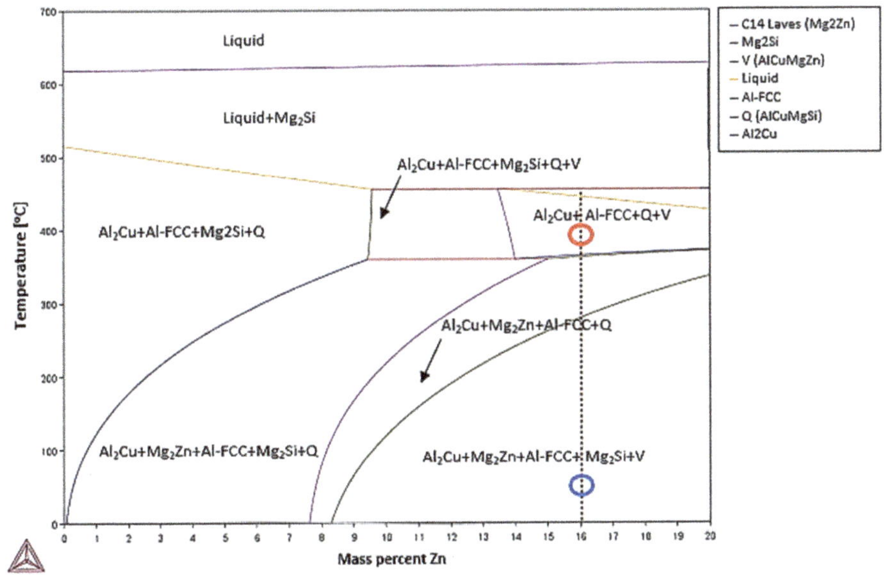

(b)

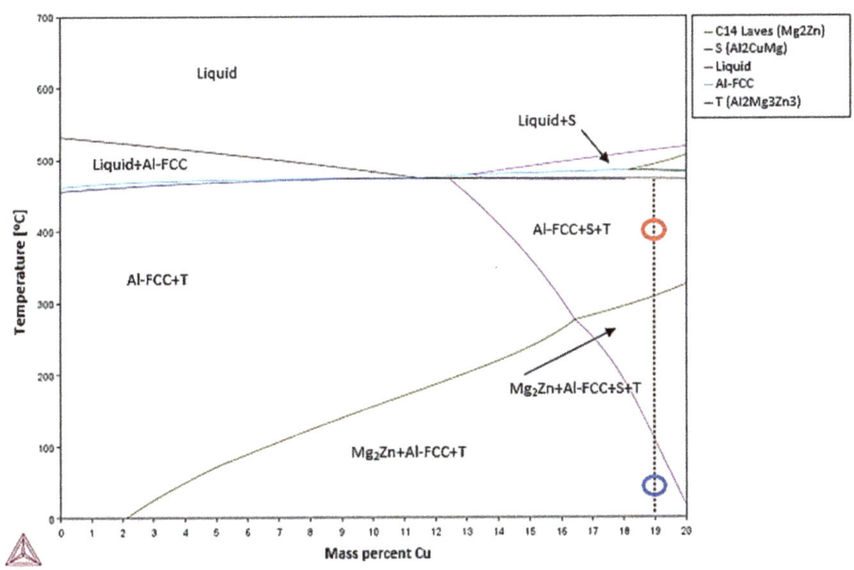

Figure 2. *Cont.*

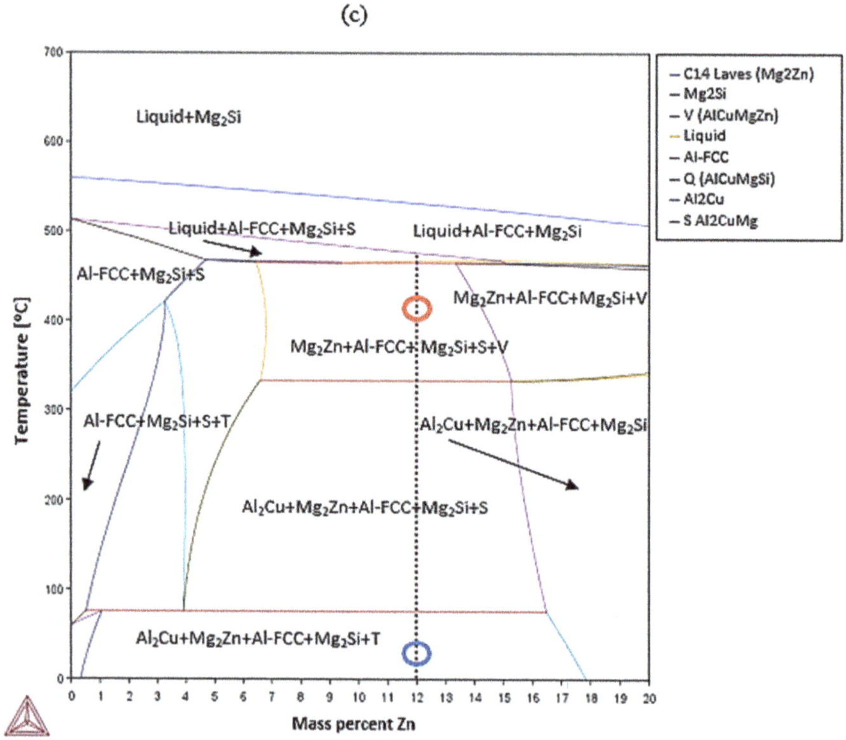

Figure 2. Al-Zn isopleth pseudo-binary phase diagram for (**a**) alloy A and (**c**) alloy AM and Al-Cu isopleth pseudo-binary phase diagram for (**b**) alloy B. The arrows represent the areas of thermodynamic stability for the various phases. The red circle represents the heat-treatment temperature and the blue circle the room temperature.

A brief description of the aforementioned phases is provided below. The θ (Al_2Cu) phase is widely known to enhance the strength of Al-Cu alloys after aging [54] and to form during solidification as well along the grain boundaries in the form of veins or skeletal shapes and after metallic mold casting, they can only be revealed after detailed and meticulous microanalysis [30]. Precipitation of the Q-$Al_4Cu_2Mg_8Si_7$ phase (also known as the $Al_5Cu_2Mg_8Si_6$ or $Al_3Cu_2Mg_9Si_7$ phase) in Al-Cu-Mg-Si quaternary alloys has been demonstrated to improve creep resistance and tensile strength, both at room temperature and at higher temperatures [55]. It was reported that the addition of copper, even with impurities, will promote the formation of the Q phase, which appears as inclusions with sharp and irregularly shaped edges [30]. The C_14 Laves ($MgZn_2$) phase is the most commonly encountered intermetallic phase in all HEAs [56]. It plays an important role in the excellent mechanical properties of the precipitation-hardenable 7xxx series [57]. The Mg_2Si phase, as mentioned previously, is a desirable phase in various Al-based alloys to improve the mechanical properties through the formation of intermetallic compounds. It is usually found in plate shapes [58], as octahedron [59] or truncated octahedron blocks [60], and as nano-particles [57]. Finally, the S (Al_2CuMg) and T ($Mg_{32}(Al,Zn)_{49}$ or $Al_2Mg_3Zn_3$) phases in the 7XXX series are responsible for the increased hardness after aging [61]. The S (Al_2CuMg) phase has been reported to provide a strong increase in the hardness of mold-casted alloys [30].

In Table 4, the volume fraction of the formed phases was estimated for the thermodynamic equilibrium conditions.

Table 4. Volume fraction of phases in the developed alloys at room temperature.

Alloy	FCC_A1	Q_AlCuMgSi	Al2Cu_C16	V	C14_Laves	Mg2Si	T	S
A	45	30	14	6	5	-	-	-
B	22	-	-	-	17	-	61	-
AM	39	-	17	-	10	38	16	-

3.3. Microstructural Characterization

3.3.1. Microstructural Characterization of As-Cast Alloys

By studying the as-cast sample A, its dendritic morphology and intense porosity was distinguished (Figure 3), which was more obvious in the upper part of the ingot (part that was in contact with the atmospheric air). Nevertheless, the existence of minor shrinkage porosity did not influence the outcome of the present study, since it can be easily addressed at the industrial scale.

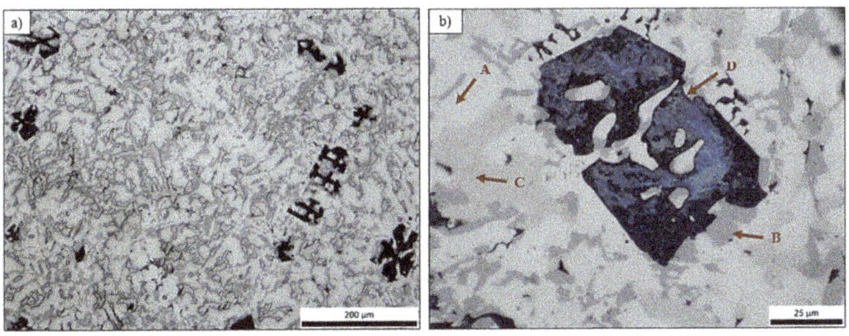

Figure 3. Optical micrographs of specimen A at (**a**) center of the middle part of the ingot and (**b**) edge of the middle part of the ingot. A, B, C, D.

The microstructure of the alloy revealed four distinct phases: a white-colored dendritic phase, which is probably the aluminum base phase (A), a dark grey phase (B), a light grey phase (C), and large dark intermetallic phases (D).

The microstructure of alloy B revealed a dendritic morphology and shrinkage porosity, mainly in the upper part of the cast (Figure 4). At low magnification, it was easy to distinguish two phases: a polygonal phase (A) and a eutectic structure. The eutectic structure consisted of a light-colored dendritic phase (B) and phase (A). The thickness of the B phase was greater than that of A, but the B phase was present in a smaller volume fraction in the microstructure. At higher magnification, a dark angular phase (C) and gray dendritic phase were evident.

An interesting observation was that the A phases appeared to be larger in the center of the ingot than at the edge. This phenomenon can be attributed to temperature gradients and the cooling rate. At the edge of the mold, the cooling rate is faster than in the center; thus, there is limited time for the phases to grow.

Similar to all the previous castings, alloy AM was characterized by a noticeable dendritic morphology and porosity (greater in the upper part of the ingot) (Figure 5). The microstructure of this alloy revealed three distinct phases: a white-colored dendritic phase, which is probably the aluminum base phase (A); a large dark angular phase (B); and a eutectic structure. At higher magnification, it can be seen that the eutectic structure consisted of small dendrites of phase (A) and a small ecru phase, designated as (C). It is worth mentioning that the dark phases (B) were slightly larger in the center of the ingot than at the edge.

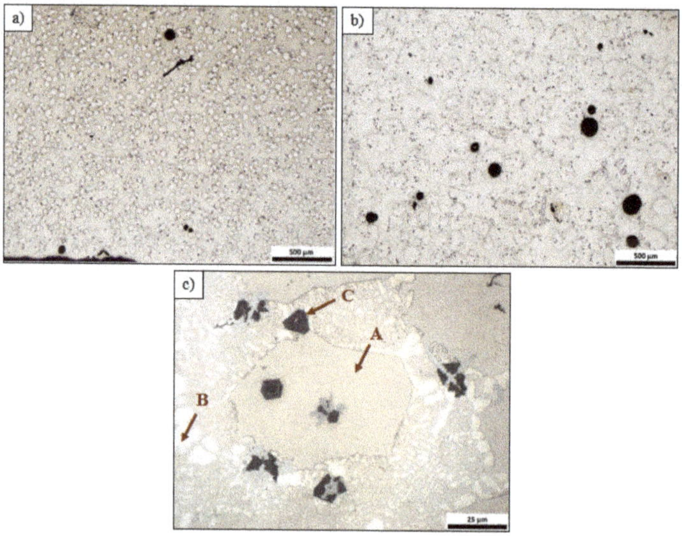

Figure 4. Optical micrographs of specimen B at (**a**) edge of the bottom of the ingot and (**b**,**c**) center of the bottom of the ingot.

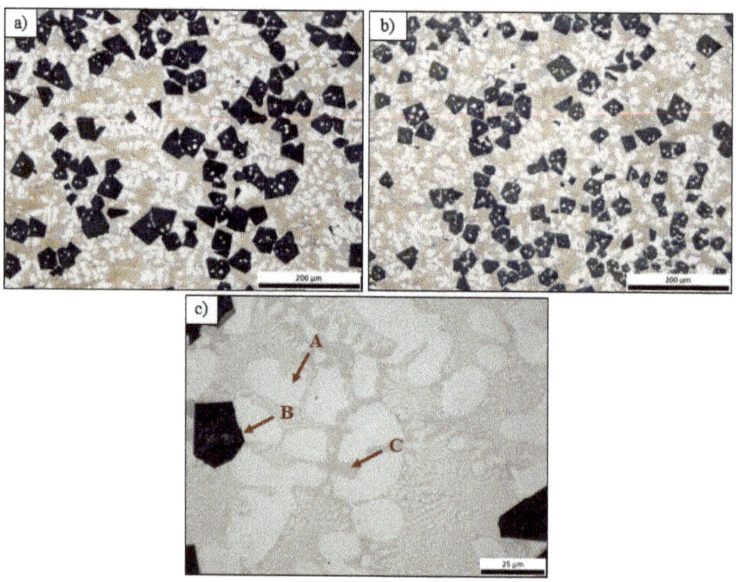

Figure 5. Optical micrographs of specimen AM from (**a**) center of the upper part of the ingot and (**b**,**c**) edge of the upper part of the ingot.

In all the heat-treated samples (A4, A24, B4, B24, AM4 and AM24), no substantial structure alteration was perceived compared to the corresponding as-cast specimens. Hence, the same main phases as the as-cast samples were identified.

The overall composition of the alloys was estimated using scanning electron microscopy (S.E.M.) and energy-dispersive X-ray spectrometry (E.D.S.) on large areas. At least three random measurements were made, and the overall values are presented in Table 5. The obtained results showed good approximations to the target compositions of the alloys (Table 2). Any deviations could be attributed to (i) partial oxidation of the components during the melting due to the absence of a protective atmosphere and (ii) E.D.S. analyses only providing semi-quantitative information. Therefore, the manufacturing process that was followed, in order to achieve the target compositions, was considered successful.

Table 5. Indicative chemical composition in wt.% and at.% of the manufactured alloys obtained by EDS.

Alloy	Al (w.t.%)	Cu (w.t.%)	Mg (w.t.%)	Si (w.t.%)	Zn (w.t.%)	Al (a.t.%)	Cu (a.t.%)	Mg (a.t.%)	Si (a.t.%)	Zn (a.t.%)
A	51.2	11.8	12.4	10.6	13.4	59.8	5.8	16	11.9	6.4
B	49	15.5	16.4	-	18.6	60.2	8.1	22.4	-	9.4
AM	51.6	9.5	19.3	8.3	11.4	57.5	4.5	24.9	8.8	5.2

Table 6 summarizes the phases in alloy A and their chemical compositions, as measured by the EDS analyses. For alloy A, the EDS analysis at spot 1 indicated that this phase is a primary Mg_2Si intermetallic compound. Based on the analysis of spot 2, the light white phase in Figure 6 is an Al_2Cu phase. Both phases were in good agreement with the Thermocalc predictions at room temperature (RT). The analysis of spot 3 verified the formation of an Al-rich solid-solution matrix with a small amount of Zn. It is interesting to note that there was very good agreement between the composition of this phase and the Al-FCC phase predicted by Thermocalc near the solidus temperature of the alloy, although the composition of the Al-FCC phase at room temperature was significantly different from the one predicted by the EDS analysis. The composition was predicted by Thermocalc under thermodynamic equilibrium conditions, whereas in reality, the cooling rates were faster. This phenomenon needs further investigation and it is worth studying whether it appears in the remaining phases. The analysis of spot 4 showed that this phase consisted of all the components in similar quantities. The composition of this phase was consistent with what Thermocalc predicted as the Q-$Al_4Cu_2Mg_8Si_7$ phase at room temperature and near the solidus temperature. Finally, the analysis performed on spot 5 detected a high content of Fe (11 w.t.%). As can be seen, the main impurity was Fe, which is commonly found in secondary aluminum alloys [62]. The presence of Fe was probably due to impurities, which were included in the industrial purity raw materials. This phase was not predicted during the alloy design and it an AlFeSi-type phase, which is well known in 3xx series alloys with a low Mg content such as Al_5FeSi [30]. It is known that all industrial aluminum alloys contain iron-bearing constituents. Such iron-bearing phases tend to not undergo significant changes during heat treatments prior to quenching [30].

Table 6. Elemental composition of the manufactured alloy in sample A obtained by EDS.

Element	Spot 1 (w.t.%)	Spot 1 (a.t.%)	Spot 2 (w.t.%)	Spot 2 (a.t.%)	Spot 3 (w.t.%)	Spot 3 (a.t.%)	Spot 4 (w.t.%)	Spot 4 (a.t.%)	Spot 5 (w.t.%)	Spot 5 (a.t.%)
Al	1.0	1.0	45.4	64.8	81.0	89.4	18.0	22.0	72.3	77.2
Mg	52.7	56.9	1.3	2.1	1.6	2.0	24.1	32.6	0.3	0.3
Cu	0.7	0.3	48.9	29.7	3.5	1.6	21.2	11.7	0	0
Zn	1.4	0.6	3.3	1.9	12.8	5.8	14.1	7.1	0	0
Si	44.2	41.3	1.1	1.5	1.1	1.2	22.7	26.6	16.2	16.7
Fe	-	-	-	-	-	-	-	-	11.1	5.7

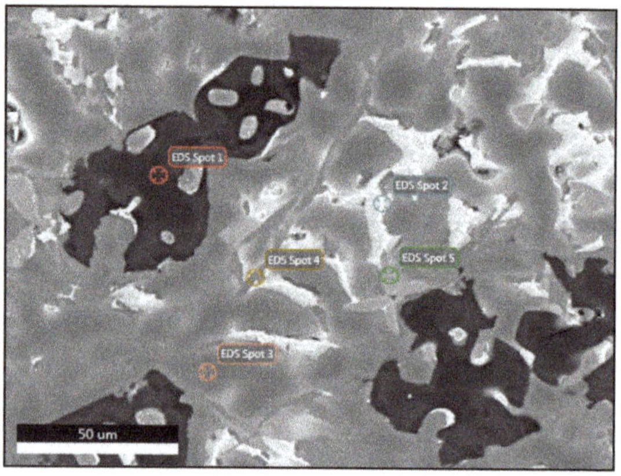

Figure 6. FESEM backscatter electron microstructure with EDS analysis of selected phases in specimen A.

Qualitative elemental mapping was conducted using EDS to identify the distribution of elements in the observed regions (Figure 7). For a better understanding of the maps, oxygen was ignored due to its insignificant amount in the composition of the alloys. The microstructure showed a matrix composed of an Al-rich area. This region corresponded to the phase predicted by Thermocalc as Al-FCC. The elements Mg and Si were located in the same areas and in places where the Mg_2Si phases were found. Si, however, appeared to be found in small amounts in other areas of the cast, possibly in phase C. Finally, Cu appeared to be located in places where the light gray phase B appeared in the OM, which was designated as Al_2Cu, while Zn was dispersed throughout the ingot.

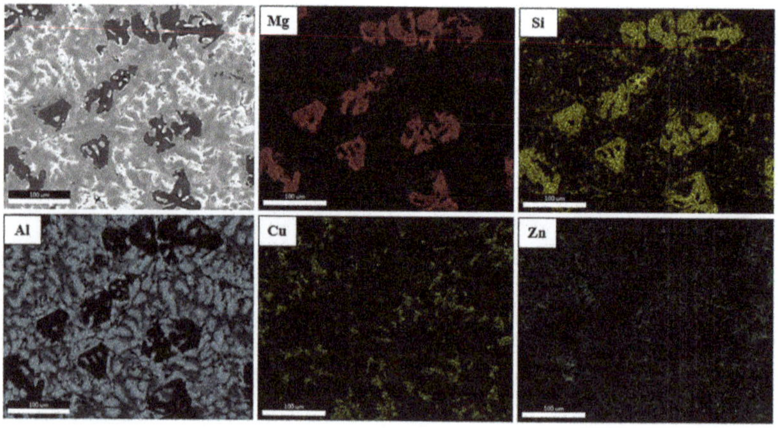

Figure 7. EDS map of specimen A.

Regarding alloy B, the EDS analysis at area 1 showed that this phase consisted of all the components that constitute the alloy in relatively equal amounts (Figure 8). Table 7 summarizes the phases in alloy B and their chemical compositions, as measured by the EDS analyses. The composition of this phase was highly consistent with that of the phase calculated by Thermocalc near the solidus temperature and reported as the T-$Mg_{32}(Al,Zn)_{49}$ phase. However, the composition calculated by the software at room temperature deviated significantly from the point analysis results. This phenomenon was also observed in the Al-FCC phase of alloy A. The analysis of spot 1 found that it was enriched in Mg and Si.

For alloy B, silicon was not included in the alloy design and it may have derived from the industrial-grade raw materials. It appears that its minimal presence in the melt led to the formation of several Mg_2Si phases, much smaller in size than those in alloy A. As for spot 2, it corresponded to an Al-rich phase. This phase is the Al-FCC phase since the Al-FCC composition predicted by Thermocalc near the solidus temperature was predominantly in agreement with that of the spot analysis. In the OM, this phase was visualized as a white color and was designated as phase B (Figure 4). The analysis carried out in the neighboring phase of α (Al-FCC) in the eutectic area showed a composition close to that of the Q-$Al_4Cu_2Mg_8Si_7$ phase. However, due to the relatively small lamellae of the eutectic phase, which were less than 0.5 μm, the measured phase was probably influenced by the neighboring phase so the displayed percentages may deviate from the actual values. Neither the C14 Laves ($MgZn_2$) phase that was predicted at room temperature or the S phase near the solidus temperature were observed.

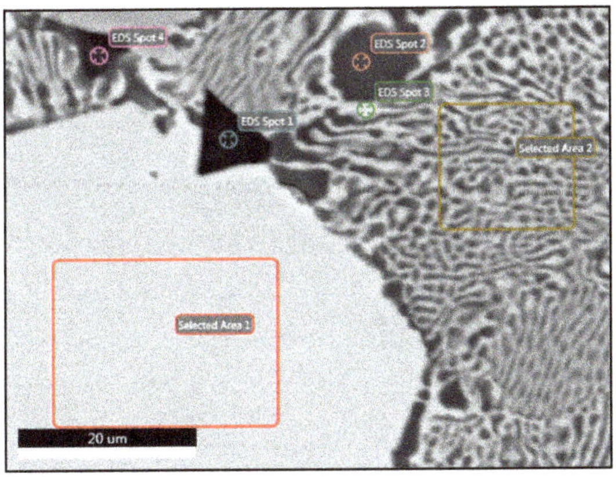

Figure 8. FESEM backscatter electron microstructure with EDS analysis of selected phases in specimen B.

Table 7. Elemental composition of the manufactured alloy in sample B obtained by EDS.

Element	Area 1 (w.t.%)	Area 1 (a.t.%)	Spot 1 (w.t.%)	Spot 1 (a.t.%)	Spot 2 (w.t.%)	Spot 2 (a.t.%)	Area 2 (w.t.%)	Area 2 (a.t.%)	Spot 3 (w.t.%)	Spot 3 (a.t.%)
Al	30.3	40.4	2.0	2.0	87.3	90.4	54.4	64.7	36.4	46.4
Mg	22.3	33.0	43.4	47.7	5.8	6.7	15.3	20.2	22.4	31.7
Cu	23.4	13.3	1.4	0.6	2.0	0.9	12.9	6.5	14.8	8.0
Zn	24.0	13.3	1.8	0.8	4.9	2.1	17.5	8.6	26.4	13.9
Si	-		51.4	49.0	-	-	-	-	-	-

Qualitative elemental mapping was conducted using EDS to identify the distribution of elements in the observed regions (Figure 9). It can be seen that Al was mainly distributed in the B phases, identified as α (Al-FCC) in the EDS analysis, and there was a stronger enrichment in the T phase's grain boundaries. Mg seemed to be more concentrated in the places where there was a presence of Si, which was probably trapped in the melt from the industrial purity raw materials and lead to the formation of Mg_2Si phases. As for Cu and Zn, there seems to be a greater concentration of them in the T phase, but in general, they were distributed throughout the cast.

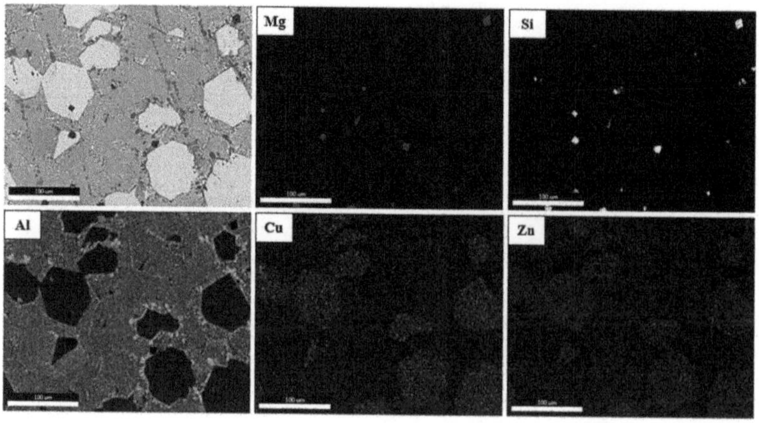

Figure 9. EDS map of specimen B.

For alloy AM, Table 8 summarizes the phases and their chemical compositions, as measured by the EDS analyses. The analysis at spot 1 (EDS Spot 1) proved that the dark angular phases were Mg_2Si phases, as they were overwhelmingly composed of Mg and Si as shown in Figure 10. This confirmed that this was an Mg_2Si phase. An expected alteration resulting from the further addition of Mg to the composition of alloy A was the formation of more Mg_2Si phases, which were much bigger and angular than those in alloy A. The results from the analysis at spot 2 revealed that the dendrites of phase A were an Al-FCC phase since its composition was in close agreement with that predicted by Thermocalc as an Al-FCC phase near the solidus temperature of the alloy. The analysis at spot 3 indicated that this phase, located in the grain boundaries of the Al-FCC dendrites, showed very high percentages of Al and Cu and a small percentage of Zn. This composition is consistent with the phase predicted by Thermocalc at room temperature as Al_2Cu, though it was not detected during the OM observations. Finally, the EDS analysis in the eutectic area (Selected Area 1) where phases A (FCC) and B alternated showed a high percentage of Al and a significant proportion of Cu, Zn, and Mg. The identification of the B phase was difficult due to its very small size (less than 0.5 μm) and therefore the analysis was probably influenced by the neighboring Al-FCC phase. It is not clear whether the phase identified as Al_2Cu, based on the analysis of spot 3, corresponded to the B phases alternating with the Al-FCC phases. However, it can be assumed that it was a phase of the T-$Mg_{32}(Al,Zn)_{49}$ or Q-$Al_4Cu_2Mg_8Si_7$ form, in which all the components that constitute the alloy were present.

Table 8. Elemental composition of the manufactured alloy in sample AM obtained by EDS.

Element	Area 1 (w.t.%)	Area 1 (a.t.%)	Spot 1 (w.t.%)	Spot 1 (a.t.%)	Spot 2 (w.t.%)	Spot 2 (a.t.%)	Spot 3 (w.t.%)	Spot 3 (a.t.%)
Al	50.0	63.7	0.9	0.9	86.2	91.0	47.2	2.5
Mg	10.5	14.9	61.0	64.9	3.6	4.3	1.6	66.4
Cu	18.0	9.8	0.5	0.2	2.3	1.1	47.5	28.4
Zn	20.9	11.0	0.9	0.3	7.3	3.2	3.1	1.8
Si	0.6	0.7	36.7	32.7	0.5	0.5	0.7	0.9

By studying the qualitative elemental map from the EDS analysis in Figure 11, it can be seen that Al was mainly found in the large dendrites of the A phase, identified as Al-FCC, and to a lesser extent in the smaller dendrites throughout the cast. The elements Mg and Si were located in the same areas where the presence of Mg_2Si phases was evident. Zn and

Cu appeared to be uniformly distributed throughout the ingot, in places far from the Mg_2Si and Al-FCC phases.

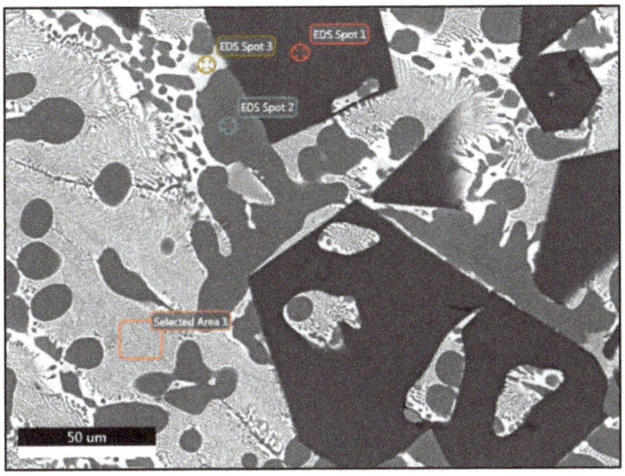

Figure 10. FESEM backscatter electron microstructure with EDS analysis of selected phases in specimen AM.

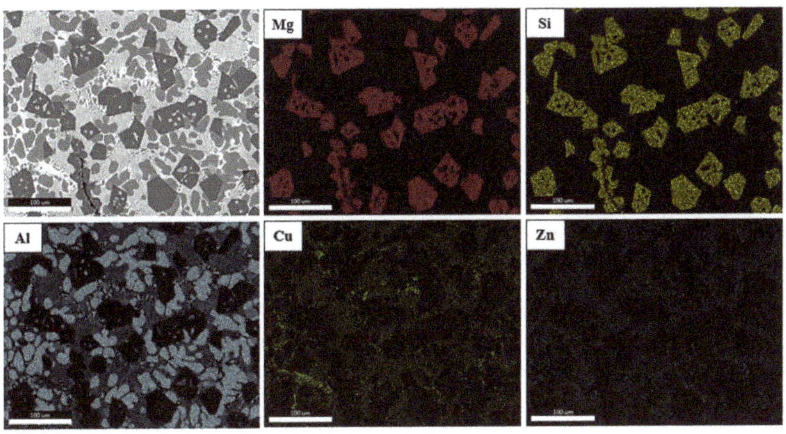

Figure 11. EDS map of specimen AM.

3.3.2. Microstructural Characterization of Heat-Treated Alloys

Despite the fact that no morphological changes were observed after the heat treatments in alloy A, there were changes in the compositions of its phases, according to the spot analyses which are summarized in Table 9. In particular, the Al-FCC phase displayed a significant decrease in Al (10%), while it was significantly enriched in Zn (13%) as shown in Figure 12. Comparing the spot analysis of the Q phase in the as-cast and the heat-treated samples, a major reduction in Zn (by 11%) was observed. Therefore, it can be assumed that this is where the enrichment of Zn in the Al-FCC phase was derived from. There was also a small drop in the percentages of Al and Cu (2% and 3%, respectively), while significant increases in Mg and Si (8% each) were observed. Mg and Si probably diffused into the Q phase from the interface of the Mg_2Si phases or from the Mg_2Si into the matrix phase and then to the Q phase, as observed by the small reduction (2%) in these percentages (EDS Spot 1). Nonetheless, distortions in the analysis from the neighboring phases should always be taken into consideration. As for the Al_2Cu phase (EDS Spot 3), its composition remained stable after the heat treatment. Regarding the remaining heat-treated samples

(B4, B24, AM4, and AM24), no alterations in the compositions of their phases were noted in comparison with the corresponding as-cast specimens.

Table 9. Elemental composition of the manufactured alloy in heat-treated sample A4 obtained by EDS.

Element	Spot 1 (w.t.%)	Spot 1 (a.t.%)	Spot 2 (w.t.%)	Spot 2 (a.t.%)	Spot 3 (w.t.%)	Spot 3 (a.t.%)	Spot 4 (w.t.%)	Spot 4 (a.t.%)
Al	0.9	0.8	16.0	17.7	47.7	67.3	69.5	83.6
Mg	50.5	54.6	32.7	40.1	0.8	1.3	1.0	1.4
Cu	0.6	0.2	17.7	8.3	47.6	28.5	3.3	1.7
Zn	1.3	0.5	3.1	1.4	2.9	1.7	25.5	12.6
Si	46.8	43.8	30.5	32.4	0.9	1.2	0.6	0.8

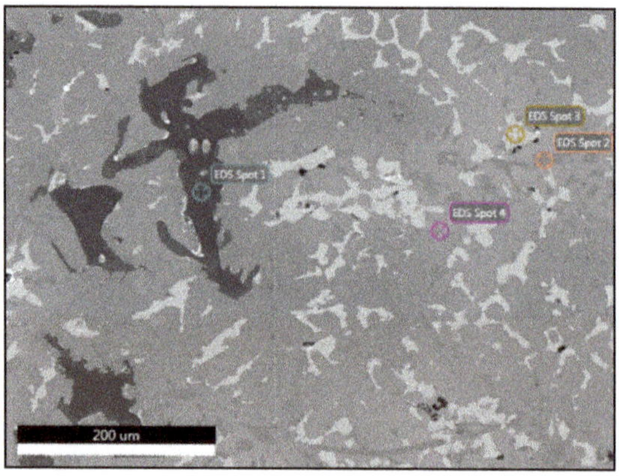

Figure 12. FESEM backscatter electron microstructure with EDS analysis of selected phases in heat-treated specimen A4.

3.4. X-Ray Diffraction Analysis

In addition, diffraction analysis was used to evaluate the produced alloys in order to verify the phases that were identified. In Figure 13, the XRD analysis results for samples A, B, and AM are presented.

In alloy A, the parent Al-base phase with an FCC structure can be observed. Moreover, the intermetallic phases of Mg_2Si (FCC), Al_2Cu (Tetragonal), and $Al_{1.9}CuMg_{4.1}Si_{3.3}$ (Hexagonal), which is the stoichiometric equivalent of the $Al_4Cu_2Mg_8Si_7$–Q phase, were noted. The $Al_9Fe_{0.84}Mn_{2.16}Si$ (Hexagonal) is the phase that appeared due to the impurities included in the industrial purity raw materials. As for alloy B, the primary $Mg_{32}(Al,Zn)_{49}$—T phase (Bergman Cubic) was recognized. Additionally, the Al-base phase (FCC) and Mg_2Si phase (FCC) were observed. For alloy AM, the XRD analysis identified the Al-base matrix phase (FCC), along with various intermetallic phases, such as Mg_2Si (FCC), Al_2Cu (Tetragonal), $Al_{1.9}CuMg_{4.1}Si_{3.3}$—Q (Hexagonal) and $MgZn_2$ (Hexagonal) phases. Thanks to the XRD analysis, it was obvious that the eutectic lamellae that were not identified in the EDS analysis was the Q phase. Thus, the XRD analysis confirmed the results from the EDS analyses and all the phases were accurately identified.

Regarding the heat-treated samples, only alloy B showed alterations. It was interesting to note that, in the case of the heat-treated B4 alloy, the XRD analysis identified a $MgZn_2$ phase. This could mean that this phase precipitated following the thermal treatment at the nanometer scale. It would be intriguing to use TEM to verify this hypothesis.

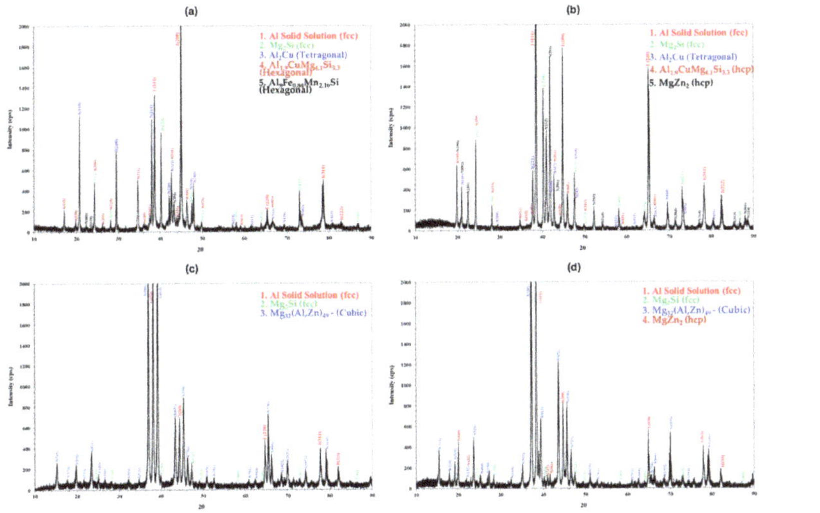

Figure 13. X-ray diffraction patterns of (**a**) specimen A, (**b**) specimen AM, (**c**) specimen B, and (**d**) heat-treated specimen B4.

3.5. Mechanical and Physical Properties

The hardness of the as-cast and heat-treated alloys at room temperature is shown in Figure 14. Additionally, a hardness/density ratio diagram is presented, with the purpose of evaluating their performance in relation to that of commercial aluminum alloys (Figure 15).

The hardness of the as-cast and heat-treated alloys was measured at room temperature and the results are shown in Table 10. In addition, the hardness to density ratio is depicted as well.

Table 10. Average hardness values of as-cast and heat-treated samples of manufactured alloys.

Specimen	Hardness (HV4)	Density (g/cm^3)	Specific Hardness
A	236	3.15	74.9
A4	256	3.15	81.3
A24	241	3.15	76.5
B	245	3.18	77
B4	245	3.18	77
B24	251	3.18	78.9
AM	246	2.73	90.1
AM4	229	2.73	83.9
AM24	219	2.73	80.2

According to hardness measurements, alloy AM had the highest hardness (246 HV_4) among the as-cast samples, while from the heat-treated ones, alloy A showed the highest value (256 HV_4). In alloy A, the hardness increased by 8.5% in the first 4 h of the heat treatment, followed by a 6% decrease. In alloy B, the hardness remained constant in the first 4 h of the heat treatment, followed by a small increase (2.5%), while in alloy AM, a reduction of 7% in hardness was observed after 4 h of soaking, followed by a reduction of 4% after 24 h of soaking in the furnace. However, comparing the hardness to density ratios, alloy AM showed the highest value both in the as-cast and heat-treated samples. The increase in hardness after the heat treatment for alloys A and B indicates that the most prominent strengthening mechanism was precipitation hardening due to the natural aging effect, while in the case of alloy AM, the decrease in hardness indicates that the main strengthening mechanism was probably solid-solution hardening.

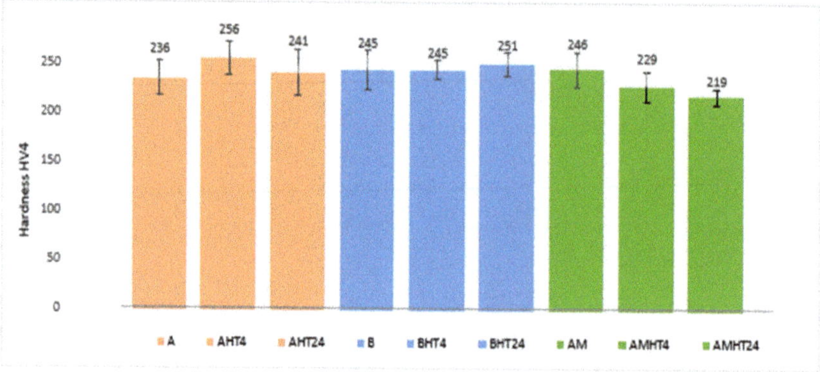

Figure 14. Hardness values of as-cast and heat-treated samples of manufactured alloys.

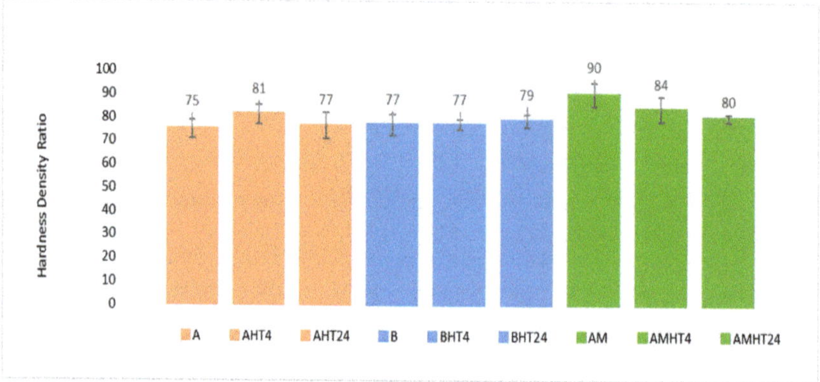

Figure 15. Hardness/density ratio values of as-cast and heat-treated samples of manufactured alloys.

Furthermore, electrical conductivity measurements were conducted in order to estimate the amount of alloying elements that were trapped in the solid solution and diffused/precipitated during the heat treatment stage, as shown in Figure 16. It is evident that after the heat treatment, the electrical conductivity of all the alloys increased, with the highest percentage increase occurring in the first 4 h of the heat treatment. Alloy A presented the highest conductivity values in both conditions (as-cast and heat-treated). Alloy B showed the lowest values, although after a 4 h heat treatment, it demonstrated the largest percentage increase (36%) compared to alloys A and AM, with percentage increases of 7% and 13%, respectively. The fact that the electrical conductivity increased in all three alloys after the heat treatment indicates that the alloying elements diffused from the solid solution, which were trapped during casting and solidification due to the limited time for diffusion, as described as the sluggish diffusion principle that defines HEAs. More importantly, since the greatest percentage increase in conductivity occurred in the first 4 h of the heat treatment, it demonstrates that for sufficient precipitation of the components trapped in the solid solution, 4 h of heat treatment is adequate.

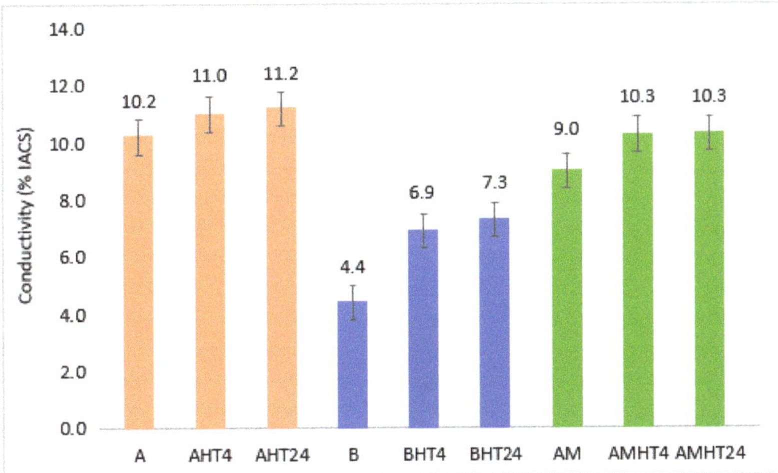

Figure 16. Conductivity measurements of as-cast and heat-treated specimens.

4. Discussion

4.1. Better Understanding of Al-Based CCA Phase Diagrams Calculated by Thermocalc

The order in which the phases solidified in each of the manufactured alloys will be mentioned bellow, since these were the phases that appeared in the final microstructure at RT.

Based on the phase diagram corresponding to alloy A, when reducing the temperature in the melt, the first phase formed was Mg_2Si at about 610 °C. As the temperature decreased to 480 °C, the Al-FCC, Q ($Al_4Cu_2Mg_8Si_7$), Al_2Cu, and V (AlCuMgZn) phases began to nucleate until the melt completely solidified at 380 °C. The only phase that was not present in the final microstructure of the alloy was the V phase.

For alloy B, by studying its phase diagram, it was clear that the first phases formed were T ($Mg_{32}(Al,Zn)_{49}$) and S (Al_2CuMg) at about 520 °C. At 480 °C, the eutectic reaction of L = Al-FCC + T ($Mg_{32}(Al,Zn)_{49}$) was taking place until the melt completely solidified. It should be noted that the S phase was not observed in the final microstructure.

Finally, in alloy AM, at around 700 °C, Mg_2Si started forming. As the temperature decreased to 550 °C, Al-FCC (pro-eutectic) began to nucleate, while below 40 °C, the eutectic reaction L = Al-FCC + Q ($Al_4Cu_2Mg_8Si_7$) started taking place until the melt completely solidified. The solidification process was completed before the C14 Laves ($MgZn_2$) and V (AlCuMgZn) phases started to nucleate, since they were not observed in the final microstructure.

This new knowledge and better interpretation of the Al-based CCA phase diagrams will help future research to be more accurate in predicting the final microstructure and consequently will considerably improve CALPHAD alloy design strategies.

4.2. Alloy Design Outcome

The design strategy for alloy A was based on the restriction of the Mg_2Si phase to a controlled quantity. Nevertheless, this phase formed to a large extent. Thanks to a better understanding of the phase diagrams due to this research, it was concluded that it is challenging to limit this phase to a controlled quantity, since, according to the phase diagrams (Figure 2), the Mg_2Si phase was the first to form and it was found that it remained stable until room temperature was reached.

Therefore, one approach to reduce the percentage of this phase, as demonstrated in the case of alloy B, is to minimize the percentage of Si. However, it was found that even the minimum amount of Si led to a few small-sized Mg_2Si phases with a particularly angular morphology, the edges of which may generate crack initiation points and thus make the material brittle [29]. Such an approach did not result in the expected improvement in mechanical properties and brittleness. Another intriguing design direction would be to add an appropriate modifier to the melt in order to convert the angular morphology of the brittle Mg_2Si phases into a more spherical one [28]. Lastly, the addition of industrial grain refining, e.g., Ti-B, could be considered as an alternative to controlling the brittle Mg_2Si phases. As for alloy AM, it was successfully designed with a major Mg addition, without ignition occurring in the melt pool, and thus a lightweight HEA with a very low density of 2.73 g/cm^3 was produced. Nonetheless, Mg_2Si is an intermetallic compound, which results in high-hardness alloys and since it has a high melting point, its formation has potential in automotive components where thermal loads are a challenge.

It is important to highlight that the presence of impurities is very likely to lead to the appearance of unexpected phases, which are not predicted by Thermocalc. Therefore, any research intended to produce industry-like Al-based CCAs with a more industrially applicable approach should expect that the utilization of industrial-grade raw materials will create unanticipated phases, which must be taken into account during the alloying design.

For thermal stability, composition design should incorporate an element that will provide coherent precipitates that do not dissolve or coarsen at the operating temperature, e.g., something like Al_3X where X = Sc, Zr, Ti, V, etc.

It is evident that the proposed alloys showed high hardness–density ratios, specifically when compared to other cast aluminum alloys, as shown in Figure 17. These results are very promising, since the newly manufactured alloys have been shown to be an appealing alternative for applications where commercial cast aluminum alloys are used, since the hardness-to-density ratios are higher. Furthermore, this new system can be directly compared with some major LWHEAs from the literature. From an economic point of view, taking into consideration production and cost limitations, all three alloys offer a friendly alternative since they were manufactured with industrial purity raw materials using a common furnace. Although there is still a lot of room for exploration in the field of CCAs, the microstructures and the superior strength-to-density ratios of the investigated alloys in combination with the manufacturing method that focuses on industrial practices can broaden the horizon for various applications in the structural, automotive and energy industries. The assessment of particular properties based on the exact needs can confirm the suitability of alloys for each specific application.

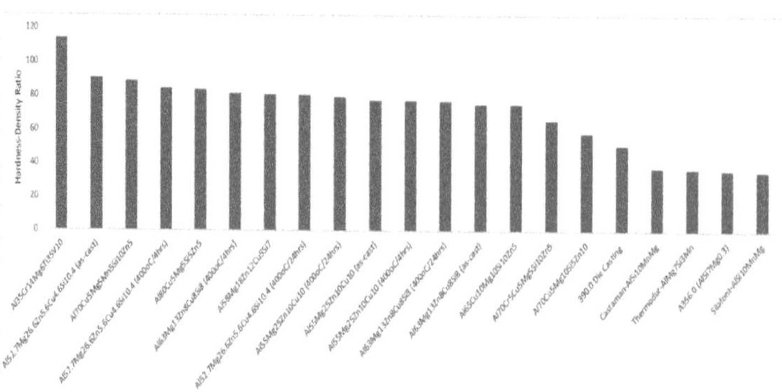

Figure 17. Comparative diagram of compressive strength hardnesses and density ratios from current study and of various alloys found in commercial applications and in the literature [13,24,29,63–65].

5. Conclusions

In this work, three new lightweight aluminum-based CCAs were successfully produced using an industrially relevant approach: $Al_{52}Mg_{9.6}Zn_{16}Cu_{15.5}Si_{6.9}$ w.t.% or $Al_{63}Mg_{13}Zn_{8}Cu_{8}Si_{8}$ a.t.% (alloy A), $Al_{44}Mg_{18}Zn_{19}Cu_{19}$ w.t.% or $Al_{55}Mg_{25}Zn_{10}Cu_{10}$ a.t.% (alloy B), and $Al_{47}Mg_{21.4}Zn_{12}Cu_{9.7}Si_{9.7}$ w.t.% or $Al_{52.7}Mg_{26.6}Zn_{5.6}Cu_{4.6}Si_{10.4}$ a.t.% (alloy AM), with low densities of 3.15 g/cm^3, 3.18 g/cm^3, and 2.73 g/cm^3, respectively. They were manufactured using industrial-grade raw materials, a standard furnace, and a pour mold casting process.

- The Thermocalc software predictions showed good agreement with the experimental results as it successfully predicted the majority of phases that appeared in the alloys. Therefore, Thermocalc can be a useful tool to guide alloy design.
- During the alloy design process, the presence of impurities in industrial-grade raw materials should be taken into consideration since it could lead to the formation of unforeseen phases, though such phases do not undermine the alloy's properties.
- Alloy AM had the highest hardness (246 HV4) from among the as-cast samples, while from the heat-treated ones, alloy A showed the highest value (256 HV4). By comparing the hardness/density ratios of the alloys of this work with other alloys in the HEA literature and commercial cast Al alloys that are used for automotive pistons, it was evident that they could offer alternatives with a high hardness strength for automotive Al cast pistons.
- Heat treatment was found to induce alterations in the elemental composition of the constitutional phases along with changes in hardness. It was found that natural aging can be used in these alloys to improve their hardness. Alloy A underwent an 8% increase in hardness.

Author Contributions: Conceptualization, S.C., E.G. and S.P.; methodology, S.C., C.T. and E.G.; software, S.C., C.T., M.B. and F.T.; resources, E.G., M.B. and F.T.; writing—original draft preparation, S.C., C.T., M.B. and F.T.; writing—review and editing, S.C., C.T., E.G., M.B., F.T. and S.P.; visualization, S.C.; supervision, S.C. and S.P.; project administration, S.P. All authors have read and agreed to the published version of the manuscript.

Funding: This research received no external funding.

Data Availability Statement: Data are contained within the article.

Acknowledgments: The authors express their gratitude to the Hellenic Research Centre for Metals ELKEME S.A. for supporting this research. We thank E. Stachouli for contributing to the metallography and electron microscopy analyses and A. Flampouri for the design and realization of the casting trials on behalf of ELKEME S.A., V. Loukadakis on behalf of National Technical University of Athens.

Conflicts of Interest: Author Spyridon Chaskis was employed by the company ELVALHALCOR S.A. and authors Dr. Evangelos Gavalas, Dr. Marianthi Bouzouni, Fotis Tsiolis were employed by the company ELKEME S.A. The remaining authors declare that the research was conducted in the absence of any commercial or financial relationships that could be construed as a potential conflict of interest.

References

1. Poznak, A.; Freiberg, D.; Sanders, P. Automotive Wrought Aluminium Alloys. In *Fundamentals of Aluminium Metallurgy*; Woodhead Publishing: Sawston, UK, 2018; pp. 333–386. [CrossRef]
2. Ducker Frontier. Aluminum Content In European Passenger Cars. Available online: www.duckerfrontier.com (accessed on 2 May 2022).
3. European Aluminum. Aluminium Content in Cars Up 20% in the Last Three Years | EURACTIV PR. Available online: https://pr.euractiv.com/pr/aluminium-content-cars-20-last-three-years-195339 (accessed on 2 May 2022).
4. Gondhaleka, A.A. Design and Development of Light Weight High Entropy Alloys. Master's Thesis, JÖNKÖPING University, Jönköping, Sweden, 2019.

5. Aluminium Automotive Manual—European Aluminium. Available online: https://european-aluminium.eu/blog/aluminium-automotive-manual/ (accessed on 5 June 2024).
6. Koos, R. Correlation Between 3D Microstructure and Thermo-Mechanical Behavior of Near Eutectic Piston Alloys. Ph.D. Thesis, Technische Universität Wien, Vienna, Austria, 2015. [CrossRef]
7. Raabe, D.; Ponge, D.; Uggowitzer, P.J.; Roscher, M.; Paolantonio, M.; Liu, C.; Antrekowitsch, H.; Kozeschnik, E.; Seidmann, D.; Gault, B.; et al. Making sustainable aluminum by recycling scrap: The science of "dirty" alloys. *Prog. Mater. Sci.* **2022**, *128*, 100947. [CrossRef]
8. Sanchez, J.M.; Vicario, I.; Albizuri, J.; Guraya, T.; Garcia, J.C. Phase prediction, microstructure and high hardness of novel light-weight high entropy alloys. *J. Mater. Res. Technol.* **2018**, *8*, 795–803. [CrossRef]
9. Maulik, O.; Kumar, D.; Kumar, S.; Dewangan, S.K.; Kumar, V. Structure and properties of lightweight high entropy alloys: A brief review. *Mater. Res. Express* **2018**, *5*, 052001. [CrossRef]
10. Yeh, J.-W.; Chen, S.K.; Lin, S.-J.; Gan, J.-Y.; Chin, T.-S.; Shun, T.-T.; Tsau, C.-H.; Chang, S.-Y. Nanostructured High-Entropy Alloys with Multiple Principal Elements: Novel Alloy Design Concepts and Outcomes. *Adv. Eng. Mater.* **2004**, *6*, 299–303. [CrossRef]
11. Pickering, E.J.; Jones, N.G. High-entropy alloys: A critical assessment of their founding principles and future prospects. *Int. Mater. Rev.* **2016**, *61*, 183–202. [CrossRef]
12. Miracle, D.B.; Senkov, O.N. A critical review of high entropy alloys and related concepts. *Acta Mater.* **2017**, *122*, 448–511. [CrossRef]
13. Sanchez, J.M.; Vicario, I.; Albizuri, J.; Guraya, T.; Acuña, E.M. Design, Microstructure and Mechanical Properties of Cast Medium Entropy Aluminium Alloys. *Sci. Rep.* **2019**, *9*, 6792. [CrossRef]
14. Oh, H.S.; Odbadrakh, K.; Ikeda, Y.; Mu, S.; Körmann, F.; Sun, C.-J.; Ahn, H.S.; Yoon, K.N.; Ma, D.; Tasan, C.C.; et al. Element-resolved local lattice distortion in complex concentrated alloys: An observable signature of electronic effects. *Acta Mater.* **2021**, *216*, 117135. [CrossRef]
15. Gao, M.C.; Liaw, P.K.; Yeh, J.W.; Zhang, Y. *High-Entropy Alloys: Fundamentals and Applications*, 1st ed.; Springer: Berlin/Heidelberg, Germany, 2016.
16. Kumar, A.; Gupta, M. An Insight into Evolution of Light Weight High Entropy Alloys: A Review. *Metals* **2016**, *6*, 199. [CrossRef]
17. Tsai, M.-H. Three Strategies for the Design of Advanced High-Entropy Alloys. *Entropy* **2016**, *18*, 252. [CrossRef]
18. Barnett, M.; Senadeera, M.; Fabijanic, D.; Shamlaye, K.; Joseph, J.; Kada, S.; Rana, S.; Gupta, S.; Venkatesh, S. A scrap-tolerant alloying concept based on high entropy alloys. *Acta Mater.* **2020**, *200*, 735–744. [CrossRef]
19. Mitrica, D.; Badea, I.C.; Serban, B.A.; Olaru, M.T.; Vonica, D.; Burada, M.; Piticescu, R.-R.; Popov, V.V. Complex Concentrated Alloys for Substitution of Critical Raw Materials in Applications for Extreme Conditions. *Materials* **2021**, *14*, 1197. [CrossRef] [PubMed]
20. Chaskis, S.; Bouzouni, M.; Gavalas, E.; Loukadakis, V.; Papaefthymiou, S. Development of Complex Concentrated Alloys (CCAs) Utilizing Scrap to Preserve Critical Raw Materials. *Mater. Proc.* **2022**, *5*, 109. [CrossRef]
21. Shao, L.; Zhang, T.; Li, L.; Zhao, Y.; Huang, J.; Liaw, P.K.; Zhang, Y. A Low-Cost Lightweight Entropic Alloy with High Strength. *J. Mater. Eng. Perform.* **2018**, *27*, 6648–6656. [CrossRef]
22. Sanchez, J.M.; Vicario, I.; Albizuri, J.; Guraya, T.; Koval, N.E.; Garcia, J.C. Compound Formation and Microstructure of As-Cast High Entropy Aluminums. *Metals* **2018**, *8*, 167. [CrossRef]
23. Sanchez, J.M.; Pascual, A.; Vicario, I.; Albizuri, J.; Guraya, T.; Galarraga, H. Microstructure and Phase Formation of Novel $Al_{80}Mg_5Sn_5Zn_5X_5$ Light-Weight Complex Concentrated Aluminum Alloys. *Metals* **2021**, *11*, 1944. [CrossRef]
24. Sanchez, J.M.; Galarraga, H.; Del Molino, E.; Albizuri, J.; Guraya, T.; Hudson, S.W. Microstructure and mechanical properties of two novel scrap tolerant Al65Cu10Mg10Si10Zn5 and Al80Cu5Mg5Si5Zn5 high entropy aluminum alloys. *Intermetallics* **2023**, *162*, 108023. [CrossRef]
25. Tun, K.; Murugan, P.; Srivatsan, T.; Gupta, M. Synthesis and Characterization of aluminium based multicomponent alloys. *Mater. Today Proc.* **2021**, *46*, 1210–1214. [CrossRef]
26. Sudha, P.; Tun, K.S.; Gupta, M.; Mourad, A.I.; Vincent, S. Electrochemical characterization of a novel multicomponent $Al_{75}Mg_5Li_{10}Zn_5Cu_5$ low entropy alloy in different pH environments. *Mater. Corros.* **2022**, *73*, 2071–2083. [CrossRef]
27. Cui, L.; Zhang, Z.; Chen, X.-G. Lightweight Al-based entropy alloys: Overview and future trend. *Sci. China Mater.* **2023**, *67*, 31–46. [CrossRef]
28. Dong, H.; Xiang, S.; Lv, J.; Wang, Y.; Li, L.; Yu, W. Modification of Mg_2Si Phase Morphology in Mg-4Si Alloy by Sb and Nd Additions. *J. Mater. Eng. Perform.* **2020**, *29*, 3678–3687. [CrossRef]
29. Chaskis, S.; Stachouli, E.; Gavalas, E.; Bouzouni, M.; Papaefthymiou, S. Microstructure, Phase Formation and Heat-Treating of Novel Cast Al-Mg-Zn-Cu-Si Lightweight Complex Concentrated Aluminum Based Alloy. *Materials* **2022**, *15*, 3169. [CrossRef] [PubMed]
30. Zolotorevsky, V.S.; Belov, N.A.; Glazoff, M.V. *Casting Aluminum Alloys*; Elsevier: Amsterdam, NX, The Netherlands, 2007; pp. 1–530. [CrossRef]

31. Zhu, X.; Yang, H.; Dong, X.; Ji, S. The effects of varying Mg and Si levels on the microstructural inhomogeneity and eutectic Mg2Si morphology in die-cast Al–Mg–Si alloys. *J. Mater. Sci.* **2018**, *54*, 5773–5787. [CrossRef]
32. Srivatsan, T.S.; Gupta, M. (Eds.) *High Entropy Alloys: Innovations, Advances, and Applications*; CRC Press: Boca Raton, FL, USA, 2020.
33. Subedi, U.; Kunwar, A.; Coutinho, Y.A.; Gyanwali, K. pyMPEALab Toolkit for Accelerating Phase Design in Multi-principal Element Alloys. *Met. Mater. Int.* **2021**, *28*, 269–281. [CrossRef]
34. Thermo-Calc Software. *TCS Al-Based Alloy Database (TCAL7)*; Thermo-Calc Software, Inc.: McMurray, PA, USA, 2020; Available online: www.thermocalc.com (accessed on 14 June 2022).
35. Computational Materials Engineering—Thermo-Calc Software. Available online: https://thermocalc.com/ (accessed on 11 October 2022).
36. Kuryntsev, S. A Review: Laser Welding of Dissimilar Materials (Al/Fe, Al/Ti, Al/Cu)—Methods and Techniques, Microstructure and Properties. *Materials* **2021**, *15*, 122. [CrossRef]
37. Electronegativity | Periodic Table of Elements—PubChem. Available online: https://pubchem.ncbi.nlm.nih.gov/ptable/electronegativity/ (accessed on 7 January 2025).
38. WebElements Periodic Table 'Magnesium' Electronegativity. Available online: https://www.webelements.com/magnesium/electronegativity.html (accessed on 7 January 2025).
39. Magnesium—Crystal Structure. Available online: https://www.periodic-table.org/magnesium-crystal-structure/?utm_content=cmp-true (accessed on 7 January 2025).
40. Magnesium—Atomic Radius—Mg. Available online: https://www.periodic-table.org/magnesium-atomic-radius/ (accessed on 7 January 2025).
41. Lattice Constants for All the Elements in the Periodic Table. Available online: https://periodictable.com/Properties/A/LatticeConstants.html (accessed on 7 January 2025).
42. Zinc—Periodic Table and Atomic Properties. Available online: https://material-properties.org/zinc-periodic-table-atomic-number-mass-radius-density/ (accessed on 7 January 2025).
43. Zinc (Zn). Available online: https://periodictable.chemicalaid.com/element.php/Zn?lang=en (accessed on 7 January 2025).
44. Silicon—Atomic Radius—Si. Available online: https://www.periodic-table.org/silicon-atomic-radius/ (accessed on 7 January 2025).
45. Silicon—Crystal Structure. Available online: https://www.periodic-table.org/Silicon-crystal-structure/ (accessed on 7 January 2025).
46. Biran, A.; López-Pulido, R. Basic Ship Hydrostatics. *Sh. Hydrostatics Stab.* **2014**, 23–75. [CrossRef]
47. Heydari, H.; Tajally, M.; Habibolahzadeh, A. Calculations to introduce some light high entropy alloys based on phase formation rules. *J. Alloys Compd.* **2022**, *912*, 165222. [CrossRef]
48. Yang, X.; Chen, S.Y.; Cotton, J.D.; Zhang, Y. Phase Stability of Low-Density, Multiprincipal Component Alloys Containing Aluminum, Magnesium, and Lithium. *JOM* **2014**, *66*, 2009–2020. [CrossRef]
49. Mitrica, D.; Badea, I.C.; Olaru, M.T.; Serban, B.A.; Vonica, D.; Burada, M.; Geanta, V.; Rotariu, A.N.; Stoiciu, F.; Badilita, V.; et al. Modeling and Experimental Results of Selected Lightweight Complex Concentrated Alloys, before and after Heat Treatment. *Materials* **2020**, *13*, 4330. [CrossRef]
50. Sahin, H.; Zengin, H. Microstructure, Mechanical and Wear Properties of Low-Density Cast Medium and High Entropy Aluminium Alloys. *Int. J. Met.* **2022**, *16*, 1976–1984. [CrossRef]
51. Tsai, M.-H.; Yeh, J.-W. High-Entropy Alloys: A Critical Review. *Mater. Res. Let.* **2014**, *2*, 107–123. [CrossRef]
52. Gao, M.; Zhang, C.; Gao, P.; Zhang, F.; Ouyang, L.; Widom, M.; Hawk, J. Thermodynamics of concentrated solid solution alloys. *Curr. Opin. Solid State Mater. Sci.* **2017**, *21*, 238–251. [CrossRef]
53. Cui, L.; Zhang, Z.; Chen, X.-G. Development of lightweight Al-based entropy alloys for elevated temperature applications. *J. Alloys Compd.* **2023**, *938*, 168619. [CrossRef]
54. Zhang, Y.; Li, R.; Chen, P.; Li, X.; Liu, Z. Microstructural evolution of Al2Cu phase and mechanical properties of the large-scale Al alloy components under different consecutive manufacturing processes. *J. Alloys Compd.* **2019**, *808*, 151634. [CrossRef]
55. Kim, K.; Bobel, A.; Baik, S.-I.; Walker, M.; Voorhees, P.; Olson, G. Enhanced Coarsening Resistance of Q-phase in Aluminum alloys by the addition of Slow Diffusing Solutes. *Mater. Sci. Eng. A* **2018**, *735*, 318–323. [CrossRef]
56. Tsai, M.-H.; Tsai, R.-C.; Chang, T.; Huang, W.-F. Intermetallic Phases in High-Entropy Alloys: Statistical Analysis of their Prevalence and Structural Inheritance. *Metals* **2019**, *9*, 247. [CrossRef]
57. Lervik, A.; Marioara, C.; Kadanik, M.; Walmsley, J.; Milkereit, B.; Holmestad, R. Precipitation in an extruded AA7003 aluminium alloy: Observations of 6xxx-type hardening phases. *Mater. Des.* **2019**, *186*, 108204. [CrossRef]
58. Ding, L.; Jia, Z.; Nie, J.-F.; Weng, Y.; Cao, L.; Chen, H.; Wu, X.; Liu, Q. The structural and compositional evolution of precipitates in Al-Mg-Si-Cu alloy. *Acta Mater.* **2018**, *145*, 437–450. [CrossRef]
59. Mohamed, A.; Samuel, F.; Al Kahtani, S. Microstructure, tensile properties and fracture behavior of high temperature Al–Si–Mg–Cu cast alloys. *Mater. Sci. Eng. A* **2013**, *577*, 64–72. [CrossRef]

60. Li, C.; Wu, Y.; Li, H.; Liu, X. Morphological evolution and growth mechanism of primary Mg2Si phase in Al–Mg2Si alloys. *Acta Mater.* **2011**, *59*, 1058–1067. [CrossRef]
61. Mondal, C.; Mukhopadhyay, A. On the nature of T(Al2Mg3Zn3) and S(Al2CuMg) phases present in as-cast and annealed 7055 aluminum alloy. *Mater. Sci. Eng. A* **2005**, *391*, 367–376. [CrossRef]
62. Zhang, L.; Gao, J.; Damoah, L.N.W.; Robertson, D.G. Removal of Iron From Aluminum: A Review. *Miner. Process. Extr. Met. Rev.* **2012**, *33*, 99–157. [CrossRef]
63. Huang, Y.; Wen, J.; Liu, Y.; Zhao, Y. Effects of electromagnetic frequency on the microstructure and mechanical properties of Al70Zn10Mg10Cu5Si5 medium entropy alloy. *J. Mater. Res. Technol.* **2022**, *17*, 3105–3117. [CrossRef]
64. Chauhan, P.; Yebaji, S.; Nadakuduru, V.N.; Shanmugasundaram, T. Development of a novel light weight Al35Cr14Mg6Ti35V10 high entropy alloy using mechanical alloying and spark plasma sintering. *J. Alloys Compd.* **2020**, *820*, 153367. [CrossRef]
65. Aluminum Alloys Wrought and Cast Property Data. Available online: https://www.matweb.com/reference/aluminum.aspx (accessed on 7 January 2025).

crystals

MDPI

Article

Ductility Index for Refractory High Entropy Alloys

Ottó K. Temesi [1,2,3], Lajos K. Varga [4,*], Nguyen Quang Chinh [3] and Levente Vitos [4,5]

1 H-ION Kft., Konkoly-Thege Miklós út, 29-33, H-1121 Budapest, Hungary; otto.temesi@h-ion.hu
2 SMARTUS Zrt., Gyár utca 2, 2040 Budaörs, Hungary
3 Department of Materials Physics, Eötvös Loránd University, Pázmány Péter Sétány 1/A, H-1117 Budapest, Hungary; chinh@metal.elte.hu
4 HUN-REN Wigner Research Center for Physics, P.O. Box 49, H-1525 Budapest, Hungary; leveute@kth.se
5 Department of Materials Science and Engineering, Royal Institute of Technology, SE-100 44 Stockholm, Sweden
* Correspondence: vlk@h-ion.hu or varga.lajos@wigner.hun-ren.hu

Abstract: The big advantage of refractory high entropy alloys (RHEAs) is their strength at high temperatures, but their big disadvantage is their brittleness at room temperature, which prevents their machining. There is a great need to classify the alloys in terms of brittle-ductile (B-D) properties, with easily obtainable ductility indices (DIs) ready to help design these refractory alloys. Usually, the DIs are checked by representing them as a function of fraction strain, ε. The critical values of DI and ε divide the DI—ε area into four squares. In the case of a successful DI, the points representing the alloys are located in the two diagonal opposite squares, well separating the alloys with (B-D) properties. However, due to the scatter of the data, the B-D separation is not perfect, and it is difficult to establish the critical value of DI. In this paper, we solve this problem by replacing the fracture strain parameter with new DIs that scale with the old DIs. These new DIs are based on the force constant and amplitude of thermal vibration around the Debye temperature. All of them are easily available and can be calculated from tabulated data.

Keywords: ductility; elastic anisotropy; Pugh and Cauchy criteria; refractory high entropy alloys

1. Introduction

The enlarged compositional freedom afforded by the extended solid solution structure of refractory high-entropy alloys (RHEAs), based on the nine early transition elements, Ti, Zr, Hf, V, Nb, Ta, Cr, Mo, and W), opens up the possibility for the design of alloys for different applications. However, casting the RHEAs for high strength at high temperature applications reveals again the age-old problem of metallurgy, namely, the higher the strength, the lower the ductility and malleability [1,2].

The strength–ductility trade-off is represented as a hyperboloid function relationship [3] of the two parameters: the higher the strength, the lower the ductility. After stress, the ductility can be characterized by the following measurable parameters: elongation at break (ε (%)), reduction in cross-sectional area at break, energy absorbed up to break, and strain hardening exponent (n) determined from the true stress–strain diagram. Elongation to fracture seems to be the easiest to apply in practice, but since the majority of the samples in the literature have been tested for compression rather than tension, it is better to discuss the fracture strain to characterize the degree of deformation until fracture. This, however, is very contingent in terms of its extent, so it is only a qualitative characteristic, and it was necessary to find its quantitative substitute in this work.

Predicting the ductility or brittle behavior before stress can be conducted with the ductility index based on the existing and newly invented criteria. These criteria are connected to the strengthening mechanism of the RHEAs. When looking for ductile behavior, one has to avoid the strengthening mechanisms, which are as follows:

Grain refinement: Smaller grains mean more grain boundary, that is, more discontinuities in the path of dislocations, that is, less ductility. The increase in strength can be expressed by the Hall–Petch relation [4], which expresses the increase in yield stress (σ_y) due to the decrease in grain size (d):

$$\sigma_y = \sigma_o + \frac{k}{d^{1/2}} \tag{1}$$

where σ_o is the yield stress for large grains and k is a constant.

Strain hardening, which happens during the plastic deformation when the continuous increase in dislocation densities (ρ) causes the increase in stress (σ_y) necessary to move the dislocations. This mechanism [5] is described by:

$$\sigma_y = \sigma_o + \frac{G \cdot b}{2}\sqrt{\rho} \tag{2}$$

where G is the shear modulus, b is the Burger's vector, and σ_o is the yield stress for the sample state before plastic deformation.

Solid solution strengthening [6] originates from the interaction of the dislocations with the stress field of the solute atoms, which increases together with the increasing atomic size difference (δR), and depends on the nature of the solute atom: the small interstitials (B, C, and N) produce tensile stress, whereas the large substitutional solutes produce compressive stress around them.

Precipitation hardening stems from the interaction of the stress field of dislocation with the large elastic distortion around the coherent precipitate particles. This elastic distortion is missing around incoherently precipitated particles, and will not contribute to the hardening (it is just a neutral inclusion).

Dispersion hardening happens when fine precipitate particles (a second phase such as oxides, borides, carbides, nitrides) extend all over the sample, and the increase in the yield stress originates [7] from the stress necessary to move a dislocation of length ℓ (equal to the mean spacing between the particles) pinned at both ends with Burger's vectors of b, in a matrix with shear modulus G:

$$\Delta\sigma_y = \frac{G \cdot b}{\ell} \tag{3}$$

Phase transformation strengthening happens when the phase born in transformation has lower symmetry then the original matrix, like the hardening caused by austenite–martensite transformation in ball bearing steels, or FCC-BCC and FCC-B2 transformation in FCC Cantor HEAs alloyed with Al [8].

Based on this strengthening mechanism, one can make predictions concerning the strength of the alloy, but no clear predictive correlations have yet been established between the increase in strength and decrease in ductility for a particular alloy composition. For example, we cannot handle the strength–ductility trade-off in the case of the best refractory alloys with the highest yield stress and brittle at room temperature, presenting a ductile elongation as law as 4% [9]. It is annoying that when we are in the midst of such great discoveries in nuclear physics and space exploration, we cannot reconcile an everyday phenomenon such as tensile strength with the formability of the sample, so the millennia-old task remains of how to make the material tough while preserving the great tensile strength. At the moment, we can only make suggestions about the presence or absence of ductility. For example, how to transform the so called Senkov alloys, NbTaMoW (yield stress, YS = 1000 MPa) and MoWNbTaV (YS = 1250 MPa), with a law elongation strain (<5%) into a more malleable alloy at room temperature, thus preserving the excellent tensile stress values [10].

2. Review of the Most Relevant Ductility Criteria

The Pugh criterion [11] is based on the model that the yield stress (a measure of the resistance to plastic deformation) scales with the shear modulus, whereas the fracture stress scales with the bulk modulus. A ductile behavior is expected when the ratio of shear and bulk moduli is $G/B < 0.575$, above this value, brittle behavior is predicted. This prediction is based on two elastic moduli, however, this is not always possible. For example, the elastic moduli (E,G,B) of the elements Zn and Zr did not differ too much, although they behaved differently in a visible way under stress or bending.

The Pettifor criterion [12] is based on the Cauchy pressure: $C'' = C_{12} - C_{44}$. The positive value of C'' corresponds to the isotropic metallic bond, which characterizes an intrinsically ductile material. Materials with negative Cauchy pressure possess, at least partially, direction (covalent) bonding and are intrinsically brittle.

The Rice and Thomson criterion [13] is where the model predicts an approximate condition for sharp cleavage cracks (i.e., brittle behavior):

$$Gb/\gamma_s > 7.5\text{–}10 \quad (4)$$

where γs is the surface energy, G is the shear modulus, and b is the Burgers vector.

It is worth mentioning that for iron, the $Gb/\gamma s \sim 7.5\text{–}8$. This value is at the limit of ductile behavior. This means that the ductile-to-brittle transition temperature of iron is not much below room temperature [14].

The Rice criterion [15] is the plasticity resulting from the competition between dislocation emissions and cleavage fracture propagation. This competition can be expressed [16,17] as a ratio of the critical stress intensities responsible for the dislocation emission (K1e) and cleavage (K1c), respectively:

$$D = K1e/K1c \quad (5)$$

The parameter D can be used as ductility meters. Intrinsically ductile behavior is expected when the stress intensity necessary for dislocation emission is smaller than the cleavage stress $D < 1$. The ratio of critical stresses can only be obtained by ab initio calculations. For example, when conducting density-functional theory simulations [18] to investigate the competition between cleavage decohesion and dislocation emissions from the crack tip, it turned out that the dislocation plasticity was found to be unfavorable for five representant compositions of refractory RHEAs: HfNbTiZr, MoNbTaVW, MoNbTaV, MoNbTiV, and NbTiVZr, which are all brittle at room temperature.

Theoretically, it is important that the ratio of critical stress intensities can be replaced [19] by the surface energy (γ_{surf}) of cleavage and the energy of the unstable stacking fault, γ_{usf}, [19]:

$$D = \gamma surf/\gamma usf \quad (6)$$

The higher the ratio $\gamma_{surf}/\gamma_{usf}$, the higher the fracture deformation, ε_p (the ductility is improved).

The VEC~4.2 criterion: Qi and Chrzan [20] proposed that the intrinsic ductility of a bcc refractory alloy could be estimated based on VEC. They showed that Mo- and W-based alloys could become intrinsically ductile if their average valence electron numbers were decreased by alloying. Moreover, Sheikh et al. [21] demonstrated a ductility around VEC = 4.2, proposing the idea of metastability engineering. Changing the composition of the alloy, the brittleness increased as the VEC value increased from 4 to 6 [21]. Exploiting the transformation induced plasticity (TRIP), the ductilization of BCC–RHEAs was obtained by alloying, reducing the VEC to 4.2 [22,23].

Local lattice distortion (LLD) criterion: Theoretical calculations [24] revealed that the local lattice distortion of the refractory HEAs was much more significant than that of the 3d HEAs.

Recently, a ductility index was presented [25] based on the ab initio calculation of the ratio of the average local lattice displacement and vector norm of the lattice displacement of local lattice distortions (LLDs). A ratio smaller then 0.3 is ductile, and that larger than 0.3 indicates brittle behavior.

We have to keep in mind that all predictions concerning the ductility behavior refers to single-phase alloys Usually, all of these ductility meters are checked by representing them as a function of fraction strain, (ε). The check is valid if the points of the alloys are sorted in two opposing quadrants only separated by the critical values of the DI under investigation and taking, rather arbitrarily, (ε) = 10% for the brittle–ductile limit. The problem is with the set of data containing samples prepared with different methods, different phase compositions, and (ε) measured with different accuracies. The scope of this work was to find a proper way for sorting the alloys into brittle and ductile quadrants. This scope was realized by introducing new DI parameters that scale linearly with the old ones. First, we represented the elemental metal data and decided the critical values of the parameters at the borderline of B-D behavior. Then, we checked the alloys available in different published datasets.

To sum up, each ductility criterion defines a ductility index with a characteristic critical value separating the ductile and brittle behavior. Two classes of the ductility indices were used in the application of the ductility criteria. In the phenomenological class, we used those based on the tabulated data of constituent elements of the alloys (Pugh, Pettifor, and Rice and Thomson). The LLD (local lattice distortion) and D parameter ($D = \gamma_{surf}/\gamma_{usf}$) belong to the theoretical ductility meters only obtainable by ab initio calculations. The scope of the present work is twofold. First, to renew the existent phenomenological criteria by introducing new parameters that are available from tabulated data. Second, the new DI parameter will be not checked with the fraction strain (ε) parameter with uncertain numerical value, but with an old DI parameter with which it can be scaled (i.e., it has a linear relationship).

3. Results

In the following four paragraphs, we present how to classify the alloys of a dataset into ranges B-D with the "old" DI parameters discussed in the introduction. The following nine figures are based on the dataset (about 50 RHEAs) collected by Hu et al. [19].

3.1. Application of Fracture Strain as Ductility Index

At first, it seemed that the most obvious ductility index is the deformation until fracture, expressed as a percentage of the original size. The values of the fracture deformation, ε, obtained from compression measurements, are very contingent and can differ by as much as 50–100%, especially if the faces of the square column-shaped sample are not plan–parallel. Furthermore, our ductility criteria are valid for single phase homogenous alloys. The single-phase structure of the samples with ε data published in the literature was not documented carefully, which is why we were looking for other decision-maker ductility meters. Considering the fracture strain as just a qualitative parameter only, we defined, rather arbitrarily, the ductile behavior as $\varepsilon > 10\%$ and brittle for $\varepsilon < 10\%$. Furthermore, we defined a derivative of this ductility meter, the effective fracture energy FE as the product of yield stress and fracture strain:

$$FE = \sigma_y{}^{\times}\varepsilon \tag{7}$$

In Figure 1, we present the fracture energy as a function of fracture strain for a set of 50 refractory HEAs with a BCC structure. The data were collected by Hu et al. [19], together with the $D = \gamma_{surf}/\gamma_{usf}$ calculated parameters.

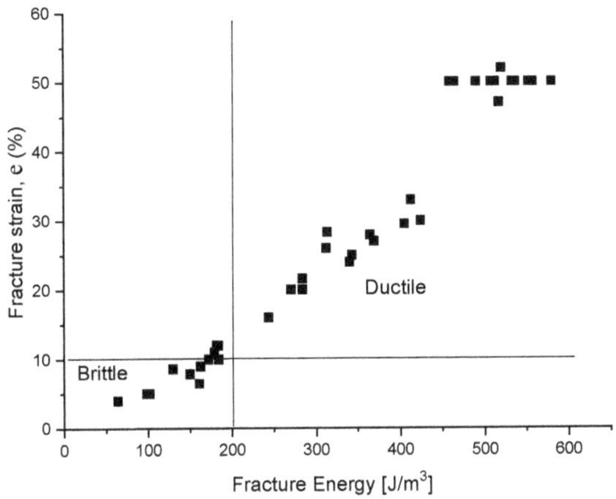

Figure 1. Fracture energy versus fracture strain. Two opposite quadrants separately contain the brittle and ductile alloys, B and D. The other two quarters are empty if the critical values are chosen correctly. ε_c = 10% and E_c = 200 MJ/m^3.

3.2. Application of Reduced Valence Electron Counts Parameter, R-VEC

It is known that a number of physical properties depend monotonously from the total valence electron counts (VEC). Furthermore, we demonstrated in a recent paper [26] that a correlation exists with the number of unpaired d electrons. To emphasize the role of unpaired d electrons, here, we introduced the so called "reduced VEC":

$$\text{R-VEC} = (\text{VEC} - 2)/4.5 \quad (8)$$

valid for the RHEAs only. This expression gives the relative number of d electrons, starting from Ti with 2 d electrons (VEC = 4) and ending at VEC = 6.5, where the maximum occurs for those properties that show strong VEC dependence (hardness, cohesion energy, elastic moduli, see [26]).

In Figure 2 the fracture strain as a function of R-VEC is presented. Accepting 10% fracture strain as the limit of B-D behavior the critical value for R-VEC will be 0.72.

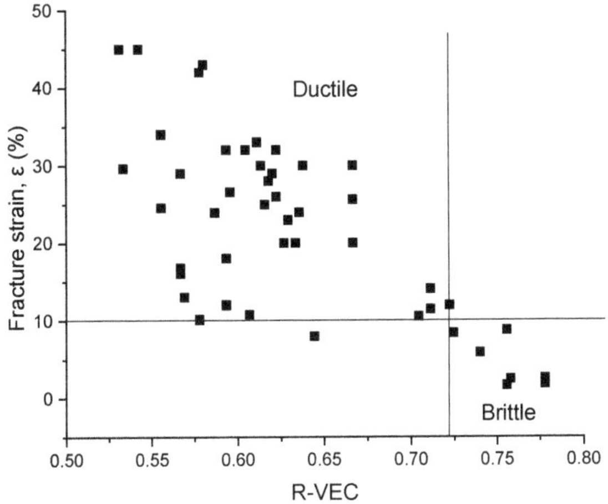

Figure 2. Fracture strain ε as a function of reduced VEC. The critical values are ε_c = 10%, R-VEC$_c$ = 0.72, and VEC$_{cr}$ = 5.4.

3.3. *Application of Poisson Ratio and Its Derivative, Parameter D* of Christensen*

If there are no available ductility meters, then we have to create one. The most promising parameter looks like to be the *D** parameter of Christensen because out of the four elastic parameters (B, E, G, and ν), the Poisson ratio, ν, is the most sensitive to the brittleness of deformations. Having already determined the average moduli values for the given alloy, G_{ave} and B_{ave}, with the rule of mixture (ROM), the Poisson ratio can be calculated with Formula (9):

$$\nu = \frac{3B - 2G}{6B + 2G} \tag{9}$$

The rather limited interval of variation of the ν parameter (between 0.2 and 0.5) was overcome by Christensen [27], defining a ductility index parameter, D*, as:

$$D^* = \left(\frac{3\nu}{\nu + 1}\right)^2 \tag{10}$$

In practice, D* varies between 0.25 and 1 while ν varies between 0.2 and 0.5. D* = 1 corresponds to perfect ductility and D = 0 to total brittleness. The critical value for the In general, Poisson ratio ν > 0.28 and D* > 043 means ductile. Representing the fracture strain as a function of D* (see Figure 3), the critical value for D* = 0.562. It should be mentioned that Formula (9) is valid for isotropic alloys only.

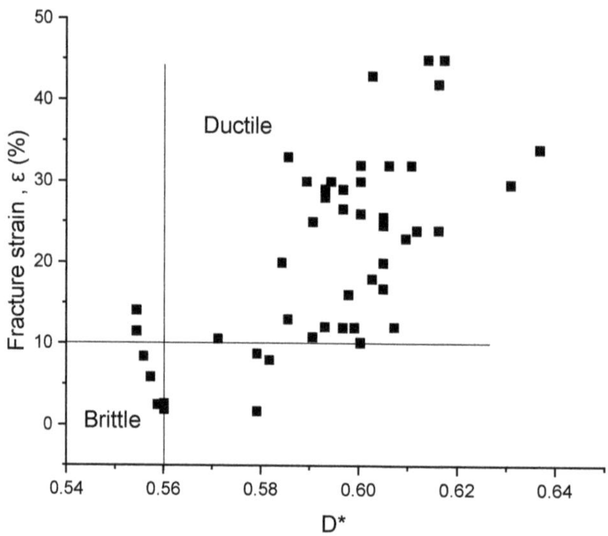

Figure 3. Fracture strain c as a function of modified Poisson ratio parameter D*. The critical values are ε_c = 10% and D_c^* = 0.562.

3.4. *Application of Modified Rice Ratio, the Parameter D, (See Equation (6)) as Ductility Index*

First, we report on a very important observation. The D parameter of Rice obtainable by theoretical calculations only, scales with the phenomenological obtainable D* parameter (see Figure 4). This observation makes it possible to obtain the D parameter of an alloy through the role of mixture calculations based on the tabulated D values of the constituent elements. Representing in Figure 5 the fracture strain as a function of D parameter helps determining the critical value for D = 2.62.

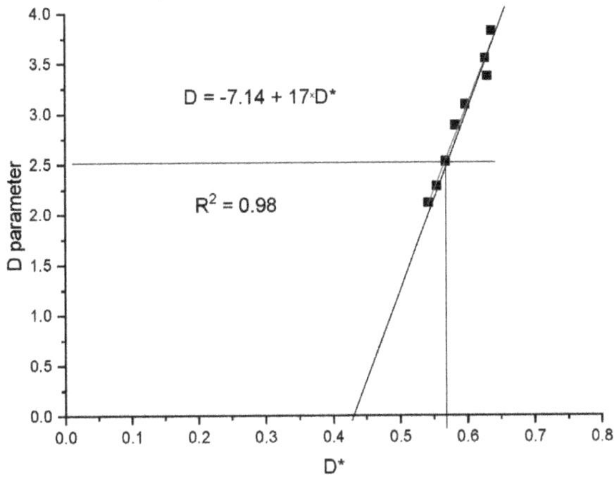

Figure 4. The D parameter of Rice scales with the D* of Christensen.

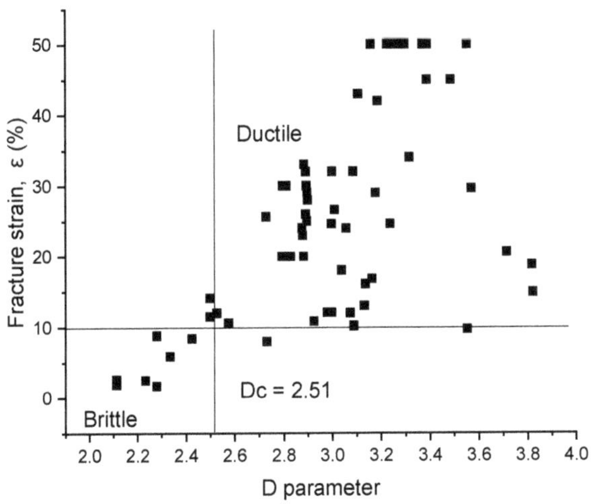

Figure 5. Fracture strain ε as a function of the D parameter. The critical values are ε_c = 10% and D_c = 2.51.

3.5. Application of Cauchy Pressure and Pugh Ratio as Ductility Meters

According to the Senkov and Miracle calculations, the Cauchy pressure $C'' = C_{12} - C_{44}$ can also be expressed in terms of the elastic moduli (see Equation (6) in Ref. [28]):

$$\begin{gathered} C'' = \left(\frac{C12-C44}{B}\right) = \left(1 - f(A)\frac{G}{B}\right) \\ f(A) = \frac{5}{3} \cdot \frac{(2A+3)(2+3A)}{3A^2+19A+3} \end{gathered} \tag{11}$$

The critical value of Cauchy pressure, $C'' = 0$, determines the critical values of the Pugh ratio, G/B, as

$$(G/B)_{cr} = 1/f(A). \tag{12}$$

This critical value of the Pugh ratio varies between 0.568 for A = 2 and 0.3 for an infinite value of A.

The critical term is the Zener anisotropy [29], defined as:

$$A = 2C_{44}/(C_{11} - C_{12}) \tag{13}$$

For isotropic polycrystalline materials, A = 1 signals perfect disorder in grain orientation. This is difficult to realize experimentally because casting produces inherent thermal cooling anisotropy. However, the largest effect on elastic anisotropy is produced by the metastable structure around $VEC = 4.2$ when the material is at the limit of BCC and HCP phase formation, which means that C' ($C' = (C_{11} - C_{12})/2$) becomes zero, and anisotropy A has a singular point presenting a large value of the order of ten.

This singular point was around VEC = 4.1–4.2, both for elemental metals and alloys (see Figure 6). It turns out that for the elements at the beginnings of rows 3d–4d and 5d (VEC around 4), the anisotropy presented a large, singular point, and the critical G/B value was minimal, at 0.30. For VECs out of the singularity region, the C' values differed from zero, and the critical G/B value achieved their normal values between 0.57 and 0.3. It should be mentioned that because of this VEC dependence of Zener anisotropy, it cannot be defined as a unique anisotropy factor (f(A)) for a set of samples with different VEC values. We recommend choosing, rather arbitrarily, the interval 0.26–0.28 for the minimal values of the Poisson ratio for the BCC–RHEAs. The corresponding interval or the critical value for G/B is 0.568–0.515, and for the anisotropy factor $f(A)$, the interval is 1.76–1.941.

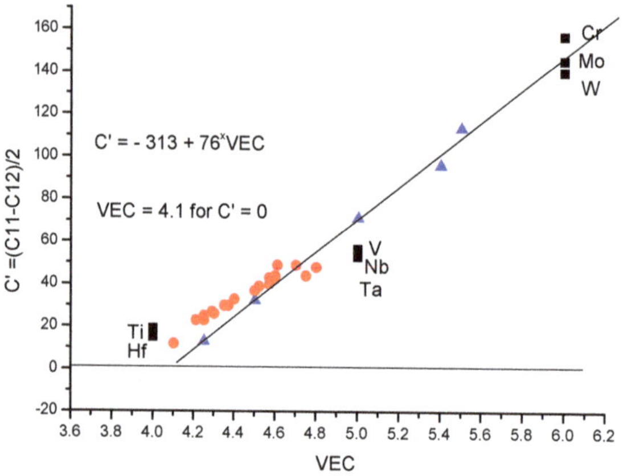

Figure 6. Valence electron counts dependent of the C' component of the shear modulus for refractory metals and their alloys. Linear correlation was found between the valence electron counts and the C' component of shear modulus for refractory metals and alloys. For alloys the data from [18,30] and for pure metals the tabulated data have been used.

The expression of Cauchy pressure as a function of Pugh ratio and anisotropy factor serves as a new ductility index, Parameter PSM (Pettifor–Senkov–Miracle):

$$P = 1 - f(A)\frac{G}{B} \tag{14}$$

where $P > 0$ means ductile, and $P < 0$ means brittle.

The anisotropy factor, f(A), was determined as follows:

i. From the tabulated data, we collected the G and B values of the alloying elements;
ii. With the rule of mixture, we calculated the average values for the given composition, G_{ave} and B_{ave};
iii. With Formula (15), we calculated the critical value of G/B by selecting the critical value ν_c of the Poisson ratio.

$$\left(\frac{G}{B}\right)_{Cr} = \frac{3}{2}\frac{1-2\nu}{1+\nu} \quad (15)$$

Usually, ν_c = 0.28 and $(G/B)_{Cr}$ = 0.515. The anisotropy factor will be the inverse of $(G/B)_{Cr}$, f(A) = 1.939.

With this, the P parameter of Equation (14) can be calculated for the dataset and can be represented as a function of accepted ductility indices (fracture strain, LLD, and D parameters).

The important values of these parameters are collected in Table 1.

Table 1. The critical values for the PSM parameters: Poisson ratio.

ν	D*	(G/B)cr	f(A)	A
0.2	0.25	0.75	1.33	0.8
0.25	0.36	0.6	1.66	1
0.28	0.43	0.515	1.9417	3.35
0.31	0.5039	0.435	2.298	6.94
0.33	0.5625	0.375	2.66	14.4
0.5	1	0.3	3.33	inf.

Although we have tried different DI pairs for the PSM parameter (fracture strain in Figure 7a, reduced VEC in Figure 7b. and parameter D in Figure 8), the strong dependence of Cauchy pressure from the Zener anisotropy makes it less applicable for the delimitation of B-D regions.

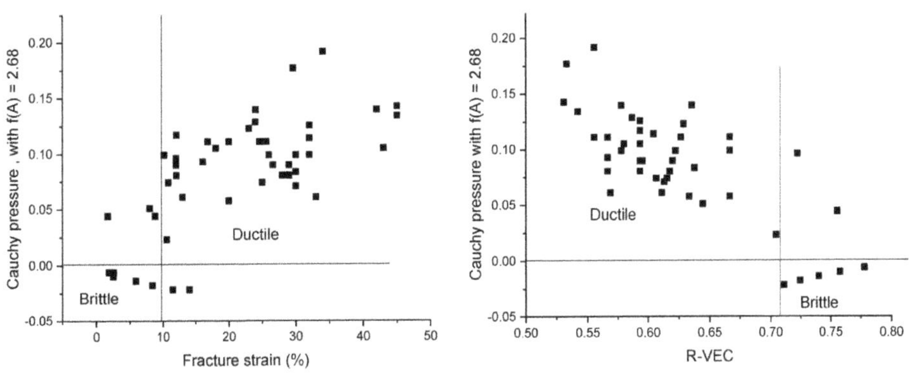

Figure 7. Delimitation of the B-D regions using Cauchy pressure parameter P = 1 − 2.66^x (*G/B*) versus (**a**) fracture strain and (**b**) *R-VEC*.

Unfortunately, the application of the oldest Pugh criterion also does not give better results (see Figure 9) in terms of classification according to B-D properties because the Pugh ratio is also dependent on anisotropy factor A, which can vary arbitrarily from sample to sample. Applying the Voigt–Reuss–Hill (VRH) approximation [31], we obtained the following expressions for the shear and bulk moduli as a function of the C_{11}, C_{12}, and C_{44} elastic constants and Zener anisotropy A (see Equation (13)):

$$G = \frac{3C_{44}(A2+12A+2)}{5A(2A+6)}$$
$$B = \frac{2C_{44}}{3A} + C_{12} \quad (16)$$

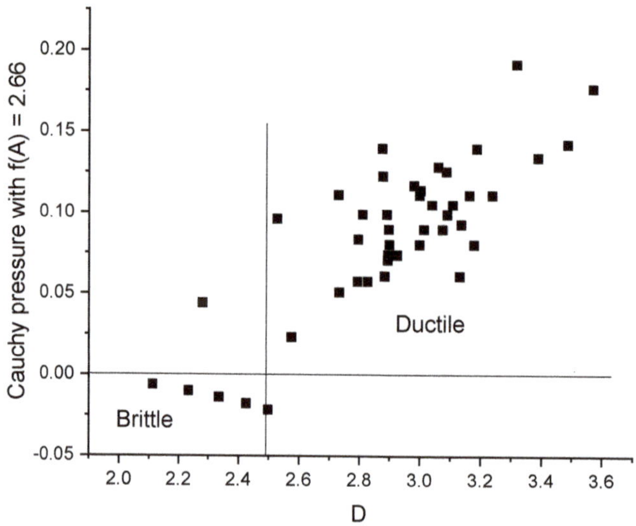

Figure 8. Cauchy pressure $C'' = \text{PSM} = 1 - 2.66\ \text{G/B}$ as a function of the D parameter. Again, we obtained the critical value $D_c = 2.5$.

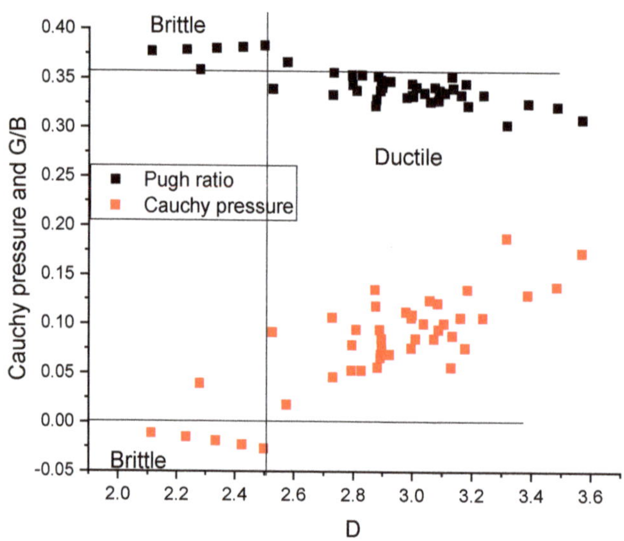

Figure 9. Delimitation of B-D regions using Cauchy pressure, P = 1 − 2.66^{x}*G/B*), and the Pugh ratio versus *D* parameter.

It is worth noting that while applying the Pugh criterion there are problems due to the singularities of the anisotropy around the value of VEC = 4.2, until then, this does not affect the application of the Rice and Thomson criterion.

Changing the pairs of ductility indices, no improvement in the delimitation of B-D regions could be detected. The only advantage of these figures is that it makes clear the constancy of critical values, irrespective of the chosen pair. However, it seems necessary to search for new DIs to improve the classification of alloys in the D-B range.

3.6. Possible New Physical Parameters for New Ductility Indexes

Parameter k (renewed Pugh criterion)

In order to renew the existing ductility meters with new parameters, we had to demonstrate that the new parameters scaled with the old ones and were eventually easier to access through tabulated values. We considered the following new parameters: the force constant, k, the amplitude of thermal vibration, u, the melting temperature, and the surface energy, γ_S.

The force constant, k, can be calculated from the tabulated atomic mass, A, and Debye temperature (θ_D[K]) data as follows [32]:

$$k = \left(\frac{k_B}{\hbar}\right)^2 \cdot A \cdot \theta_D^2 = 2.86 \cdot 10^{-5} \cdot A \cdot \theta_D^2 \tag{17}$$

The unit of k is N/m (J/m^2). In order to obtain the elastic moduli unit, we have to divide it with a length parameter, which we arbitrarily selected as the amplitude of thermal oscillations at the Debye temperature. This selection was motivated by the fact that it varies in the opposite sense with the force constant, so the larger the force constant, the smaller the thermal amplitude.

The mean square amplitude at T = θ_D is given in the textbook [33] as:

$$\begin{aligned} &< u^2 >= \frac{9\hbar^2}{k_B} \cdot \frac{1}{A \cdot \theta_D} = 436.86 \cdot 10^{-20} \cdot \frac{1}{A \cdot \theta_D} \\ &< u >= \sqrt{\frac{436.86}{A \cdot \theta_D}} \left[\overset{\circ}{A} \right] \end{aligned} \tag{18}$$

where A and θ_D are the tabulated atomic mass and Debye temperature, respectively.

Finally, the new DI parameter, k/<u>, has the same unit, N/m^2, as the elastic moduli:

$$\frac{k}{< u >} = \frac{2.86 \cdot 10^{-5} A \theta_D^2}{\sqrt{\frac{436.86}{A \cdot \theta_D}}} = 0.137 \cdot A^{3/2} \cdot \theta_D^{5/2} \tag{19}$$

In Figures 10 and 11, we demonstrate that both the force constant and the ratio k/<u> can be scaled with the shear modulus, so can consequently replace it in the Pugh criterion.

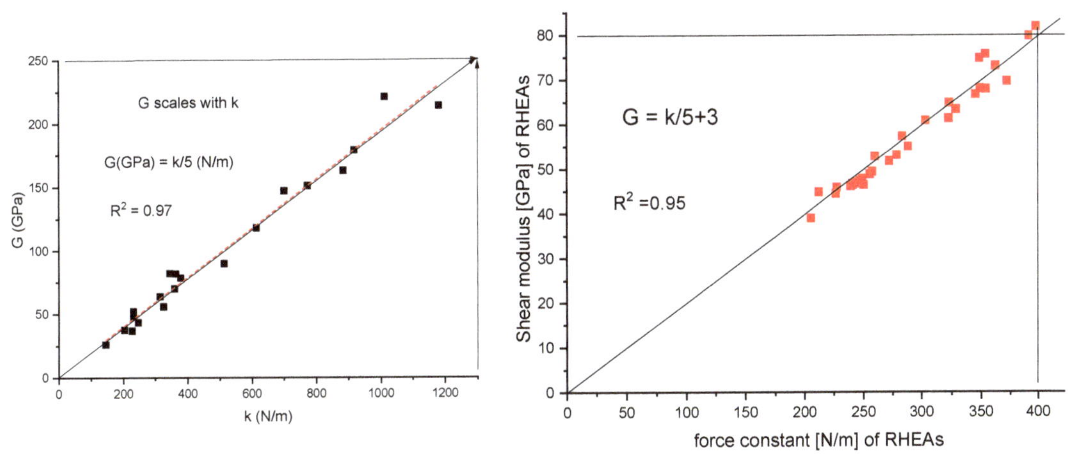

(a) (b)

Figure 10. Scaling of the shear moduli with the force constant for elemental metals (**a**) and alloys (**b**).

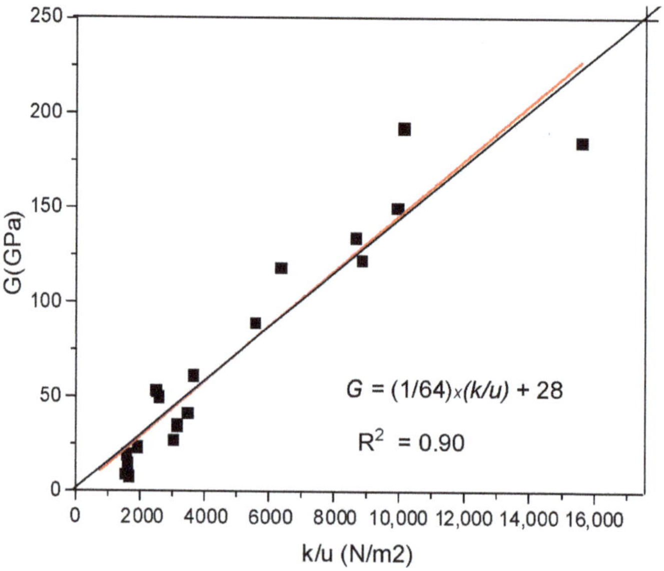

Figure 11. Scaling of the shear moduli with the ratio of the force constant and thermal amplitude for elemental metals.

The shear moduli was scaled with both (a) the force constant, G = k/5, and with (b) the ratio of k/u, G = (1/64)k/u + 28. Furthermore, we had to find the appropriate substitute for the bulk modulus, which is the second term of the Pugh ratio. We remind the reader of the work by Wacke et al. [33], who showed that the bulk moduli scaled with the density of cohesion energy. Here we will use the scaling relationship between the bulk moduli and melting temperature (see Figure 12), because Kaptay [34] have demonstrated a strong linear correlation between the cohesion energy and melting temperature. Therefore, we found the new Pugh ratio as the ratio of *k/u* and melting point T_m.

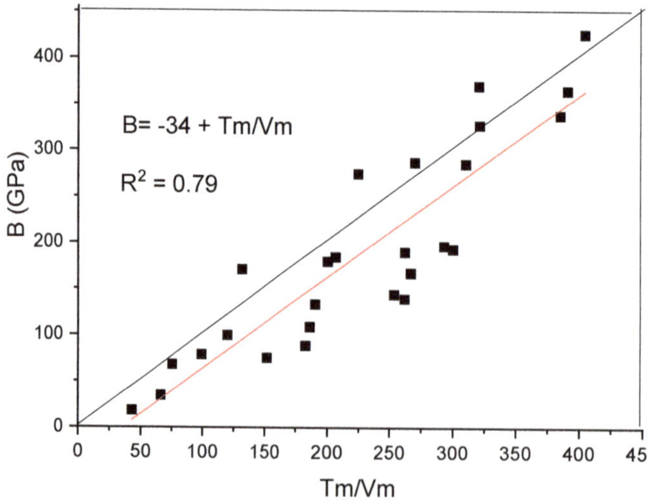

Figure 12. Scaling of the bulk modulus with the ratio of melting point and molar volume.

The dimensionless expression for the new Pugh (NP) ductility index, NP, will be:

$$NP = \frac{k}{< u >}\frac{V_m}{RT_m} \quad (20)$$

where k/u stands for G and RT_m/V_m stands for B.

Using the tabulated molar volume in $cm^3/mole$, the melting temperature in K, and the amplitude in Å, we can extract a numerical factor:

$$NP = \frac{k}{< u >}\frac{V_m}{T_m} \cdot \frac{10^4}{8.31}$$

obtaining the reduced new Pugh (RNP) index as:

$$RNP = \frac{k}{< u >}\frac{V_m}{T_m} \quad (21)$$

In practice, we have found that a better relationship can be obtained with the square root of RNP as a function of the "old" Pugh ratio, G/B. The result is shown in Figure 13, where the known brittle elemental metals are clearly delimitated from the ductile ones. A similar good delimitation was obtained for the refractory alloys dataset collected by Borg [35] and Gorsee [36,37].

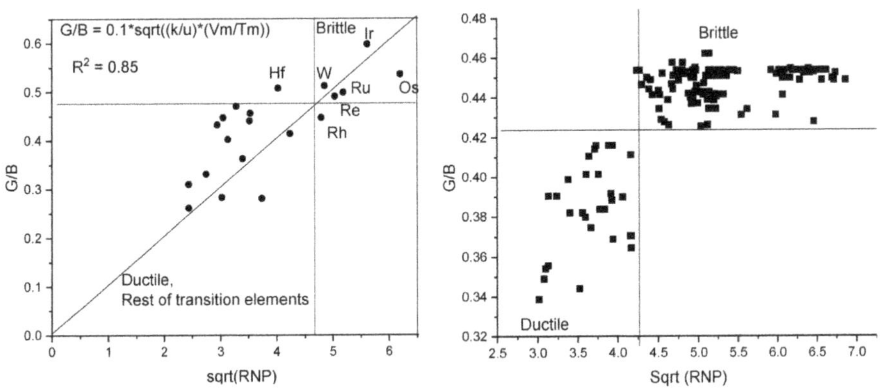

(a) (b)

Figure 13. Clear delimitation of D and B quadrants in G/B versus sqrt (RNP) representations for (**a**) elemental metals and (**b**) for RHEAs.

3.7. Easy Access Surface Energy and Renewed Rice and Thomson Meter

Considering that all of the cohesion related properties were expected to scale with the melting temperature, it was not out of the ordinary that the surface energy scaled with the melting temperature (see Figure 14), even though the units of the two energy correlated properties were not the same. An extremely simple numerical expression helps obtain the surface energy by dividing the melting temperature in Kelvin with 1000, and the unit of γ_{surf} will be J/m^2. As we do not have the space here to enter into an extensive analysis of this simple relationship, we just employed it to renew the Rice–Thomson expression. Instead of $G^x b/\gamma_{surf}$, we can apply the two dimensionless expression k/γ_{surf} or $2R^x(k/u)/\gamma_{surf}$, where 2R is the near-neighbor distance, u is the thermal amplitude at the Debye temperature, and γ_{surf} is the tabulated surface energy. The scaling of the new and old Rice and Thomson parameters are presented in Figures 15 and 16. It turns out that the brittle elemental metals were those with the highest melting point. The linear correlation was also preserved for the collection of alloys, although none of them were really brittle, having a R-T parameter less than 10.

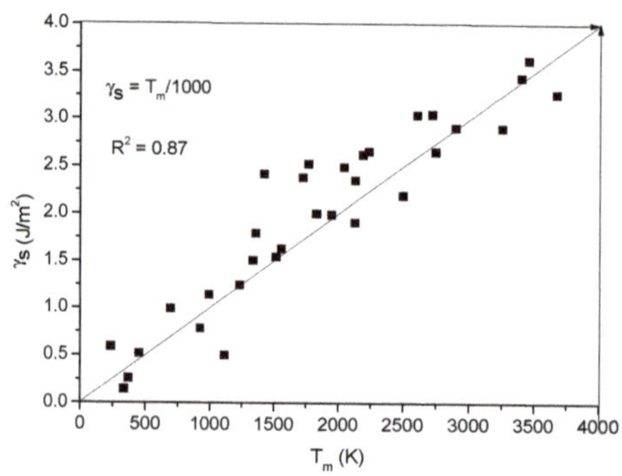

Figure 14. Surface energy of metals scaled with the melting temperature, similar to the cohesion energy.

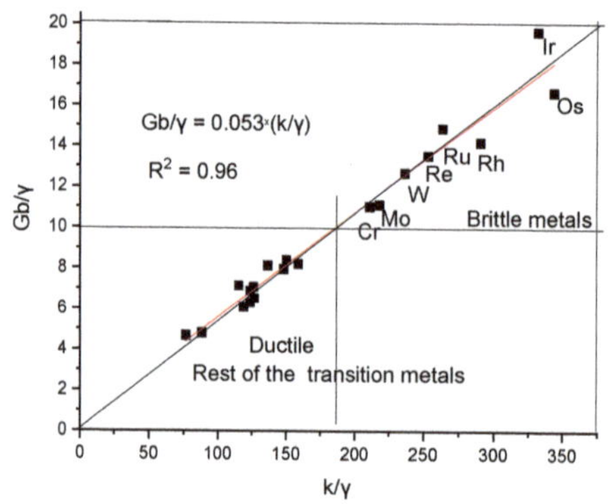

Figure 15. Scaling the R-T parameter with $k/\gamma_{surf,}$ for elemental metals. The critical value $(k/\gamma)_{cr}$~185, whereas $(Gb/\gamma)_{cr} = 10$.

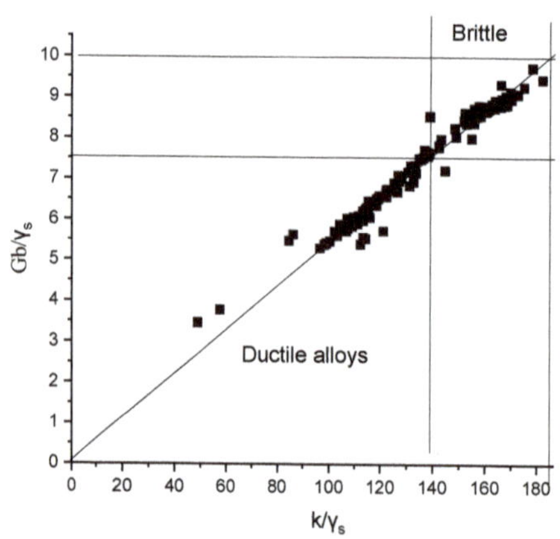

Figure 16. Assignment in B-D regions in the representation of old and new R-T parameters. The alloys were taken from the [35] dataset.

Concerning the application of B-D criteria and the corresponding DIs, we observed that the application of the Pugh criterion is complicated because of the anisotropy factor, which showed a singularity around VEC = 4.2, whereas the Rice and Thomson parameter was free of this problem.

Finally, we present a calculation example of how to apply the old–new Pugh and the old–new Rice and Thomson criteria based only on the tabulated data. First, we collected the necessary tabulated data, as shown in Table 2:

Table 2. Collection of the tabulated data for the application of the Pugh and R-T criteria.

Metal	A	NN-Dist	θ_D	Tm	Dens	Vm	γ_{surf}	k	ν	G	B
	g/mol	A	K	K	g/cm^3	cm^3/mol	J/m^2	J/m^2		GPa	GPa
Al	27		433	933	2.7	10.00	1.143	145.0	0.35	26.10	78.10
Fe	55.85	2.48	478	1808	7.87	7.097	2.417	363.0	0.29	81.90	179.5
Ti	47.9	2.89	425	1943	4.51	10.62	1.989	246.0	0.32	43.40	88.60
Zr	91.22	3.17	296	2127	6.51	14.01	1.909	227.0	0.34	36.80	75.80
Hf	178.49	3.13	253	2500	13.31	13.41	2.193	325.0	0.26	55.80	109.0
V	50.94	2.62	399	2190	6.09	8.365	2.622	231.0	0.37	48.10	139.4
Nb	92.91	3.3	277	2741	8.58	10.83	2.635	203.0	0.40	38.00	170.0
Ta	180.9	2.86	264	3269	16.68	10.85	2.902	359.0	0.334	70.00	193.0
Cr	51.99	2.91	589	2133	7.14	7.282	2.354	513.0	0.21	90.00	196.5
Mo	95.84	2.72	474	2890	10.28	9.323	2.907	613.0	0.29	118.0	285.0
W	183.8	2.74	384	3683	19.26	9.546	3.265	772	0.28	151	338

With the composition of the alloy, we can calculate the average value of each characteristic with the rule of mixture (ROM). At present, the ROM means a concentration weighted average of a given, x, characteristic, $x_{ave} = \Sigma Ci^x x_i$. The average values are collected in Table 3.

Table 3. Calculated characteristics for the alloys taken as examples.

Alloy	RVEC	D*	G	B	k	u	Vm	Tm	γ_{surf}	b
			GPa	GPa	N/m	A	cm^3	K	J/m^2	A
		ave	ave	ave	ave	ave	ave	ave	ave	ave
TiZrHfNbMo	0.57778	0.5453	58.4	145.5	322.8	0.12	11.63	2440	2.326	3.042
--	--									
VNbTaMoW	0.75556	0.5808	85.02	125	435.6	0.109	9.782	2955	2.866	2.846
	--									
NbTaMoW	0.77778	0.5477	94.25	246.5	486.75	0.1006	10.13	3145	2.927	2.905

Finally, as shown in Table 4, we calculated the ductility indices for three alloys where from 2 were at the border of the B-D properties. A perusal of Table 4 helped with the selection of the most suitable DIs, comparing the calculated ones with the above enumerated critical values. Furthermore, we can involve in the discussion above the applicability of the DIs of ductility meters that are experimentally (fracture strain) or theoretically (local lattice distortion, LLD, and the ratio of surface energy and unstable stacking energy, D) obtained.

Table 4. Comparison of ductility indices for three refractory alloys.

Alloy	Frac.strain ε	D	Gb/γ	k/γ	G/B	sqrt(k/u^{x}Vm/Tm)	k/u
	%			/10		/10	N/m^2
	DI	DI	DI	DI	DI	DI	
1. TiZrHfNbMo	10, 10.12	3.088	7.63	13.8779	0.401	0.35807	2690
				--		--	
2. VNbTaMoW	1.7, 2, 8.8	2.28	8.4415	15.19888	0.377	0.36372	4137
				--		--	
3. NBTaMoW	2, 1.9, 2.1	2.397	9.376	16.62852	0.382	0.39487	4838

Considering the criterion based on D_{cr} = 2.5, alloys 2 and 3 are brittle and alloy no. 1 is ductile. Considering the fracture strain criterion, alloy no. 1 was just at the border of the sB-D properties, but the brittleness of the other two alloy was confirmed. Both the old and new R-T criterion had values in the transition zone, with no firm decision about the B-D behavior. Both the old and new Pugh criterion predicted the ductile behavior for all three alloys, which was not the case. At least the RNT parameter varied monotonously with the brittleness of the samples while the G/B parameter varied up and down, and unless the anisotropic effect is taken into account, the G/b parameter by itself is useless. Surprisingly, when we took only half of the new Pugh index (i.e., only the k/u parameter), we obtained a much more sensitive DI, where we achieved a 50% change in the parameter in this transitional range in question. However, it cannot be considered a DI because it is not dimensionless.

The brittle–ductile behavior of single phase RHEAs can be further improved by adding elements from the fourth column of the periodic table (Ti, Zr, and Hf), facilitating the formation of two-phase (HCP + BCC) alloys. The ratio of phases permit improvement and control of the B-D nature of the alloy [38,39], which will be the subject of our next paper.

4. Conclusions

The survey of the most important ductility indices (DI) showed that no perfect B-D distinction can be made when these Dis are represented as a function of fraction strain, ε. Although ε is considered as the most direct DI, its value shows big variations as a function of the preparation and testing details of the given composition. The standard deviation of the ε data is significant and can reach 100%. Therefore, in the case of a large number of samples, data pairs also end up in the "forbidden" quadrants. In order to facilitate the D-B distinction, we recommend that the fraction strain, ε, be used only for qualitative checks to verify the correctness of the sorting. However, the sorting itself is conducted with DIs set in pairs, preferably so that the pairs consist of mutually scalable parameters.

In order to promote this way of classifying samples, a number of ductility indices have been adopted or created from the existing mechanical characteristics:

i. Based on the experimental tests, the fracture energy approximated by the product of yield stress with fracture strain, FE = σ_{YS}*ε, scales with fracture strain can be considered as a new ductility index, DI;
ii. Based on valence electron counts (VEC), there were reduced valence electron counts, which revealed the number of unpaired d electrons as RVEC = (VEC − 2)/4.5;
iii. Based on the Poisson ratio ν, the new DI will be the parameter of Christensen: $D^* = (3\nu/(1 + \nu))^2$;
iv. Based on Cauchy pressure, the Pettifor, Senkov, and Miracle (PSM) parameter = 1 − f(A)*G/B.

Furthermore, new DIs were introduced based on the thermal vibration characteristics (force constant and thermal amplitude). Special pairs were introduced consisting of an old

DI (Rice–Thomson parameter, Pugh ratio) and its new scalable version, which correlated with each other:

i. The parameters of Pugh's ratio, G/B, were replaced by reduced new Pugh parameters: RNP = (k/u)/(Tm/Vm), where the force constant, k, over the amplitude, u, scales with G and the melting point, Tm, over the molar volume, Vm, scales with B.
ii. The parameters of the Rice and Thomson ratio, Gb/γ, were replaced by k/γ, where the two parameters showed linear correlation: $Gb/\gamma = 0.053^{x}(k/\gamma)$.

When creating the parameters, we made sure that they had scalable partner parameters and were dimensionless and accessible by simple calculations based on tabulated data.

The critical values of the old and newly introduced ductility indices were established as follows:

Larger DI values than the critical one indicate a brittle property: (VEC)cr = 5.4, (RVEC)cr = 7.2, (G/B)cr = 0.48~042 and its pair (sqrt RNP)cr = 4.8–4.2, (Gb/γ)cr = 7.5–10 and its pair (k/γ)cr = 140–180.

Larger DI values than the critical one indicate a ductile property: (ε)cr = 10%, (FE)cr = 200 MJ/m^3, ν depends on the anisotropy (ν)cr = 0.28~0.33 and (D*)cr = 0.43~0.562, (D)cr = 2.41~2.5.

Aside from the new DI parameters, we emphasize the importance of some linear scaling correlations that were found between a number of parameters:

I. Surface energy, γ_s, scales with Tm (γ = Tm/1000), where the unit of γ is J/m^2 and that of Tm is K.
II. Surface energy scales with the parameter D.
III. The Poisson ratio related D* parameter scales with the D parameter: $D = -7.14 + 17^{x}D^{*}$.
IV. Due to these scaling correlations, the theoretically calculated D parameter can be obtained phenomenologically as well as using the rule of mixture (ROM) calculations based only on tabulated data. The ratio of surface energy to unstable stacking energy is the D parameter, which is preferred by the community of theoretical physicists working on the theory of the B-D property. Until now, D has only been available by ab initio theoretical calculations. Its calculability by the ROM needs further explanation.

Author Contributions: O.K.T.: Writing—original draft, data curation, L.K.V.: Conceptualization, writing—original draft, review and editing, N.Q.C.: Formal analysis, writing—review editing, L.V.: Supervision, validation, writing—editing. The link for defining the role of authors and contributors: https://www.icmje.org/recommendations/browse/roles-and-responsibilities/defining-the-role-of-authors-and-contributors.html, accessed on 22 September 2024. All authors have read and agreed to the published version of the manuscript.

Funding: This research was funded by the Hungarian Scientific Research Fund OTKA, grant number: 128229.

Data Availability Statement: The data used are from the literature and their sources are properly cited. Requests to access the datasets should be directed to vlk@h-ion.hu.

Acknowledgments: We thank the creative scientific atmosphere and encouragement provided by Smartus, who dealt with refractory high-entropy alloys in the framework of the project Smartus Zrt. 2020-1.1.2-PIACI-KFI-2020-00025.

Conflicts of Interest: Author Ottó K. Temesi was employed by the company H-ION Kft and company company SMARTUS Zrt. The remaining authors declare that the research was conducted in the absence of any commercial or financial relationships that could be construed as a potential conflict of interest.

References

1. Ritchie, R.O. The conflicts between strength and toughness. *Nat. Mater.* **2011**, *10*, 817–822. [CrossRef] [PubMed]
2. Wei, Y.; Li, Y.; Zhu, L.; Liu, Y.; Lei, X.; Wang, G.; Wu, Y.; Mi, Z.; Liu, J.; Wang, H.; et al. Evading the strength–ductility trade-off dilemma in steel through gradient hierarchical nanotwins. *Nat. Commun.* **2014**, *5*, 3580. [CrossRef] [PubMed]

3. Gao, Y.F.; Zhang, W.; Shi, P.J.; Ren, W.L.; Zhong, Y.B. A mechanistic interpretation of the strength-ductility trade-off and synergy in lamellar microstructures. *Mater. Adv.* **2020**, *8*, 100103. [CrossRef]
4. Ronald, W. Armstrong, Engineering science aspects of the Hall–Petch relation. *Acta Mech.* **2014**, *22*, 1013–1028. [CrossRef]
5. Bowen, A.W.; Partridge, P.G. Limitations of the Hollomon strain-hardening equation. *J. Phys. D Appl. Phys.* **1974**, *7*, 969. [CrossRef]
6. Ondicho, I.; Alunda, B.; Kamau, K. Solid Solution Strengthening in High-Entropy Alloys. In *High Entropy Materials—Microstructures and Properties*; IntechOpen: London, UK, 2023. [CrossRef]
7. Martin, J.W. *Precipitation Hardening: Theory and Applications*; Butterworth-Heinemann: Oxford, UK, 1998.
8. Zhang, M.; Zhou, X.; Zhu, W.; Li, J. Influence of Annealing on Microstructure and Mechanical Properties of Refractory $CoCrMoNbTi_{0.4}$ High-Entropy Alloy. *Metall. Mater. Trans. A* **2018**, *49*, 1313–1327. [CrossRef]
9. Senkov, O.N.; Wilks, G.B.; Scott, J.M.; Miracle, D.B. Mechanical properties of $Nb_{25}Mo_{25}Ta_{25}W_{25}$ and $V_{20}Nb_{20}Mo_{20}Ta_{20}W_{20}$ refractory high entropy alloys. *Intermetallics* **2011**, *19*, 698–706. [CrossRef]
10. Coury, F.G.; Kaufman, M.; Clarke, A.J. Solid-solution strengthening in refractory high-entropy alloys. *Acta Mater.* **2019**, *175*, 66–81. [CrossRef]
11. Pugh, S.F. Relations between the elastic moduli and the plastic properties of polycrystalline pure metals. *Lond. Edinb. Dublin Philos. Mag. J. Sci.* **1954**, *45*, 823–843. [CrossRef]
12. Pettifor, D.G. Theoretical predictions of structure and related properties of intermetallics. *Mater. Sci. Technol.* **1992**, *8*, 345. [CrossRef]
13. Rice, J.R.; Thomson, R. Ductile versus brittle behaviour of crystals. *Philos. Mag. A J. Theor. Exp. Appl. Phys.* **1974**, *29*, 73–97. [CrossRef]
14. Cottrell, A. Surprises in materials science. *Interdiscip. Sci. Rev.* **1997**, *22*, 318–324. [CrossRef]
15. Rice, J. Dislocation nucleation from a crack tip: An analysis based on the Peierls concept. *J. Mech. Phys. Solids* **1992**, *40*, 239–271. [CrossRef]
16. Zhou, S.J.; Carlsson, A.E.; Thomson, R. Crack blunting effects on dislocation emission from cracks. *Phys. Rev. Lett.* **1994**, *72*, 852–855. [CrossRef] [PubMed]
17. Mak, E.; Yin, B.; Curtin, W.A. A ductility criterion for bcc high entropy alloys. *J. Mech. Phys. Solids* **2021**, *152*, 104389. [CrossRef]
18. Li, X.; Li, W.; Irving, D.L.; Varga, L.K.; Vitos, L.; Schönecker, S. Ductile and brittle crack-tip response in equimolar refractory high-entropy alloys. *Acta Mater.* **2020**, *189*, 174–187. [CrossRef]
19. Hu, Y.J.; Sundar, A.; Ogata, S.; Qi, L. Screening of generalized stacking fault energies, surface energies and intrinsic ductile potency of refractory multicomponent alloys. *Acta Mater.* **2021**, *210*, 116800. [CrossRef]
20. Qi, L.; Chrzan, D.C. Tuning ideal tensile strengths and intrinsic ductility of bcc refractory alloys. *Phys. Rev. Lett.* **2014**, *112*, 115503. [CrossRef]
21. Sheikh, S.; Shafeie, S.; Hu, Q.; Ahlström, J.; Persson, C.; Veselý, J.; Zyka, J.; Klement, U.; Guo, S. Alloy design for intrinsically ductile refractory high-entropy alloys. *J. Appl. Phys.* **2016**, *120*, 164902. [CrossRef]
22. Lilensten, L.; Couzinié, J.P.; Bourgon, J.; Perrière, L.; Dirras, G.; Prima, F.; Guillot, I. Design and tensile properties of a bcc Ti-rich high-entropy alloy with transformation-induced plasticity. *Mater. Res. Lett.* **2017**, *5*, 110–116. [CrossRef]
23. Huang, H.; Wu, Y.; He, J.; Wang, H.; Liu, X.; An, K.; Wu, W.; Lu, Z. Phase-transformation ductilization of brittle high-entropy alloys via metastability engineering. *Adv. Mater.* **2017**, *29*, 1701678. [CrossRef] [PubMed]
24. Song, H.; Tian, F.; Hu, Q.M.; Vitos, L.; Wang, Y.; Shen, J.; Chen, N. Local lattice distortion in high-entropy alloys. *Phys. Rev. Mater.* **2017**, *1*, 023404. [CrossRef]
25. Singh, P.; Vela, B.; Ouyang, G.; Argibay, N.; Cui, J.; Arroyave, R.; Johnson, D.D. A ductility metric for refractory-based multi-principal-element alloys. *Acta Mater.* **2023**, *257*, 119104. [CrossRef]
26. Tian, F.; Varga, L.K.; Chen, N.; Shen, J.; Vitos, L. Empirical design of single phase high-entropy alloys with high hardness. *Intermetallics* **2015**, *58*, 1–6. [CrossRef]
27. Christensen, R.M. Failure Mechanics—Part I: The Coordination between Elasticity Theory and Failure Theory for all Isotropic Materials. *J. Appl. Mech.* **2014**, *81*, 081001. [CrossRef]
28. Senkov, O.N.; Miracle, D.B. Generalization of intrinsic ductile-to-brittle criteria by Pugh and Pettifor for materials with a cubic crystal structure. *Sci. Rep.* **2021**, *11*, 4531. [CrossRef]
29. Zener, C. *Elasticity and Anelasticity of Metals*; University of Chicago Press: Chicago, IL, USA, 1948; p. 16.
30. Schönecker, S.; Li, X.; Wei, D.; Nozaki, S.; Kato, H.; Vitos, L.; Li, X. Harnessing elastic anisotropy to achieve low-modulus refractory high-entropy alloys for biomedical applications. *Mater. Des.* **2022**, *215*, 110430. [CrossRef]
31. Chen, Q.; Sundman, B. Calculation of Debye temperature for crystalline structures—A case study on Ti, Zr and Hf. *Acta mater.* **2001**, *49*, 947–961. [CrossRef]
32. Ashcroft, N.W.; Mermin, N.D. *Solid State Physics*; Saunders College Publishing: Philadelphia, PA, USA, 1976; p. 426.
33. Wacke, S.; Górecki, T.; Górecki, C.; Książek, K. Relations between the cohesive energy, atomic volume, bulk modulus and sound velocity in metals. *J. Phys. Conf. Ser.* **2011**, *289*, 012020. [CrossRef]
34. Kaptay, G.; Csicsovszki, G.; Yaghmaee, M.S. An Absolute Scale for the Cohesion Energy of Pure Metals. *Mater. Sci. Forum* **2003**, *414–415*, 235–240. [CrossRef]

35. Borg, C.K.; Frey, C.; Moh, J.; Pollock, T.M.; Gorsse, S.; Miracle, D.B.; Senkov, O.N.; Meredig, B.; Saal, J.E. Expanded dataset of mechanical properties and observed phases of multi-principal element alloys. *Sci. Data* **2020**, *7*, 430. [CrossRef] [PubMed]
36. Gorsse, S.; Miracle, D.B.; Senkov, O.N. Mapping the world of complex concentrated alloys. *Acta Mater.* **2017**, *135*, 177–187. [CrossRef]
37. Gorsse, S.; Nguyen, M.H.; Senkov, O.N.; Miracle, D.B. Database on the mechanical properties of high entropy alloys and complex concentrated alloys. *Data Brief* **2018**, *21*, 2664–2678. [CrossRef] [PubMed]
38. Tsuru, T.; Han, S.; Matsuura, S.; Chen, Z.; Kishida, K.; Iobzenko, I.; Rao, S.I.; Woodward, C.; George, E.P.; Inui, H. Intrinsic factors responsible for brittle versus ductile nature of refractory high-entropy alloysi. *Nat. Commun.* **2024**, *15*, 1706. [CrossRef]
39. Zhang, C.; Wang, H.; Wang, X.; Tang, Y.T.; Yu, Q.; Zhu, C.; Xu, M.; Zhao, S.; Kou, R.; Wang, X.; et al. Strong and ductile refractory high-entropy alloys with super formability. *Acta Mater.* **2023**, *245*, 118602. [CrossRef]

crystals

MDPI

Article

Molecular Dynamics Analysis of Collison Cascade in Graphite: Insights from Multiple Irradiation Scenarios at Low Temperature

Marzoqa M. Alnairi [1,*] and Mosab Jaser Banisalman [2,3,*]

1 Department of Physics, Umm Al-Qura University, Makkah 2438224382, Saudi Arabia
2 Virtual Lab Inc., Wangsimni-ro, Seongdong-gu, Seoul 04779, Republic of Korea
3 EN2CORE Technology, 77, Jukdong-ro, Yuseong-gu, Daejeon 34127, Republic of Korea
* Correspondence: mmneairi@uqu.edu.sa (M.M.A.); mosab123@snu.ac.kr (M.J.B.)

Abstract: In our study, we utilize molecular dynamics simulations, specifically through the Reactive Empirical Bond Order, to unravel atomic-scale dynamics in graphite, an essential component in many advanced technologies, under varying irradiation scenarios. We shed light on the behavior of graphite when exposed to Primary Knock-on Atom (PKA) energies of 10, 20, 40, and 80 keV. The findings highlight the radiation vulnerability of graphite, especially when influenced by hydride inclusion. Both pristine graphite and its hydride variant exhibited an increase in Frenkel pairs (FPs) with escalating PKA energies. Notably, carbon PKAs manifested significant FP effects, whereas hydrogen PKAs influenced defect formation through variable diffusivity. In tested radiation scenarios, particularly in Mode C and the R1 region, cascade patterns identified distinct defect forms of diamond-like and elongated-diamond-like shapes, distinct from the typical PKA collision clusters. Furthermore, our cascade findings emphasize the formation of three-coordinated graphite rings, particularly as PKA energies increase. The graphite population statistics reveal a decline in threefold-coordinated atoms and an increase in other types of defects, with 7-carbon atom rings being the most common. Our research highlights the significance of understanding three-coordinated graphite rings, especially as PKA energies rise. Graphite population statistics reveal a decline in threefold-coordinated atoms and a rise in other defects. Notably, 7-carbon atom rings are the most common. From a clustering perspective, self-interstitial atom (SIA) clusters predominated in pristine graphite, while this trend balanced in the hydride variant. Our research highlights the importance of understanding atomic behaviors in graphite under several radiation scenarios. This knowledge is needed for advancing reliable and efficient technological applications, particularly in the field of nuclear technology. Our research underscores the need to understand atomic behaviors in graphite under radiation, paving the way for detailed study on reliable efficient technological applications.

Keywords: primary radiation damage defects; graphite; hydride; multiple cascade; molecular dynamics; collision cascade

1. Introduction

Graphite has always been at the forefront of materials suitable for nuclear applications, owing to its unique physicochemical properties. Its legacy dates to 1942 with the inception of the Chicago Pile 1, and it has been considered for advanced Generation IV reactor concepts [1]. Such persistent reliance on graphite is attributed to its remarkable radiation responses, which induce significant changes in its thermal, mechanical, and creep properties [2]. Neutron-induced interstitials in graphite can raise its energy levels, leading to phenomena like Wigner energy release if not effectively managed.

Graphite's behavior under irradiation, especially the intriguing radiation-induced dimensional change involving shrinkage, swelling, and cross-over at elevated damage levels, is complex yet crucial for reactor performance [3]. Pre-existing Mrozowski cracks and porosity in reactor-grade graphite interact with defects, causing atomic-scale changes in structure [4].

Traditional methods, like Scanning Tunneling Microscopy and Rutherford backscattering, have been employed for investigating atomic-scale radiation damage in carbon [5]. STM, although adept at surface characterization, offers limited scope in area analysis. Similarly, Rutherford backscattering often struggles with bulk materials due to the inherent challenges posed by polycrystalline structures like HOPG. In this landscape, Raman spectroscopy has emerged as a promising technique. It indirectly pinpoints defects in carbon materials, with the D to G Raman mode intensities being indicative of defect populations. The modified Tuinstra–Koenig (TK) model has further elucidated this, revealing that at higher damage levels, the D mode intensity decreases as defects become more proximate to the sixfold ring size.

Yet, it is the advent of Molecular Dynamics (MD) simulations that has truly revolutionized our understanding; offering unparalleled insights at the atomic level, MD simulations highlight pivotal factors like primary knock-on atom (PKA) energy [6–10]. Nevertheless, the selection of an appropriate interatomic potential remains contentious [11]. Historical attempts, such as using the Tersoff potential, have provided valuable findings on the sputtering of graphite [12]. Later efforts incorporated potentials like the Brenner, renowned for more properly capturing the C60 structure [13]. However, these potentials often falter when faced with long-range interactions. A significant breakthrough was the modified Tersoff potential with a long-range extension, shedding light on previously unnoticed hillock formations on graphite surfaces [14].

Despite these advancements, challenges persist. For instance, while density function theory (DFT) has offered a glimpse into early-stage defect evolution, it grapples with the intricacies of weak van der Waals interactions [15]. Yet, with the introduction of the AIREBO potential, accounting for both inter-layer covalent bonding and van der Waals forces, there is renewed hope [16]. However, values like threshold displacement energy were successfully evaluated with MD, with values spanning from 10 eV to 70 eV, as discussed by Banhart and Zinkle [17] As we move forward, understanding irradiation schemes on graphite, especially in high-energy cascades, becomes paramount.

While graphite, in its pure form (i.e., pristine), has been the subject of extensive studies, irradiation-induced impurity effects such as the hydrogen production in reactors can lead to its accumulation within graphite, morphing it into a graphite hydride [18]. Hydrogen can influence graphite's thermal stability, mechanical resilience, and radiation resistance [19]. With the critical roles of graphite in reactors like gas-cooled reactors and very high-temperature reactors (VHTRs), any change in its structure and properties, due to impurities or otherwise, is worthy of consideration. Furthermore, variant effects such as irradiation patterns can be still searchable with MD simulations.

In this work, we extend the ongoing discourse by leveraging MD simulations to closely examine radiation-induced collision cascades in graphite. Our study spans a range of primary knock-on atom (PKA) energies, by varying displacement directions, and even delves into multi-PKA irradiation scenarios for a more realistic representation.

In our study, we conducted a series of MD simulations to explore the collision cascades in both pristine graphite and graphite hydride. The investigation was carried out over a range of PKA energies, specifically at 10 keV, 20 keV, 40 keV, and 80 keV, to provide a comprehensive understanding of the energy-dependent response of the materials. To further deepen our insights, we not only examined the effect of the collision displacement direction and vibration effects but also considered different scenarios of irradiation. In addition to the traditional single PKA approach, we explored complex irradiation scenarios where the same total amount of PKA energy was distributed among several PKAs at different locations within the graphite structure, simultaneously. This multi-PKA approach allowed us to mimic more realistic irradiation conditions, potentially offering a more nuanced view of how graphite responds to radiation. By systematically varying these parameters, our work paints a wider picture of the irradiation behavior of these materials, contributing valuable insights that are relevant to their application in nuclear environments.

2. Simulation Methods

In this research, molecular dynamics (MD) simulations were performed using the Large-scale Atomic/Molecular Massively Parallel Simulator (LAMMPS) package [20].

The carbon interatomic potential was described by the modified Adaptive Intermolecular Reactive Empirical Bond Order potential (AIREBO) and used to characterize the interatomic interactions for graphite and graphite hydride [21], which was developed from the original AIREBO potential by J. Stuart [16]. The energy potential parameters for carbon–carbon (C-C) formation and migration, as well as hydrogen–hydrogen (H-H) interactions, were established through alignment with a mix of experimental data and first-principles calculations. Additionally, the cross-interaction parameters for carbon–hydrogen (C-H) were refined to conform to a specific set of first-principles data.

We conducted MD simulations by subjecting a simulation cell structure to initial conditions of 30 K and 0 Pa in order to equilibrate it. For the recoil simulations, the starting temperature was established at 30 K, but for some simulations, it was reduced to 0 K to obtain observations unaffected by the thermal movements of atoms. These simulations were categorized as the 30 K recoil simulations and the 0 K simulations, respectively.

The objective of conducting simulations at 30 K, termed 30 K recoil simulations, was to align the simulations closely with experimental values noted at low temperatures [22]. To mitigate the impact of atomic vibrations on the accuracy of our collision cascade predictions at this temperature [23], we performed several iterations for each directional displacement. This entailed conducting four distinct simulations for assessing the collision cascade in every direction, with the time gap between successive recoil incidents fixed at 50 fs. These simulations commenced at varying instances: 0, 50, 100, and 150 fs [23]. Averaging the outcomes from these runs helped in reducing the effects of thermal vibrations and temporal correlations on the cascade data.

The pure graphite structure, formed by replication rectangular graphite unit cell shown in (Figure S1a), is constructed by stacking graphene layers placed in periodic boundary conditions at all boundaries in a cell size measuring 116 Å × 118 Å. Approximately 40 of these layers are stacked together, resulting in a final structure dimension of 116 Å × 200 Å × 118 Å. This assembly comprises approximately 400,000 carbon atoms. In this configuration, the hydrogen atom substitutes for a carbon atom in each graphene unit cell, leading to the formation of C-H of the hydride structure.

The H atoms within the graphite structure were stabilized and optimized using the DFT calculation with Quantum Espresso-implemented Materials-Square Platform [24]. The selected unit cell after DFT calculation was added to the Supplementary Materials as the most stable structure (Figure S1b). The PKA was predominantly selected to be a carbon atom, with hydrogen atoms used in specific instances; both were positioned at the simulation box's center. To evaluate the collision cascade, we averaged the defect evolution measurements across the chosen dataset. The impact of vibrations was considered by testing four different vibration timings for each direction of displacement, ensuring the statistical reliability of our findings [23]. The investigation included conducting simulations for both pristine graphite and graphite hydride systems, with results compiled from 20 simulations for each structure. The simulation cell size was determined to maintain the system's average temperature below 200–300 K post-collision, preventing the displacement cascades from extending beyond the cell boundaries, even at the highest PKA energies.

An adaptive timestep was employed, setting the upper limit for displacement per step (xmax) at 0.01 Å and the maximum timestep (tmax) at 0.02 ps. These parameters proved to be effective for evaluating defect formation across various xmax and tmax values [25]. At the start of each MD simulation of recoil events, the PKA was imparted with recoil energy through its velocity components. Defect formation was assessed using Voronoi analysis [26]: a site was deemed to contain a self-interstitial atom (SIA) if it held more than one atom, considered a vacancy if it was empty, and regarded as undamaged if neither condition was met, for each level of PKA energy. Throughout the displacement cascade triggered by irradiation, the system's temperature, volume, and number of atoms were kept constant (NVT ensemble) over 30 ps to ensure the observation of all three ballistic phases was feasible. To calculate the number of atomic displacements leading to defect creation in materials under irradiation, the Norgett–Robinson–Torrens (NRT) formula is widely utilized [27].

$$\text{NRT displacement} = \frac{0.8E_{pka}}{2E_d} \tag{1}$$

Here, E_{pka} represents the energy deposited in the material due to nuclear interactions during a collision, which is approximately equal to the energy possessed by PKA as it initiates a cascade of secondary atom collisions. E_d signifies the material's threshold displacement energy (TDE), identified as 35 eV for pure pristine graphite, derived from applying the AIREBO potential within the computational framework outlined in the referenced study [28].

Table 1 details the modeling parameters utilized within this research, including a summary of the standard NRT displacement calculations. However, the NRT approach presumes a uniform impact of atomic displacements on FP's formation, an assumption that oversimplifies the actual process. The formation of FPs is influenced by numerous variables, including the displacement's direction and energy, the types of atoms involved, and the material's local structure. Consequently, the NRT model might not accurately reflect the quantity of FPs generated, as it overlooks these critical factors. In contrast, MD simulations offer an intricate and accurate method for examining FP formation and dynamics under various irradiation scenarios, aiming to accurately quantify FP formation. The analysis of the final atomic arrangements was conducted using OVITO, a tool designed for visualization and post-processing [29]. The characteristics and dimensions of the cascade were identified using the analytical features available in OVITO.

Table 1. Outline of the parameters used to simulate collision cascades. A structure measuring 116 Å by 200 Å by 118 Å was constructed. To cover various scenarios, sixteen unique simulation scenarios were devised, encompassing four vibration timing options and four displacement directions. The energy levels examined ranged between 10 keV and 80 keV.

PKA Energy (keV)	Number of Atoms in Structure	Number of Independent MD Runs	Displacements Counts (NRT)	Simulation Duration (ps)
10	400,000	20	114	30
20	400,000	20	228	30
40	400,000	20	457	30
80	400,000	20	914	30

3. Results and Discussion

3.1. Collison Cascade Dynamics

3.1.1. Frenkel Pairs (FPs) Evolution under Different PKA Energies for Graphite

This study utilized the standard error of the mean (SEM) to quantify the variations in the creation of Frenkel Pairs (FPs). The SEM was computed as the standard deviation of the average number of FPs from 20 simulations, divided by the square root of n. The damage evolution patterns (Figure 1) for both (a) pristine graphite and (b) graphite hydride at PKA energies of 10, 20, 40, and 80 keV. The patterns are quite consistent across these energy levels; higher PKA energies yielded higher peaks and prolonged times to stable recombination, whereas lower PKA energies led to shorter peaks and faster recombination. Approximately 60% of the maximum FPs survived in the peak across the range of the evaluated PKA energies. During the ballistic phase, the kinetic energy of the PKA triggers a rapid rise in atomic displacements, reaching the highest concentration of defects before most revert to their original lattice positions in a process known as recombination or stabilization. Interestingly, the peak is narrower in pristine and wider in the hydride structures. This behavior can be attributed to the influence of hydrogen in the graphite lattice. In pristine, the atomic interactions and bonds are more consistent, leading to a quicker recombination resulting in a narrower peak. Hydrogen atoms can affect the carbon–carbon bonds in the graphite lattice, which are attributed to hydrogen's ability to facilitate bond breaking, potentially hindering the recombination process and leading to a wider peak. Also, the lighter mass of H compared to carbon atoms leads to more structural deformation.

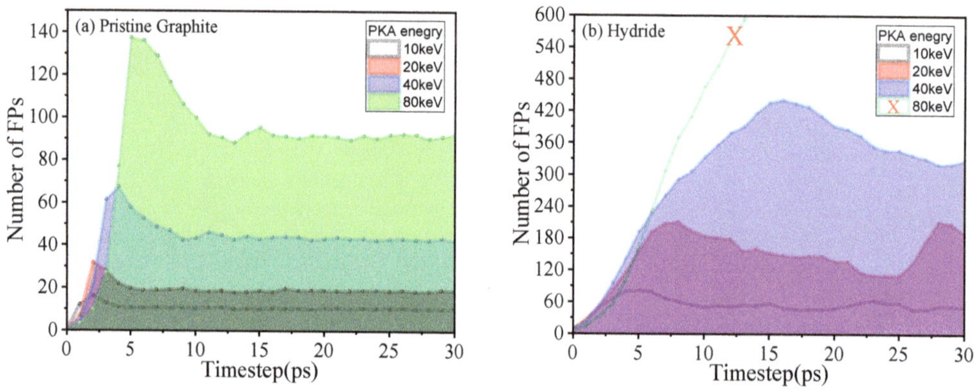

Figure 1. Displays the quantity of Frenkel pairs (FPs) generated in collision cascades at PKA energies of 10, 20, 40, and 80 keV for both pure carbon (**a**) and carbon–hydrogen systems (**b**). Each point averages the outcomes from 20 distinct simulations, factoring in five displacement directions and four vibration timing scenarios. The red "X" marks indicate the occurrence of structural failure at specific PKA energies, measured in eV.

Different PKA energy levels result in varying cascade core volumes and defect amount counts. Higher PKA energies generate results in more primary collision events and increase the probability of secondary PKAs and subsequent mini cascades [30]. This also can contribute a contribution to the broader peak observed in hydride compared to the narrower peak in pristine. It explains why an 80 keV PKA resulted in amplified structural failure in graphite hydride but remained stable and had lower counts in pristine graphite.

In Figure 1b, the occurrence of two distinct peaks (at 7 ps and 28 ps) at 20 keV requires a closer examination of the defect and structural phases. These peaks arose from interactions mainly with H atoms, leading to more FP. Though these were most common in only one displacement direction, the presented results are averages. Surprisingly, despite fewer H than C atoms, this behavior underscores the unique aspects of the radiation damage environment. In hydride, the FP counts were 6–7 times higher than in pristine, attributed to hydrogen's role in reducing the TDE to below 30 eV, as has been concluded in other studies [31]. This increased defect rate suggests the hydride's reduced stability compared to pristine. Additionally, hydrogen atoms can penetrate the graphene structure when exceeding 200 eV [32], whereas other recent studies have stated that the penetration threshold energy can be penetrated at much lower energies [33]. This information is important in irradiated graphene where passivating the carbon dangling bonds with hydrogen atoms in irradiated graphene delocalizes the charge density, reduces the density of states at the Fermi level, and increases the band dispersion, thereby tuning the properties of graphene for additional electronic functionality [34].

3.1.2. The Impact of Displacement Direction on the Frenkel Pairs Kinetics

Primary radiation damage is influenced by the crystal structure and unpredictable collision events that form sub-cascades and channels [35]. The stochastic nature of the cascade mechanism makes predicting PKA energy distribution tough, so more statistics are needed. To study the effects of PKA energy and direction on radiation damage, 20 simulations were run for each PKA energy. We looked at the number of remaining FPs or defects that lasted past the relaxation time, as shown in Figure 2.

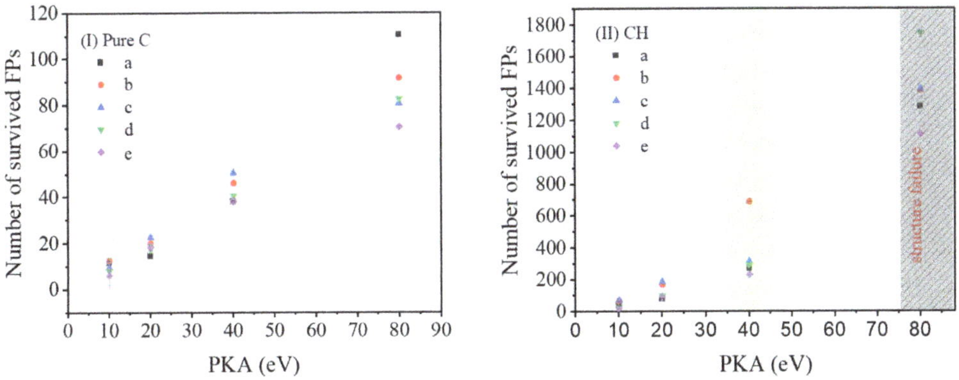

Figure 2. The number of survived FPs when different recoil directions and PKA energies are examined for (**I**) pure graphite structure and (**II**) graphite hydride structure. Displacement directions (a) [1000], (b) [1100], (c) [1120], (d) [1123], and (e) [0001] are examined.

We looked at displacement directions [1000], [1100], [1120], [1123], and [0001]. Usually, the defect count was found to follow the order: [0001] > [1000] > [1100] > [1121] > [1123]. The direction [1123] had the lowest defect count at energies like 10 keV, 20 keV, and 40 keV. At an energy level of 80 keV, the [1121] orientation exhibited a higher number of defects, potentially as a result of the "concentrated collision sequence" observed at elevated energies. In such a sequence, atomic collisions predominantly occur along a single direction, with the sequence's pattern showing minimal sensitivity to changes in atomic density. This feature elucidates the observed variation in Frenkel pairs (FPs) in the pristine [1121] orientation in relation to the energy of the primary knock-on atom (PKA).

Head-on collisions, like in the [0001] direction, often result in greater energy deposition. This direction typically has the lowest TDE energy in many crystals, including graphite and graphene. Consequently, a lower TDE means atoms are more readily displaced, as indicated in references [36,37]. As PKA energy rises, differences in defect formation between crystal directions become clearer, as seen in Figure 2a,b. For example, at 80 keV, there is a notable gap in defect counts among the five displacement directions, but this gap narrows at 10 keV. This is not unique to carbon; materials like Iron and Tungsten show similar patterns [38–40]. The amplified differentiation at higher energies might arise from inherent directional threshold variations and the effects of focused cascade collisions. In hydride, radiation causes many more defects than in pristine. Since hydrogen atoms are smaller than carbon atoms, they can fit between carbon spaces, disrupting the structure and creating more defects due to the weakened carbon–carbon bonds [41]. Our results also observed that, under the same PKA energy and collision cascade conditions, the number of broken C-H bonds was higher than that of C-C bonds (Video S0a,b). In hydride, the effect of direction on defect formation is clearer. For example, in the <1123> direction (i.e., d direction), the material remains more stable. Hydrogen atoms mainly stay within their layer, reducing the risk of hydrogen buildup. During irradiation, these atoms tend to move away from vacancies, leading to hydrogen voids. They also cluster, which can weaken the material. The presence of hydrogen, due to its weaker bond with carbon, further enhances the chances of void creation. At 40 keV PKA energy, the [1100] direction behaves differently. Irradiation here results in 700 FPs, much more than about 250 FPs in other directions (Figure 2II). This is due to the densely packed atoms in the [1100] direction, which is also graphite's c-axis where layers of graphene stack. The c-axis's weaker bonds make it more prone to radiation damage. Graphite's structure, with loosely connected graphene layers, shows layer-by-layer damage upon radiation. The top layers take most of the hit, while deeper layers, with diminishing radiation energy, have fewer defects. These layers can also slide, causing radiation to affect neighboring layers [42]. However, at 80 keV PKA energy, radiation damages multiple graphene layers, resulting in an average of 1200 to 1700 FPs across all PKA directions (Figure 2II). But, thanks to the natural strength of graphene and how radiation spreads, deeper layers might mostly stay untouched if the energy fades before reaching them.

3.1.3. Effects of Hydrogen Atoms as PKAs on Cascade Dynamics within the Hydride Structure

It is noteworthy that the employment of H atoms as PKA yields significantly fewer defects compared to C atoms. In some cases of displacement in the [1123] direction, almost no defects are observed, as illustrated in Figure S2. The trajectory of H atoms, as they navigate through the layers and rings of the CH hydride structure, varied based on the displacement direction. Complex directions presented short, twisted paths, while the more direct trajectories like the [0001] direction were linear and extended. This observation is pivotal because it accentuates the idea that the identity of the colliding atom—be it H or C—can profoundly influence the cascade and resultant damage in the hydride structure. Validating that the chances of the H or C atom being a PKA is a random and instant action in the chaotic irradiation environment. Thus, our judgment on the damage observed in Figure 1b should not be always the case.

Different displacement directions uniquely influence the PKA pathways in hydride. For example, the [1123] direction predominantly presents shorter distances between exit and re-entry points. However, the [1120] and [1000] directions typically show deeper penetration into the graphite layers. This highlights the PKA direction's impact on radiation damage severity, in accordance with studies highlighting the influence of PKA type and direction on defect formation in irradiated materials [43–45]. H atoms, being lighter than carbon in the hydride structure, inherently have lower momentum [46–48]. This makes them less likely to displace carbon atoms within the lattice. Their superior mobility, compared to carbon atoms, enhances their diffusion ability within the structure, promoting recombination with vacancies [49]. Additionally, H atoms can introduce SIA defects, which, compared to FP, exhibit less stability and a greater recombination tendency.

Notably, as we have tested, large energies of PKA (H atom) moving from 10 keV to 80 keV do not always correlate with increased FPs. Instead, the predominant factor influencing diffusion patterns mainly appears to be the displacement direction. To clarify, we present the [0001] direction as a representative case (Figure S2). Although this direction displays a relatively linear trajectory, other directions exhibit complex pathways. The visual representation, both in wrapped and unwrapped trajectory formats, highlights the defect outcomes. Videos S1–S5 further elucidate these trajectories. The H trajectory elucidates the three phases of hydrogen interaction (Figure S2d), where the initial energetic movement transitions to a zigzag pattern before the atom becomes trapped between graphene layers. In this trapped state, the hydrogen atom can marginally alter the C-C bond length before the system reverts to its original state (Figure S3e).

The timeframes for H trajectories travels across various directions have been observed as 27.3 ps ([1000]), 33.3 ps ([1100]), 24 ps ([1120]), 24 ps ([1123]), and 40 ps ([0001]). These timings indicate that more intricate directions are prone to frequent collisions, resulting in short trajectories. Conversely, the [0001] direction, which showed uninterrupted movement, hints at enhanced internal movement potential within the structure's layers.

3.2. Non-Conventional Collision Cascade Scenarios

To gain a comprehensive understanding of radiation damage in graphite structures, it is pivotal to model the unpredictable nature of irradiation environments more closely.

Heggie et al. [50] have pointed out that graphite's layered structure offers unique radiation interaction dynamics when exposed to high doses of energetic particles. In our study, to delve deeper, we explored a variety of collision cascade scenarios using a PKA energy of 80 keV in the [0001] direction. This approach is a departure from traditional methods that trigger the cascade event with just a single PKA. Our method contrasts this by referencing conventional FP outcomes to state differences in radiation damage. We considered four distinct energy deposition modes in the graphite structure:

- Mode D: A conventional single 80 keV PKA, representing the well-understood collision pattern.
- Mode C: Dual PKAs, each with 40 keV.
- Mode B: Quadruple PKAs, each carrying 20 keV.
- Mode A: Eight PKAs, each imparting 10 keV.

These modes were evaluated across three spatial regions:

- Region 1 (R1): A concentrated area, where the PKA energy—whether from a single atom or distributed across multiple PKAs—is localized.
- Region 2 (R2): A broader region, though some PKAs may still collide or interfere with one another.
- Region 3 (R3): A dispersed region where PKAs act almost independently, simulating isolated collisions.

This approach positions our simulations as an enhanced tool for understanding radiation-induced dynamics in graphite. Our findings indicate a particularly pronounced effect in Mode C of the dual PKAs, which consistently showed the highest number of survived FPs. The behavior seems contingent on the layered structure of graphite. When PKAs traverse different layers versus moving within a single layer, discrepancies in defect formation arise. This also can be weighed by the fact that the average TDE of graphite is around 35 eV [28]. Thus, each of the collided PKA already carries more energy than the TDE of the graphite. Thus, ultimately, more FPs can survive. Research and simulations alike have demonstrated that the bombardment by energetic clusters on various substrates can result in the creation of either craters or hillocks [51,52]. In the context of our study, exposing pristine surfaces to 40 keV cluster bombardments led to alterations in surface topography, as observed in MD snapshots. The peak intensity of the cluster beam, involving clusters of about 200–300 atoms, corresponds to average kinetic energies per atom in the range of 130–200 eV. These energy levels exceed the threshold displacement energy (TDE) for graphite, which is 35 eV.

However, the density of the energy deposited at the impact spot is relatively higher than that when it was a single PKA; thus, there is a higher chance for the local excavation of material [53]. Comparably, simulations have shown that cluster–surface collisions lead to the compression of material at the initial stage of impact and the pressure can locally rise to a GPa level [54]. Thus, more PKA would relatively lead to a stronger compression rate and, thus, a larger TDE due to compression strain and, finally, lower FPs, as shown in modes A and B as compared to mode C (Figure 3a).

Interestingly, the conventionally anticipated cascade from a single 80 keV PKA (Mode D) yielded the least defects. This can be attributed to the cascade mechanism—initial collisions might not fully expend the PKA's energy, causing subsequent collisions to dissipate more energy as heat, reducing defect formation as the collision process itself is an instant event and energy will be dissipated at each collision until the PKA atoms stop. In Figure 4a, Modes A, B, and C exhibit a comparable peak phase diagram, characterized by an "oblong diamond-like" shape for A and B and a "diamond-like" shape for C. This configuration indicates that multiple graphite layers are affected: the broadest sections of the shape correspond to layers subjected to the most PKA energy. For instance, in Mode A, the widest layer endures the impact of six PKAs, while the topmost and bottommost layers each experience two PKAs, mirroring the distribution in Mode B. Conversely, Mode C presents a more condensed diamond-like shape configuration, with the widths of impacted layers being more uniform. This likely reduces the melted region, leading to a higher survival rate of defects when compared to Modes A and B, where a higher likelihood of reaching melting temperatures promotes greater defect recombination. Mode D, as depicted in Figure 4a, adheres to the traditional ballistic collision cascade shape, which has been reported in several works [55,56].

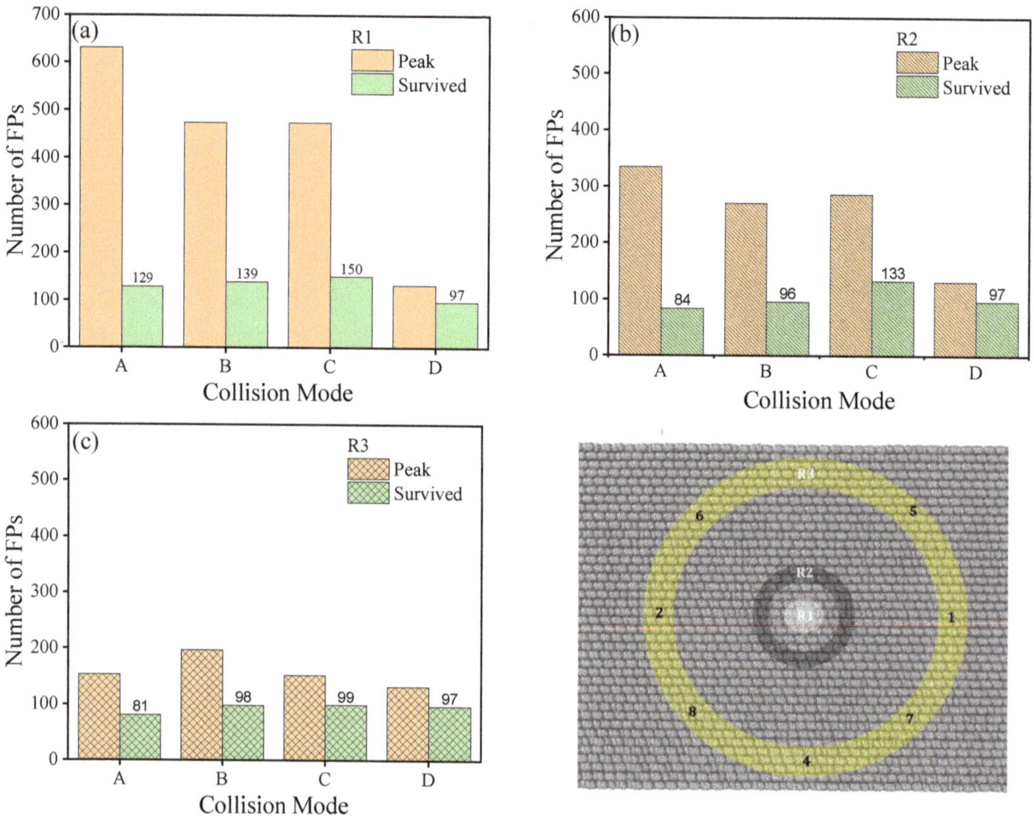

Figure 3. Delineate the results for regions (**a**) R1, (**b**) R2, and (**c**) R3, respectively, showcasing the varying number of Frenkel pairs as per the A, B, C, and D modes. Right Bottom Image explanation of the cascade location on the graphite sample.

Regarding the regions, R1 consistently resulted in more survived defects than R2 or R3 (Figure 3). This phenomenon is attributed to thermal spikes—localized temperature surges due to energetic PKA collisions. Particularly in concentrated regions like R1, multiple collisions can elevate local temperature significantly, offering two potential outcomes. Firstly, this heightened temperature can induce atomic mobility, facilitating defect annihilation or recombination. Secondly, it may prompt increased defect generation due to enhanced atomic vibrations, leading to a net increase in defects. Figure 4b highlights the isolated nature of R3, where each PKA event operates independently without overlapping with other cascade regions. This ensures that there is no compounded effect, in contrast to the pronounced interactions observed in R1 and, to a lesser extent, in R2. Our research emphasizes that in graphite, the configuration of collision cascades is influenced not only by the energy of the PKAs, the directions in which they are displaced, and the PKA atomic type of the temperature, but also by the distribution of energy and the spatial positioning of the PKAs. Assessing multiple concurrent collision scenarios provides a more precise measure of radiation damage phases compared to examining individual PKA collisions. This insight lays the foundation for enhancing graphite designs specifically tailored for radiation environments in real conditions.

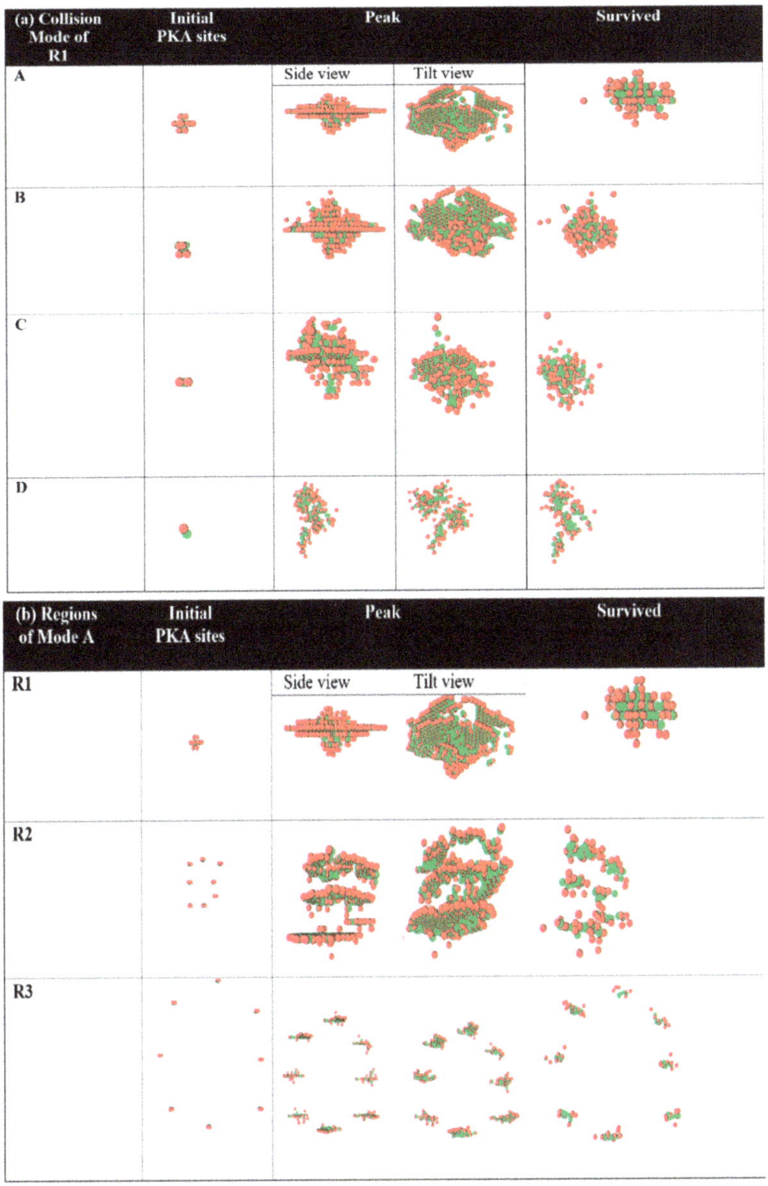

Figure 4. (**a**) Depicts the collision Modes A, B, C, and D within Region 1 (R1). (**b**) Illustrates the regions R1, R2, and R3 specific to Mode A. The figure highlights initial PKA sites, peak numbers of FPs, and the number of FPs that persisted. SIAs are represented in red, while vacancies are shown in green.

3.3. The Local-Environment Classification Scheme for Carbon Atoms Rings in Irradiated Graphite

In graphite systems, it is common to characterize atoms based on their coordination, which is generally believed to correspond to sp, sp^2, and sp^3 hybridization, indicating two, three, and four neighboring atoms, respectively [57]. Nevertheless, observation of electron-energy-loss-spectra data for irradiated graphite has underscored the significance of categorizing threefold-coordinated atoms into four subtypes [58]. The classification of these subtypes depends on the immediate surroundings such as the coordination of second neighbors, involvement in aromatic rings, and so forth. In this work, we adopt the main four categorizations with slightly different nomenclature than described originally in this work [59]. Figure 5a depicts the atom-type composition of the four structural models, showcasing the impact of increased PKA energy and disorder. The proportion of C3-Alpha atoms (indicated by Violet color) exhibits a steady decline within the targeted irradiated damaged region. In irradiated-graphite models, the C3-Beta atoms (indicated by blue symbols) are increasing, which refer to sp^2 atoms belonging to a non-hexagonal ring, and which are more prevalent than C3-Alpha at 10 keV. The observed radiation damage-induced defects also include interlayer cross-links, which are known as fourfold-coordinated atoms (C4; green symbols) that are shown to be more common, especially for 40 keV and 80 keV. Unsaturated graphene edges (C2 atoms; red symbols) are responsible for the presence of threefold-coordinated atoms bonded to fourfold- or twofold-coordinated atoms [60]. These defects play a crucial role in the stability and properties of graphene-related systems [61]. The population statistics of the graphite can be viewed as a continuation of the trends seen in the other three models (i.e., C3-Alpha, and total threefold-coordinated-atom percentage decreasing with damage and the percentage of other types increasing) (Figure 5a). The total threefold-coordinated-atom percentage of graphite is somehow akin to the percentage from previous EDIP calculations of amorphous carbon [59]. The decrease in C3-Alpha and total threefold-coordinated atoms is matched to the increased irradiation damage. On the other hand, the formation of carbon rings of specific sizes is evident, as displayed in Figure 5b. Predominantly, the 7-carbon atom ring emerges as the most prevalent, with its frequency intensifying as PKA energy increases. This is followed by the 8-carbon atom ring. On the contrary, rings comprising five carbon atoms are comparatively infrequent. Notably, at elevated PKA energies, the incidence of these 5-carbon atom rings begins to diminish.

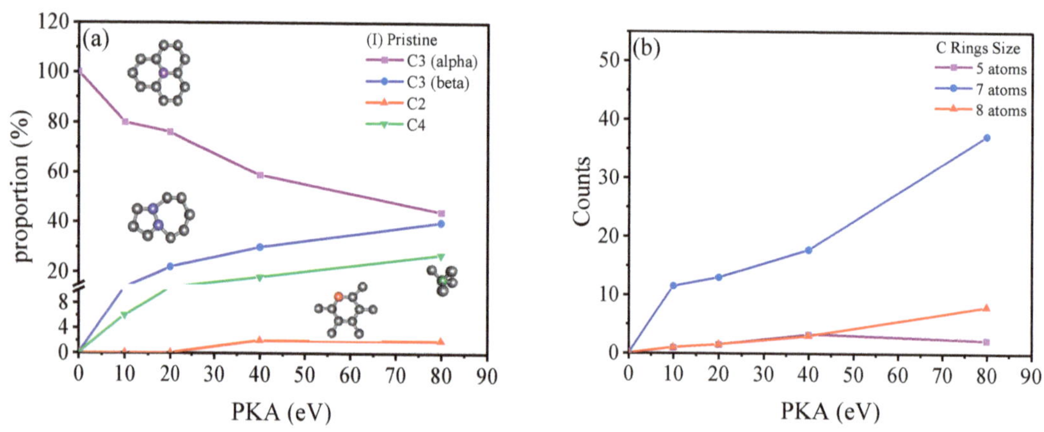

Figure 5. (**a**) Shows four structural patterns in radiation-damaged regions with increasing PKA energy, distinguished by colors: Violet for C3 alpha (threefold-coordinated carbon atoms adjacent to a fourfold-coordinated atom), Blue for C3 beta (threefold-coordinated atoms in non-hexagonal configurations), Red for C2 (twofold-coordinated in sp or sp2-radical forms), and Green for C4 (fourfold-coordinated with sp3 hybridization). (**b**) Displays rings of sizes 5, 7, and 8 within the irradiated structures. Each structure averages data from 20 simulations.

3.4. Cluster Analysis

We explored the cluster details within graphite systems, where in Figures 4b and 6a, we examine the changes in the average number of interstitial clusters, while Figure 6c,d focus on vacancy clusters. One clear trend stands out: as PKA energy surges from 10 to 80 keV, there is a commensurate rise in the cluster formation in both materials. The formation of small clusters dominated by interstitial atoms, termed SIA clusters, is notably different between the two systems. In pristine graphite, the concentration of these SIA clusters, specifically in the size range of two to four atoms, is nearly half of what is evident in the hydride structure. The disparity becomes even more pronounced for clusters larger than four atoms, where the hydride structure hosts a concentration over six times that of its pristine graphite. A closer inspection reveals that pristine predominantly favors the formation of small clusters, typically comprising less than four atoms, be it vacancies or SIAs. In marked opposition, the hydride variation inclines toward the amalgamation of more sizable clusters, particularly when the PKA energy surpasses 40 keV. This highlights a key observation: the hydride structure exhibits a distinctive characteristic of generating larger cluster formations, whereas pure graphite tends to foster smaller, compact clusters. Such variances may be attributed to the inherent lattice dynamics and bonding variations existing between the two materials. The presence of hydrogen may promote the accumulation of defects due to reasons explained in Section 3.1.1, thereby facilitating the formation of larger cluster formations. Conversely, the closely knit lattice of pristine graphite might facilitate rapid defect recombination, ultimately leading to the formation of smaller clusters.

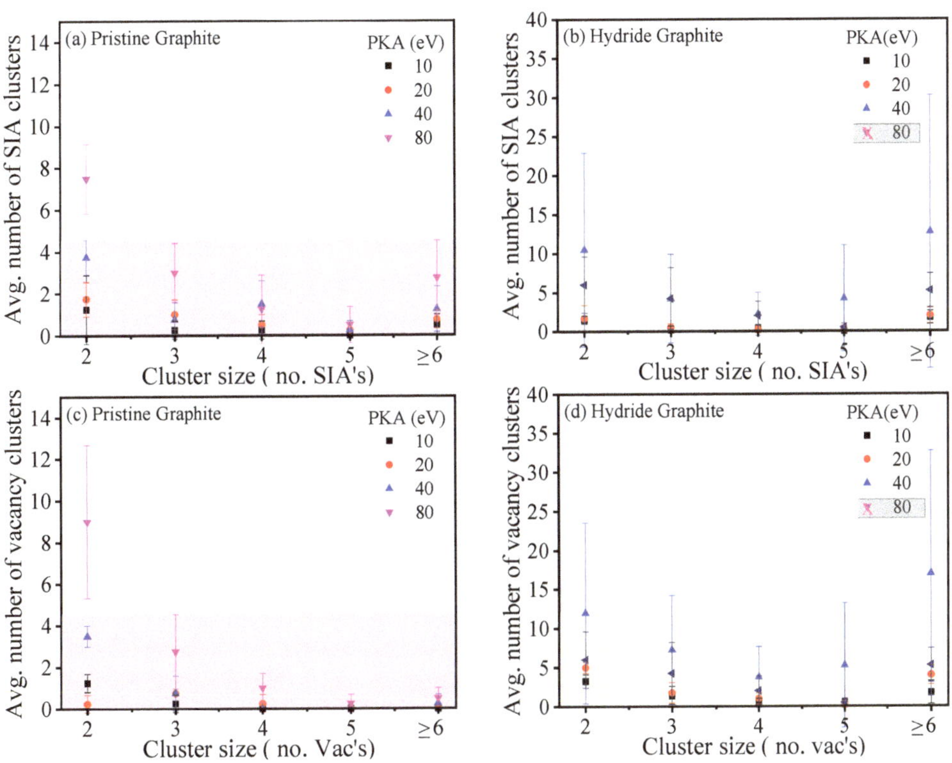

Figure 6. (**a**) Average number of cluster formations (sizes 2–6 SIAs) at various PKA energy, calculated as the mean across four displacement directions with standard deviation. Graphs (**a**,**b**) show data for SIA clusters, while (**c**,**d**) present vacancy clusters for pristine and hydride structures, respectively.

Further analysis of the data reveals a notable pattern. In pristine systems, SIA clusters are about 0.2 times more prevalent than vacancy clusters for sizes exceeding three atoms. This observation resonates with findings from other crystal studies, such as those conducted on W-Re [62] and Fe-Cr alloys [63,64]. Yet, this trend takes an unexpected turn for smaller clusters of size 2 atoms. Here, we observe that vacancy clusters outpace SIAs by a similar margin, but this reversal is specifically observed when PKA energies spike to as high as 80 keV. Such a phenomenon might be attributed to the stochastic nature of collision dynamics at elevated energies such as the focused collision sequence. The average count of vacancy clusters is relatively lower than SIA clusters, particularly for larger clusters encompassing four atoms or more. This difference can be linked to the higher formation energy of a single vacancy in graphite (7.6 eV) compared to the formation energy of an interstitial atom (6.3 eV) [65]. Specifically, the largest clusters we noticed comprised 30 SIA and 15 vacancies. Although the latter appeared more frequently, it was significantly rarer compared to other smaller-sized clusters, like groups of two or three atoms.

Upon transitioning to the hydride configuration, both the SIA and vacancy clusters exhibit comparable quantities and configurations. The inclusion of H may have an impact on lattice dynamics, thus creating an environment that facilitates the formation of both types of clusters similarly, which underscores the fact that even slight compositional alterations can have on the dynamics of defect formation in graphite.

Furthermore, we observed a distinct concave upward pattern in the formation of vacancies and SIA clusters in hydride Figures 4b and 6a. When examining the rate of interstitial cluster formation, we noted that the hydrated structures showed a faster increase than the pristine graphite structures, especially at higher PKA energies.

3.5. Extra Analysis of Layered Responses to Consecutive Collision Cascades in Graphite

The same PKA was subjected to three consecutive collision events. Each of these collisions had an energy of 80 keV and spanned over 20 picoseconds for each collision (60 ps total evolution time). The aftermath of the initial collision saw a formation of 198 defects. This number increased to 364 after the second hit, but intriguingly, the second collision, in a few ps, caused an astounding rise in FPs again, tallying up to 3890 FPs, though it later stabilized to a final count of 1300 (Figure 7 and Video S6). The layered structure of graphite provides an explanation for this behavior. Each layer of graphite, while connected to the entire structure, reacts somewhat autonomously to external disruptions. Initially, the collision's energy concentrates in the middle of the structure. But, with subsequent impacts, this energy disperses, affecting more layers but with a lessened intensity. It resembles the ripple effect in water: the initial splash is intense, but as the ripples spread out, they diminish in strength.

Post the second collision, the rapid changes observed suggest that the layers of graphite do not operate in isolation. When one-layer experiences a change, it can influence its neighboring layers, leading to a domino effect throughout the entire structure.

Interestingly, after 10 ps of the third collision, the number of FPs started to decrease. This suggests that graphite has a natural self-recovery capability. Past studies have also observed graphite's ability to restore its structure, especially when subjected to high-temperature annealing [66,67]. However, it is crucial to note that the graphite still retains some residual stress [42], which can induce other defects if observed over an extended period. After experiencing multiple collision events, the graphite's structure is no longer pristine. These defects can make the structure more vulnerable to future breakdowns, even though a significant portion of FPs revert to their initial configuration.

It is noteworthy that irradiated graphite shows a decrease in total threshold displacement energy. For instance, the threshold energy in pristine graphite reduces from 24 eV to approximately 5 eV in a fully disordered graphite structure [68]. This aligns with our observation of a larger volume of FPs in hydride compared to pristine graphite. The accumulation of defects alters the microstructure, diminishing its crystalline nature [69,70], and fostering cross-links between layers. Such alterations significantly influence physical attributes like thermal conductivity [71] and Young's modulus [58]. Moreover, neutron irradiation can cause notable size changes in the graphite due to radiation damage accumulation [72].

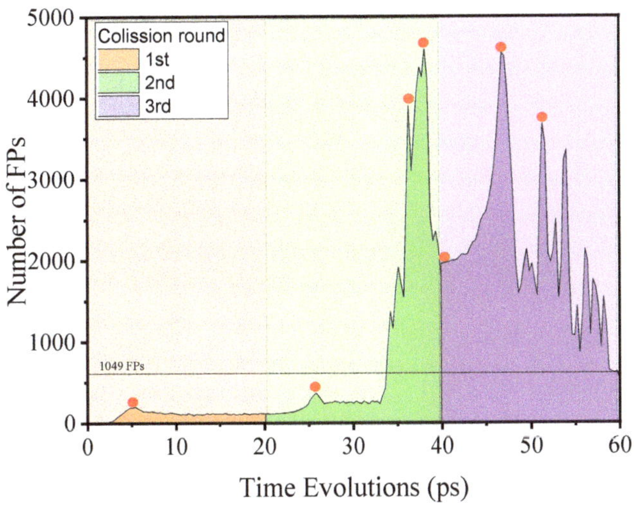

Figure 7. Time evolution of Frenkel pairs in graphite during three collision events, each with 80 keV energy and 20 ps duration. The first region shows the initiation of the primary knock-on atom (PKA) with 80 keV. The second region, in green, displays multiple peaks, with the highest peak marked by red circles. The third region is in indigo.

4. Conclusions

Graphite, pivotal in many advanced technologies, exhibits vulnerability to radiation-induced modifications, especially within nuclear reactors. Employing MD simulations through the Reactive Empirical Bond Order, we delved into the atomic-level dynamics of graphite under varied irradiation scenarios, elucidating the implications of hydride inclusion. Radiation-induced defects in graphite are not merely influenced by differing PKA energy levels (10, 20, 40, and 80 keV), unique displacement trajectories, PKA species, or thermal interactions. They also hinge on the nature of the cascade patterns—whether isolated or cumulative—and collision densities. Empirically, for both graphite structures, irradiation led to an augmented defect count in the hydride compared to pristine graphite. However, both variants displayed an escalation in FPs as PKA energy surged. PKA types—either C or H—had differential impacts, with carbon inducing pronounced effects on FPs due to its greater momentum. However, hydrogen's diffusivity varied, especially in the [0001] displacement direction, resulting in elongated diffusion times, and heightened layer penetration. Analyzing non-traditional collision modes beyond a standard single collision revealed that Mode C and the R1 region within the cascade pattern predisposed graphite to an elevated defect tally, with distinctive defect shapes such as diamond-like and oblong diamond-like morphologies in distinction to the traditional cluster-like shape as in Mode D. Our study emphasizes the importance of understanding three-coordinated graphite rings, with their nuanced atomic behaviors considering escalating PKA energies. The population statistics of graphite show a decrease in the percentage of threefold-coordinated atoms and an increase in other types of defects. The formation of carbon rings of specific sizes is also observed, with 7-carbon atom rings being the most prevalent. From a cluster perspective, SIA clusters were more prevalent in pristine graphite compared to vacancy clusters, a trend that was nearly balanced in the hydride version.

Supplementary Materials: The following supporting information can be downloaded at: https://www.mdpi.com/article/10.3390/cryst14060522/s1, Figure S1: The Graphite unit cell and the optimized structures of H atom doping in the graphite; Figure S2: the diffusion trajectory of 80 keV H atom PKA, of 0001 direction, in the Hydride structure of the graphite; Video S0a: S0a_C-first50steps; Video S0b: S0b_CH-first50steps; Video S1: Dir_a__H_atoms_stopped_at91stepof130step; Video S2: Dir_b__H_atoms_stopped_at111stepof130step; Video S3: Dir_C__H_atoms_stopped_at80stepof130step; Video S4: P3_96_130_PKA_H_atoms; Video S5: Dir_d__H_atoms_stopped_at80stepof130step; Video S6: C_Video_Multiple.

Author Contributions: M.J.B. and M.M.A.: Experiment design and methodology, M.J.B.: Software and simulation, M.J.B. and M.M.A.: collecting the literature, preparation of draft, and revision; M.J.B. and M.M.A.: editing and revision, M.J.B.: supervision. All authors have read and agreed to the published version of the manuscript.

Funding: This research received no external funding.

Data Availability Statement: The calculations were conducted using the LAMMPS code. Some of the structural design features of MatSQ were employed in this process.

Acknowledgments: We would like to extend our appreciation to Jungho Lee of Virtual Lab Inc. for his cooperation in the provision of essential codes, particularly those used in identifying the distinct graphitic rings in the irradiated structure.

Conflicts of Interest: The authors declare no conflicts of interest.

References

1. Marsden, B.J.; Jones, A.N.; Hall, G.N.; Treifi, M.; Mummery, P.M. *Graphite as a Core Material for Generation IV Nuclear Reactors;* Yvon, P., Ed.; Woodhead Publishing: Sawston, UK, 2017; pp. 495–532. [CrossRef]
2. Marsden, B.J.; Haverty, M.; Bodel, W.; Hall, G.N.; Jones, A.N.; Mummery, P.M.; Treifi, M. Dimensional change, irradiation creep and thermal/mechanical property changes in nuclear graphite. *Int. Mater. Rev.* **2015**, *61*, 155–182. [CrossRef]
3. Shibata, T.; Kunimoto, E.; Eto, M.; Shiozawa, S.; Sawa, K.; Oku, T.; Maruyama, T. Interpolation and Extrapolation Method to Analyze Irradiation-Induced Dimensional Change Data of Graphite for Design of Core Components in Very High Temperature Reactor (VHTR). *J. Nucl. Sci. Technol.* **2010**, *47*, 591–598. [CrossRef]
4. Johns, S.; He, L.; Bustillo, K.; Windes, W.E.; Ubic, R.; Karthik, C. Fullerene-like defects in high-temperature neutron-irradiated nuclear graphite. *Carbon* **2020**, *166*, 113–122. [CrossRef]
5. Atamny, F.; Spillecke, O.; Schlögl, R. On the STM imaging contrast of graphite: Towards a "true" atomic resolution. *Phys. Chem. Chem. Phys.* **1999**, *1*, 4113–4118. [CrossRef]
6. Wang, J.; Chen, D.; Chen, T.; Shao, L. Displacement of carbon atoms in few-layer graphene. *J. Appl. Phys.* **2020**, *128*, 085902. [CrossRef]
7. Trevethan, T.; Heggie, M. Molecular dynamics simulations of irradiation defects in graphite: Single crystal mechanical and thermal properties. *Comput. Mater. Sci.* **2016**, *113*, 60–65. [CrossRef]
8. Christie, H.; Robinson, M.; Roach, D.; Ross, D.; Suarez-Martinez, I.; Marks, N. Simulating radiation damage cascades in graphite. *Carbon* **2015**, *81*, 105–114. [CrossRef]
9. Krasheninnikov, A.V.; Banhart, F. Engineering of nanostructured carbon materials with electron or ion beams. *Nat. Mater.* **2007**, *6*, 723–733. [CrossRef]
10. Chen, D.; Shao, L. Using irradiation-induced defects as pinning sites to minimize self-alignment in twisted bilayer graphene. *Appl. Phys. Lett.* **2021**, *118*, 151602. [CrossRef]
11. Tersoff, J. Empirical Interatomic Potential for Carbon, with Applications to Amorphous Carbon. *Phys. Rev. Lett.* **1988**, *61*, 2879–2882. [CrossRef]
12. Smith, R. A classical dynamics study of carbon bombardment of graphite and diamond. *Proc. R. Soc. Lond. Ser. A Math. Phys. Sci.* **1990**, *431*, 143–155. [CrossRef]
13. Smith, R.; Beardmore, K. Molecular dynamics studies of particle impacts with carbon-based materials. *Thin Solid Films* **1996**, *272*, 255–270. [CrossRef]
14. Nordlund, K.; Keinonen, J.; Mattila, T. Formation of Ion Irradiation Induced Small-Scale Defects on Graphite Surfaces. *Phys. Rev. Lett.* **1996**, *77*, 699–702. [CrossRef]
15. Yazyev, O.V.; Tavernelli, I.; Rothlisberger, U.; Helm, L. Early stages of radiation damage in graphite and carbon nanostructures: A first-principles molecular dynamics study. *Phys. Rev. B* **2007**, *75*, 115418. [CrossRef]
16. Stuart, S.J.; Tutein, A.B.; Harrison, J.A. A reactive potential for hydrocarbons with intermolecular interactions. *J. Chem. Phys.* **2000**, *112*, 6472–6486. [CrossRef]
17. Banhart, F. Irradiation effects in carbon nanostructures. *Rep. Prog. Phys.* **1999**, *62*, 1181–1221. [CrossRef]

18. Kane, S.N.; Mishra, A.; Dutta, A.K. International Conference on Recent Trends in Physics 2016 (ICRTP2016). *J. Phys. Conf. Ser.* **2016**, *755*, 011001. [CrossRef]
19. Lechner, C.; Baranek, P.; Vach, H. Adsorption of atomic hydrogen on defect sites of graphite: Influence of surface reconstruction and irradiation damage. *Carbon* **2018**, *127*, 437–448. [CrossRef]
20. Plimpton, S. Fast Parallel Algorithms for Short-Range Molecular Dynamics. *J. Comput. Phys.* **1995**, *117*, 1–19. [CrossRef]
21. Miyashiro, S.; Fujita, S.; Okita, T.; Okuda, H. MD simulations to evaluate effects of applied tensile strain on irradiation-induced defect production at various PKA energies. *Fusion Eng. Des.* **2012**, *87*, 1352–1355. [CrossRef]
22. Mihaila, A.I.C.; Susi, T.; Kotakoski, J. Influence of temperature on the displacement threshold energy in graphene. *Sci. Rep.* **2019**, *9*, 12981. [CrossRef]
23. Banisalman, M.J.; Park, S.; Oda, T. Evaluation of the threshold displacement energy in tungsten by molecular dynamics calculations. *J. Nucl. Mater.* **2017**, *495*, 277–284. [CrossRef]
24. Virtual Lab. Inc. (January 01 Materials Square 2017). Available online: https://www.materialssquare.com/ (accessed on 1 December 2023).
25. Banisalman, M.J.; Oda, T. Atomistic simulation for strain effects on threshold displacement energies in refractory metals. *Comput. Mater. Sci.* **2019**, *158*, 346–352. [CrossRef]
26. Rycroft, C.H. VORO++: A three-dimensional Voronoi cell library in C++. *Chaos Interdiscip. J. Nonlinear Sci.* **2009**, *19*, 041111. [CrossRef]
27. Norgett, M.; Robinson, M.; Torrens, I. A proposed method of calculating displacement dose rates. *Nucl. Eng. Des.* **1975**, *33*, 50–54. [CrossRef]
28. Hu, Z.; Chen, D.; Kim, S.; Chauhan, R.; Li, Y.; Shao, L. Effect of Stress on Irradiation Responses of Highly Oriented Pyrolytic Graphite. *Materials* **2022**, *15*, 3415. [CrossRef]
29. Stukowski, A. Visualization and analysis of atomistic simulation data with OVITO—The Open Visualization Tool. *Model. Simul. Mater. Sci. Eng.* **2009**, *18*, 015012. [CrossRef]
30. Stoller, R.E.; Greenwood, L.R. Subcascade formation in displacement cascade simulations: Implications for fusion reactor materials. *J. Nucl. Mater.* **1999**, *271–272*, 57–62. [CrossRef]
31. Brice, D. Evidence for a single shallow hydrogen trap in hydrogen implanted graphite. *Nucl. Instrum. Methods Phys. Res. B* **1990**, *44*, 302–312. [CrossRef]
32. Sun, J.; Li, S.; Stirner, T.; Chen, J.; Wang, D. Molecular dynamics simulation of energy exchanges during hydrogen collision with graphite sheets. *J. Appl. Phys.* **2010**, *107*, 113533. [CrossRef]
33. Miao, M.; Nardelli, M.B.; Wang, Q.; Liu, Y. First principles study of the permeability of graphene to hydrogen atoms. *Phys. Chem. Chem. Phys.* **2013**, *15*, 16132–16137. [CrossRef] [PubMed]
34. Weerasinghe, A.; Ramasubramaniam, A.; Maroudas, D. Electronic structure of electron-irradiated graphene and effects of hydrogen passivation. *Mater. Res. Express* **2018**, *5*, 115603. [CrossRef]
35. Calder, A.; Bacon, D.; Barashev, A.; Osetsky, Y. On the origin of large interstitial clusters in displacement cascades. *Philos. Mag.* **2010**, *90*, 863–884. [CrossRef]
36. Krasheninnikov, A.V.; Banhart, F.; Li, J.X.; Foster, A.S.; Nieminen, R.M. Stability of carbon nanotubes under electron irradiation: Role of tube diameter and chirality. *Phys. Rev. B Condens. Matter Mater. Phys.* **2005**, *72*, 125428. [CrossRef]
37. Zobelli, A.; Ivanovskaya, V.; Wagner, P.; Suarez-Martinez, I.; Yaya, A.; Ewels, C.P. A comparative study of density functional and density functional tight binding calculations of defects in graphene. *Phys. Status Solidi B Basic Res.* **2011**, *249*, 276–282. [CrossRef]
38. Salman, M.B.; Park, M.; Banisalman, M.J. Atomistic Study for the Tantalum and Tantalum–Tungsten Alloy Threshold Displacement Energy under Local Strain. *Int. J. Mol. Sci.* **2023**, *24*, 3289. [CrossRef] [PubMed]
39. Salman, M.B.; Park, M.; Banisalman, M.J. A Molecular Dynamics Study of Tungsten's Interstitial Dislocation Loops Formation Induced by Irradiation under Local Strain. *Solids* **2022**, *3*, 219–230. [CrossRef]
40. Salman, M.B.; Park, M.; Banisalman, M.J. Revealing the Effects of Strain and Alloying on Primary Irradiation Defects Evolution in Tantalum Through Atomistic Simulations. *Met. Mater. Int.* **2023**, *29*, 3618–3629. [CrossRef]
41. Ruffieux, P.; Gröning, O.; Schwaller, P.; Schlapbach, L.; Gröning, P. Hydrogen Atoms Cause Long-Range Electronic Effects on Graphite. *Phys. Rev. Lett.* **2000**, *84*, 4910–4913. [CrossRef]
42. Bernal, J.D.; Walker, S.J.; Bernal, J.D.; Walker, S.J. The Structure of Graphite. *Proc. R. Soc. London. Ser. A Contain. Pap. A Math. Phys. Character* **1924**, *1*, 1123–1125.
43. Banisalman, M.; Oda, T. Molecular Dynamics Study for the Uniaxial Stress Effect on the Threshold Displacement Energy and Defect Formation Energies in Tungsten. In Proceedings of the KNS 2018-Transactions of the Korean Nuclear Society Autumn Meeting, Jeju, Republic of Korea, 16–18 May 2018.
44. Banisalman, M.J.; Oda, T. Iron Interstitial Defects Stability: Under the Uniaxial Stress Effect. *Cumhur. Sci. J.* **2018**, *39*, 16–22. [CrossRef]
45. Il Choi, S.; Banisalman, M.J.; Lee, G.-G.; Kwon, J.; Yoon, E.; Kim, J.H. Using rate theory to better understand the stress effect of irradiation creep in iron and its based alloy. *J. Nucl. Mater.* **2020**, *536*, 152198. [CrossRef]
46. Vos, M.; Fang, Z.; Canney, S.; Kheifets, A.; McCarthy, I.E.; Weigold, E. Energy-momentum density of graphite by(e,2e)spectroscopy. *Phys. Rev. B* **1997**, *56*, 963–966. [CrossRef]
47. Koga, T. Low- and high-momentum density functions in many-electron atoms. *Chem. Phys. Lett.* **2005**, *411*, 243–247. [CrossRef]

48. Liao, W.; Alles, M.L.; Zhang, E.X.; Fleetwood, D.M.; Reed, R.A.; Weller, R.A.; Schrimpf, R.D. Monte Carlo Simulation of Displacement Damage in Graphene. *IEEE Trans. Nucl. Sci.* **2019**, *66*, 1730–1737. [CrossRef]
49. Granja-DelRío, A.; Alonso, J.A.; López, M.J. Competition between Palladium Clusters and Hydrogen to Saturate Graphene Vacancies. *J. Phys. Chem. C* **2017**, *121*, 10843–10850. [CrossRef]
50. Heggie, M.I.; Latham, C.D. The atomic-, nano-, and meso-scale origins of graphite's response to energetic particles. In *Computational Nanoscience*; Bichoutskaia, E., Ed.; Theoretical and Computational Chemistry; The Royal Society of Chemistry: Cambridge, UK, 2011; Volume 4, pp. 378–414.
51. Samela, J.; Nordlund, K.; Keinonen, J.; Popok, V.N.; Campbell, E.E. Argon cluster impacts on layered silicon, silica, and graphite surfaces. *Eur. Phys. J. D* **2007**, *43*, 181–184. [CrossRef]
52. Popok, V. Energetic cluster ion beams: Modification of surfaces and shallow layers. *Mater. Sci. Eng. R Rep.* **2011**, *72*, 137–157. [CrossRef]
53. Liu, J.; Trautmann, C.; Müller, C.; Neumann, R. Graphite irradiated by swift heavy ions under grazing incidence. *Nucl. Instrum. Methods Phys. Res. Sect. B* **2002**, *193*, 259–264. [CrossRef]
54. Allen, L.P.; Insepov, Z.; Fenner, D.B.; Santeufemio, C.; Brooks, W.; Jones, K.S.; Yamada, I. Craters on silicon surfaces created by gas cluster ion impacts. *J. Appl. Phys.* **2002**, *92*, 3671–3678. [CrossRef]
55. Popok, V.; Samela, J.; Nordlund, K.; Popov, V. Impact of keV-energy argon clusters on diamond and graphite. *Nucl. Instrum. Methods Phys. Res. Sect. B* **2011**, *282*, 112–115. [CrossRef]
56. Egerton, R.F. The threshold energy for electron irradiation damage in single-crystal graphite. *Philos. Mag.* **1977**, *35*, 1425–1428. [CrossRef]
57. Frohs, W.; von Sturm, F.; Wege, E.; Nutsch, G.; Handl, W. Carbon, 3. Graphite. In *Ullmann's Encyclopedia of Industrial Chemistry*; John Wiley & Son Ltd.: Hoboken, NJ, USA, 2010. [CrossRef]
58. Farbos, B.; Freeman, H.; Hardcastle, T.; Da Costa, J.-P.; Brydson, R.; Scott, A.J.; Weisbecker, P.; Germain, C.; Vignoles, G.L.; Leyssale, J.-M. A time-dependent atomistic reconstruction of severe irradiation damage and associated property changes in nuclear graphite. *Carbon* **2017**, *120*, 111–120. [CrossRef]
59. Iwata, T.; Nihira, T. Atomic Displacements by Electron Irradiation in Pyrolytic Graphite. *J. Phys. Soc. Jpn.* **1971**, *31*, 1761–1783. [CrossRef]
60. Wagner, P.; Ivanovskaya, V.V.; Melle-Franco, M.; Humbert, B.; Adjizian, J.-J.; Briddon, P.R.; Ewels, C.P. Stable hydrogenated graphene edge types: Normal and reconstructed Klein edges. *Phys. Rev. B* **2013**, *88*, 094106. [CrossRef]
61. Zhong, M.; Huang, C.; Wang, G. Edge magnetism modulation of graphene nanoribbons via planar tetrahedral coordinated atoms embedding. *J. Mater. Sci.* **2017**, *52*, 12307–12313. [CrossRef]
62. Fu, J.; Chen, Y.; Fang, J.; Gao, N.; Hu, W.; Jiang, C.; Zhou, H.-B.; Lu, G.-H.; Gao, F.; Deng, H. Molecular dynamics simulations of high-energy radiation damage in W and W–Re alloys. *J. Nucl. Mater.* **2019**, *524*, 9–20. [CrossRef]
63. Malerba, L.; Terentyev, D.; Olsson, P.; Chakarova, R.; Wallenius, J. Molecular dynamics simulation of displacement cascades in Fe–Cr alloys. *J. Nucl. Mater.* **2004**, *329–333*, 1156–1160. [CrossRef]
64. Terentyev, D.A.; Malerba, L.; Chakarova, R.; Nordlund, K.; Olsson, P.; Rieth, M.; Wallenius, J. Displacement cascades in Fe–Cr: A molecular dynamics study. *J. Nucl. Mater.* **2006**, *349*, 119–132. [CrossRef]
65. Li, L.; Reich, S.; Robertson, J. Defect energies of graphite: Density-functional calculations. *Phys. Rev. B* **2005**, *72*, 184109. [CrossRef]
66. Dostov, A.I. A method of calculating the rate of release of Wigner energy in heat conduction problems for irradiated graphite. *High Temp.* **2005**, *43*, 259–265. [CrossRef]
67. Eto, M.; Oku, T. The effect of pre-stressing and annealing on the young's modulus of some nuclear graphites. *J. Nucl. Mater.* **1973**, *46*, 315–323. [CrossRef]
68. Vukovic, F.; Leyssale, J.-M.; Aurel, P.; Marks, N.A. Evolution of Threshold Displacement Energy in Irradiated Graphite. *Phys. Rev. Appl.* **2018**, *10*, 064040. [CrossRef]
69. Thrower, P.; Reynolds, W. Microstructural changes in neutron-irradiated graphite. *J. Nucl. Mater.* **1963**, *8*, 221–226. [CrossRef]
70. *IAEA-TECDOC-1154*; Irradiation Damage in Graphite Due to Fast Neutrons in Fission and Fusion Systems. IAEA: Vienna, Austria, 2000.
71. Taylor, R.; Brown, R.; Gilchrist, K.; Hall, E.; Hodds, A.; Kelly, B.; Morris, F. The mechanical properties of reactor graphite. *Carbon* **1967**, *5*, 519–531. [CrossRef]
72. Heggie, M.; Suarez-Martinez, I.; Davidson, C.; Haffenden, G. Buckle, ruck and tuck: A proposed new model for the response of graphite to neutron irradiation. *J. Nucl. Mater.* **2011**, *413*, 150–155. [CrossRef]

crystals

MDPI

Article

Effect of Ultrasonic Vibration on Tensile Mechanical Properties of Mg-Zn-Y Alloy

Wenju Yang [1], Zhichao Xu [1,2,*], Feng Xiong [1], Haolun Yang [1], Xuefeng Guo [1] and Hongshan San [1,2]

[1] School of Materials Science and Engineering, Henan Polytechnic University, Jiaozuo 454003, China
[2] Henan International Joint Research Laboratory for High-Performance Light Metallic Materials and Numerical Simulations, Henan Polytechnic University, Jiaozuo 454003, China
* Correspondence: xzc@hpu.edu.cn

Abstract: Ultrasonic vibration assisted (UVA) plastic forming technology has proven to be a highly effective processing method, particularly for materials that are challenging to deform. In this research, UVA tensile tests were carried out on $Mg_{98.5}Zn_{0.5}Y_1$ alloy at different vibration frequencies and amplitudes. The experimental results indicate that, compared with conventional tensile tests, the yield strength and tensile strength of $Mg_{98.5}Zn_{0.5}Y_1$ alloy exhibit a decrease. Furthermore, the application of ultrasonic vibration demonstrates an ability to enhance the material's elongation and plasticity. In order to further predict the stress-strain relationship in the metal tensile process, a hybrid constitutive model coupling the frequency and amplitude of ultrasonic vibration was developed based on the modified Johnson Cook model. The calculated results of the constitutive equation are in good agreement with the experimental results, indicating that the established constitutive equation can accurately predict the trend of alloy stress change at different amplitudes and frequencies. It establishes a theoretical foundation for scrutinizing the deformation mechanisms of the alloy under ultrasonic vibration. Furthermore, Abaqus finite element analysis software was employed to simulate and analyze the UVA tensile process, elucidating the impact of ultrasonic vibration on stress distribution, strain patterns, and material flow in the tensile behavior of $Mg_{98.5}Zn_{0.5}Y_1$ alloys.

Keywords: Mg-Zn-Y alloy; ultrasonic vibration; deformation mechanism; constitutive equation; simulation analysis

1. Introduction

As a light structural material, magnesium alloy has a very good application prospect in the automobile, aerospace, national defense, biomedicine and other industries. It should be noted that the Mg-Zn-Y alloy has generated great interest due to its good combination of strength and ductility [1,2]. However, the poor formability of magnesium alloys at room temperature limits their wide application. Such poor formability of Mg alloys mainly comes from their crystal structure being hcp with a limited number of deformation systems [3]. The problem concerning how to improve the plastic formability of magnesium alloys is the key to expand their application range. Previous studies [4] have shown that although it is a common practice to increase the number of active slip systems by thermal deformation, there are many shortcomings to this approach. For instance, magnesium alloy's low specific heat makes it challenging to maintain a steady temperature during processing. Additionally, the repeated annealing and high-temperature processing will result in a complicated forming process and considerable energy usage. As a result, many researchers are dedicated to developing short and environmentally friendly processing methods [5,6]. Among them, ultrasonic vibration assisted (UVA) plastic forming technology is considered as a relatively mature metal forming technology. Following the seminal work of Blaha and Longenecker [7], which demonstrated the substantial reduction of flow stress in the tensile testing of single zinc crystals under ultrasonic vibration, a series of studies have been conducted to delve into UVA plastic forming and scrutinize the influence of ultrasonic vibration on material properties. Compared with the traditional forming process, the superposition of vibration on the die or workpiece can reduce the forming load of the material, improve the plasticity and fracture limit, and effectively improve the mechanical properties of the material [8,9]. Ultrasonic vibration, known for its high frequency, rapid speed, short cycles, user-friendliness, environmental friendliness, and controllability, has emerged as a highly effective method for surface strengthening and stress relief in various metal-plastic forming processes. To date, processing technologies such as ultrasonic assisted drawing [10], ultrasonic assisted extrusion [11] and ultrasonic assisted deep drawing [12] have been developed. The advantage of UVA plastic deformation of metals is that it cannot only reduce the forming force of materials and improve the forming quality of materials, but also reduce the friction between specimens and molds and improve the surface quality of materials. It can effectively refine the grains [13,14] and promote the dislocation slip [15]. Moreover, the softening effect induced by ultrasonic vibration is not only more efficient, but also more energy-saving compared to traditional heating-induced softening [16].

Ultrasonic vibration in metal UVA plastic deformations induces two primary effects: the volume effect [17] and the surface effect [18]. The volume effect primarily signifies the impact of ultrasonic vibration on physical properties, such as the internal structure of materials. Meanwhile, the surface effect primarily reflects the influence of friction between the die and the blank during vibration. Based on various explanations of UVA plastic deformation, many scholars have proposed the constitutive model of metal deformation under ultrasonic vibration. Yao et al. [19] established acoustic softening and acoustic residual hardening models of aluminum under ultrasonic vibration based on thermal activation model and dislocation evolution theory in metal plastic deformation, which can accurately predict the stress-strain relationship of aluminum under high-frequency vibration. Prabhakar [20] expanded the constitutive model according to the acoustic softening of aluminum alloys under ultrasonic vibration, and successfully predicted the acoustic softening behavior of aluminum alloys under different vibration conditions. Sedaghat et al. [21] proposed a new constitutive model to predict the acoustic plastic behavior of materials based on the deformation behavior of aluminum alloys under ultrasonic vibration, combined with the dislocation dynamics and acoustic energy conversion mechanism of materials under ultrasonic vibration. Meng et al. [22] established the ultrasonic vibration constitutive model of materials at high temperature according to the plastic deformation behavior of Inconel 718 alloy under ultrasonic vibration and the dislocation evolution theory of materials. The outcomes reveal that the predicted results align well with the experimental findings, demonstrating the model's effectiveness in describing the material's behavior under these conditions. Lin et al. [23] established the constitutive relationship of pure titanium under ultrasonic vibration using the original Johnson cook and its modified model according to the compression deformation of pure titanium under different strain rates and different amplitudes. This model can accurately predict the material flow behavior under different amplitudes and strain rates.

In the present work, the effect of frequencies, amplitudes and strain rates on tensile properties of Mg-Zn-Y alloy was systematically studied. The stress-strain relationship of Mg-Zn-Y alloy under different conditions was compared. Additionally, the effect of ultrasonic vibration on material dislocations and dislocation motion evolution in forming process were analyzed. As a result, we developed a constitutive model that incorporates both frequencies and amplitudes to better understand and describe the material's behavior under ultrasonic vibration.

2. Experimental Procedures

2.1. Specimen Preparation

The $Mg_{98.5}Zn_{0.5}Y_1$ alloy used in the experiment was melted by pure magnesium (99.99 wt%), pure zinc (99.99 wt%) and Mg-25 wt% Y alloy. The alloy is smelted in SXL-1200 resistance furnace filled with argon. The ingots were homogenized at 500 °C for 16 h and then cooled in air. The composition of alloy ingot was measured by X-ray fluorescence spectrometer. The alloy's chemical composition is detailed in Table 1. The microstructure was examined using optical microscopy (OM, Carl Zeiss Axio Observer. 5m) (Carl Zeiss, Oberkochen, Germany). X-ray diffractometry (XRD) (Shimadzu, Kyoto, Japan) was carried out in a θ–2θ diffractometer Philips 1810 using Cu Kα radiation, voltage of 35 kV, and current of 50 mA. Figure 1 shows the microstructure of the $Mg_{98.5}Zn_{0.5}Y_1$ alloy. The average grain size, as illustrated in Figure 1a, is 146 μm. Additionally, Figure 1b indicates that the primary phase of the $Mg_{98.5}Zn_{0.5}Y_1$ alloy is $Mg_{12}YZn$ (LPSO).

Table 1. Chemical composition of $Mg_{98.5}Zn_{0.5}Y_1$ alloy.

Nominal Alloy	Actual Composition (at %)			
	Mg	Zn	Y	Y/Zn
$Mg_{98.5}Zn_{0.5}Y_1$	98.5	0.5	1	2

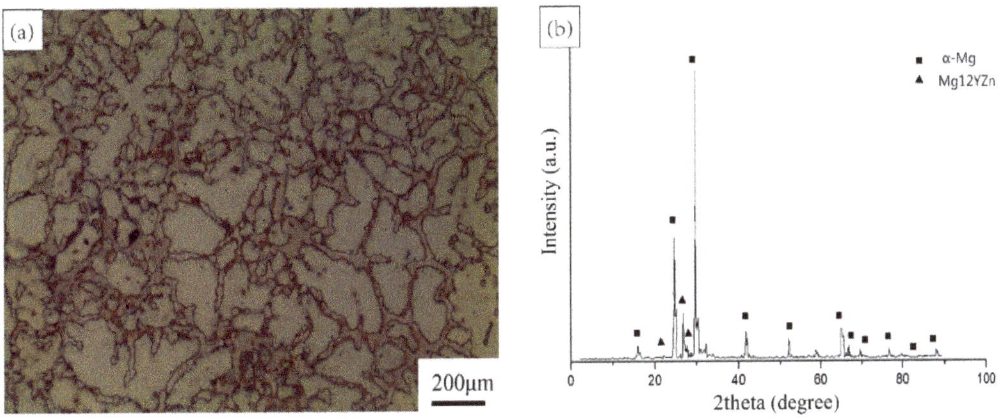

Figure 1. (**a**) OM image and (**b**) XRD pattern of $Mg_{98.5}Zn_{0.5}Y_1$ alloy.

2.2. Experimental Setup

The UVA tensile test of Mg-Zn-Y alloy was carried out on an EAST testing machine. The ultrasonic vibration system includes an ultrasonic generator, a sound transducer, and a horn. The ultrasonic vibration device was installed on the tensile testing machine, as presented in Figure 2. A 10 mm diameter hole is milled to fix the end of the specimen by a bolt. The other end of the specimen was attached to the lower clamp with bolts. Throughout the tensile tests, the horn amplified ultrasonic vibration and transmitted ultrasonic energy to the specimens. The wavelength of the ultrasonic vibration was used to calculate the length of the horn. Given the limited space available in the test system, the resonance frequencies were chosen to be 20, 25 and 30 kHz, respectively. The ultrasonic generator had the capability to adjust the ultrasonic amplitude, directly proportional to the ultrasonic energy. The sets of experimental parameters are detailed in Table 2.

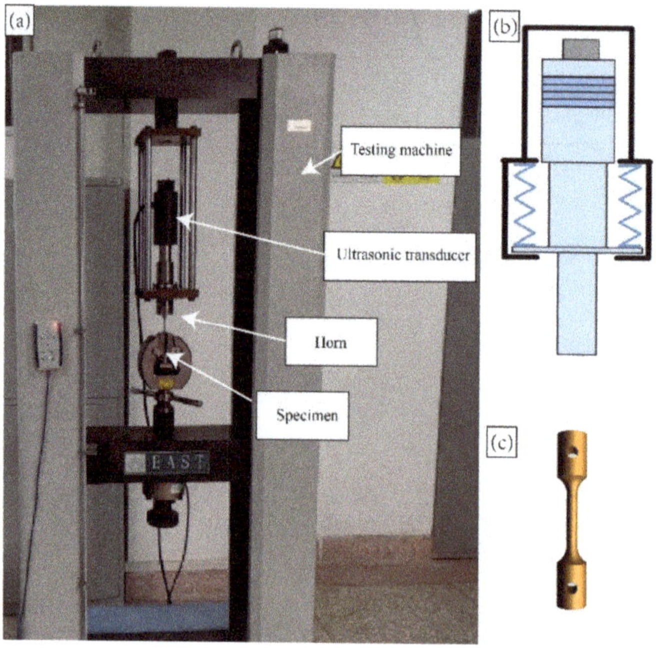

Figure 2. Ultrasonic vibration tensile testing device: (**a**) UVA tensile test system; (**b**) ultrasonic transducer; (**c**) tensile specimen.

Table 2. Experimental variables and conditions of tensile test.

Category	Frequency/kHz	Amplitude/μm	Strain Rate/s^{-1}
Tensile test	No vibration	———	0.001 0.01 0.1 1
	20 kHz/25 kHz/30 kHz	3 μm/5 μm/7 μm	0.001 0.01 0.1 1

3. Results and Discussion

3.1. Effect of the Ultrasonic Vibration on Flow Stress

The experiments encompass three distinct ultrasonic vibration amplitudes and frequencies, offering a comprehensive exploration of the coupling effects between frequency and amplitude on the material's performance. The stress-strain curves for both conventional and UVA tensile tests are illustrated in Figures 3 and 4. From the figures, one can easily observe that the flow stress falls once ultrasonic vibration is applied. To quantitatively assess the impact of vibration amplitude and frequency on flow stress, the yield stress under various conditions is presented in Table 3. It can be clearly seen that the yield stress decreases with the increase of the vibration amplitude and frequency. For instance, the most significant reduction in yield stress is 29.1 MPa when the vibration amplitude is 7 μm, the frequency is 30 kHz, and the stress rate is 0.001 s^{-1}, constituting 19% of the yield stress.

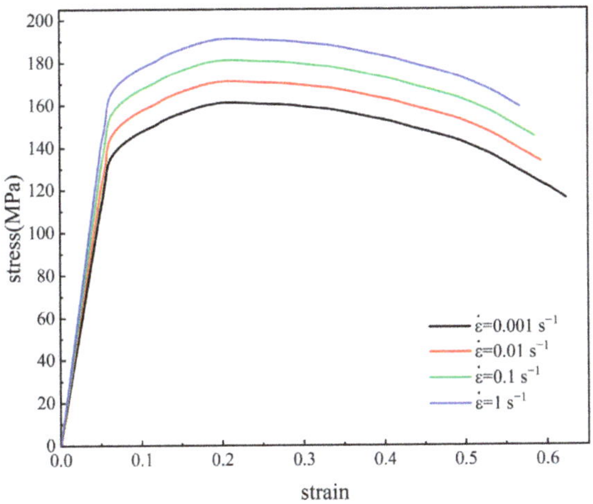

Figure 3. Stress-strain curves at different strain rates without vibration.

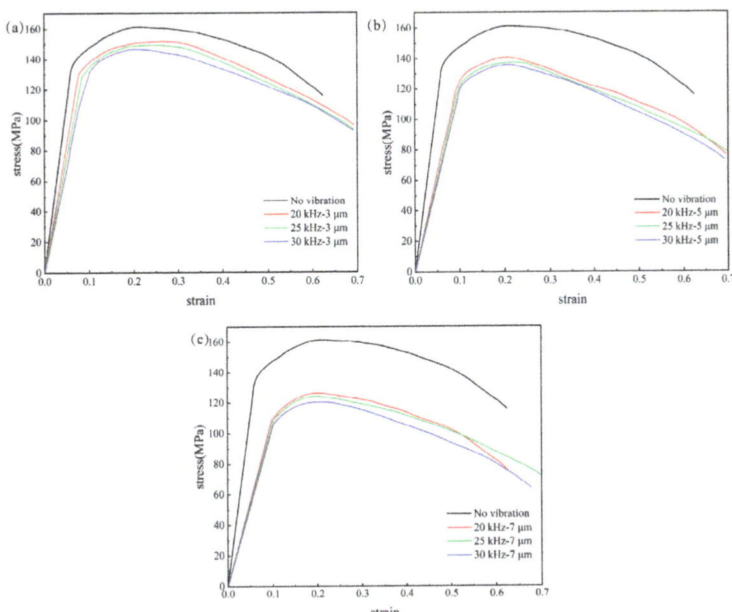

Figure 4. Stress-strain curves at different amplitude frequencies when the strain rate is 0.001 s^{-1}: (**a**) 3 μm, (**b**) 5 μm, (**c**) 7 μm.

It can be concluded that the tensile curves under different conditions exhibit almost the same trend of change. The acoustic softening and stress superposition are the main reasons for the flow stress reduction [24]. The acoustic softening means that the dislocation in the alloy will absorb enough acoustic energy through the application of high frequency ultrasonic vibrations, so that the dislocation is more likely to deviate from its equilibrium position. Stress superposition refers to the phenomenon where the oscillating stress, which initially reaches peak levels, is reduced to an average stress due to the elastic deformation of the specimens. On the other hand, there are lattice defects in the alloy, which are easier to absorb sound energy [25]. We can infer that the combined impact of the two reasons leads to a decrease in tensile stress. From the comparison of the stress-strain diagram, it can also be concluded that the application of ultrasonic vibration improves the ductility of the Mg-Zn-Y alloy. It is reported [15] that ultrasonic vibration can promote the slip and rearrangement of dislocations, which reduces the stacking of dislocations in the alloy and increases the proportion of dynamic recrystallization, thereby improving the plasticity of the alloy.

Table 3. Yield stress at different amplitudes and frequency (MPa).

Frequency	Amplitude	Initial Strain Rate/s^{-1}			
		0.001	0.01	0.1	1
No vibration	——	131.96	145.15	152.16	165.98
20 kHz	3 μm	126.78	137.77	147.55	161.40
	5 μm	119.33	131.78	142.80	151.81
	7 μm	107.56	119.37	127.68	139.89
25 kHz	3 μm	125.97	134.67	146.35	159.78
	5 μm	116.01	128.35	139.89	149.22
	7 μm	105.46	116.67	124.87	135.67
30 kHz	3 μm	123.63	132.41	145.76	157.91
	5 μm	115.33	125.31	137.24	147.82
	7 μm	105.23	115.62	123.02	133.56

3.2. *Effects of Ultrasonic Vibration on Microstructure*

3.2.1. Microstructure

Figure 5 shows the crystal orientation distribution and the orientation angle distribution of the specimens with different situation. As seen from the images, there is no significant difference between the grain size and the crystal orientation distribution. It can be calculated that the average size of the grains is 30 μm. The grain size and grain orientation distribution of alloy are not appreciably altered by ultrasonic vibration. Simultaneously, it can be seen from the figure that there are small recrystallized grains and tension twins in the specimen without UV. It is worth noting that the recrystallization and twining of the alloy after ultrasonic vibration is not obvious. Namely, after the application of ultrasonic vibration, there is a decrease in the quantity of both recrystallized and twinned grains. By comparing the statistical distribution of orientation angles, it can be easily deduced that the application of ultrasonic vibration significantly influences the formation of small-angle grain boundaries during the deformation.

In order to better understand the effect of UV on small-angle grain boundaries, we compared the composition of small-angle grain boundaries in different states of specimens, as displayed in Figure 6. By statistics, after tensile deformation, the percentage of small angle grain boundaries in the specimen is 14.8%. When ultrasonic vibration is applied, the content of small-angle grain boundaries in the alloy is reduced to 13.2%. This suggests that ultrasonic vibration contributes to a decrease in the number of small-angle grain boundaries. As we all know, fewer small-angle grain boundaries indicate lower dislocation density, which means better plasticity of the material [26]. This phenomenon is consistent with the softening of the material observed in the previous experiments. In this paper, the acoustic softening effect is considered as main reason. Ultrasonic vibrations increase the vibrational energy of atoms, making it easier for crystal defects such as dislocations to be created and move [15,16]. Ultrasonic vibration enhances the coordination and uniformity of plastic deformation in the material. The sooner ultrasonic vibration is applied, the less dislocation entanglement and plugging occur within the material. Consequently, the material exhibits a weaker hardening tendency in the subsequent molding process, marking the primary cause of the "acoustic residual softening effect" [27]. A detailed analysis will be conducted by Section 3.2.2.

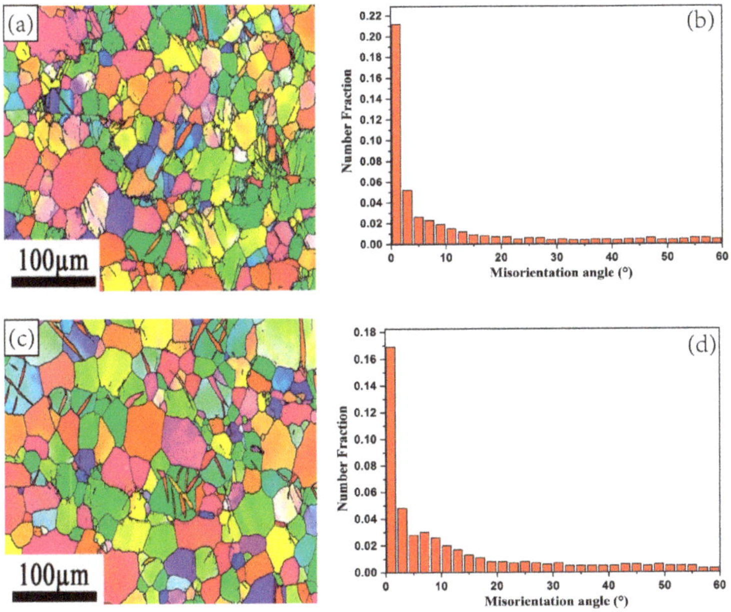

Figure 5. EBSD microstructure and misorientation angle distribution maps of Mg-Zn-Y alloy without UVA (**a**,**b**) with UVA (**c**,**d**).

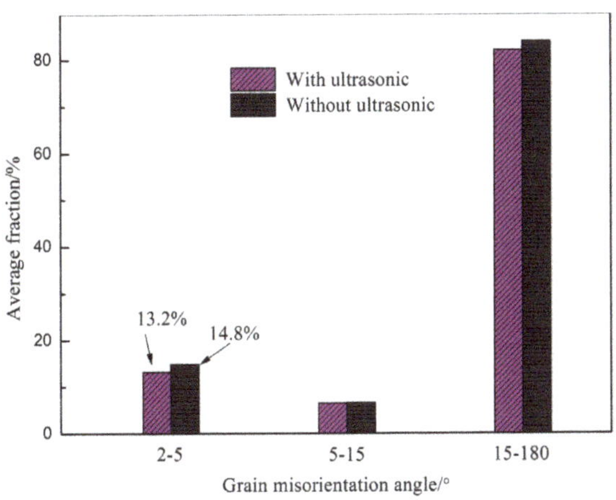

Figure 6. Comparison of average misorientation angle distribution.

KAM (Kernel average misorientation) is a common method to characterize the degree of plastic deformation of metallic materials by calculating the orientation difference between the target pixel point and its surrounding pixel points and then determining the average value of these differences, resulting in the KAM value. The KAM value can qualitatively reflect the degree of homogenization of plastic deformation. The higher value of the location indicates a higher degree of plastic deformation or a higher density of defects [28]. Figure 7 depicts the KAM maps of Mg-Zn-Y alloy during tensile deformation at room temperature.

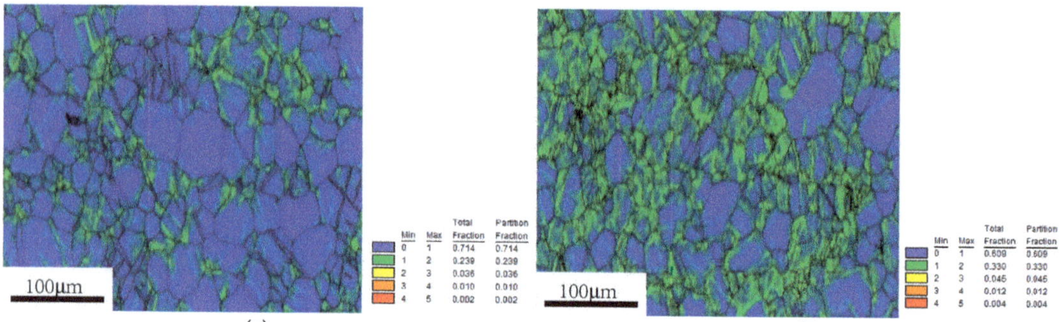

(a) (b)

Figure 7. KAM value distribution map: (**a**) without UVA; (**b**) with UVA.

From Figure 7, it can be seen that the KAM values of deformed specimens without vibration are not uniformly distributed. KAM values are generally higher at grain boundaries and lower within most individual grains. This observation suggests that, in the absence of vibration, plastic deformation in the material primarily occurs near the grain boundaries, and not all grains are involved in the deformation. With the application of UV, KAM values become higher and are uniformly distributed both at grain boundaries and within most individual grains. This observation signifies that, after introducing ultrasonic vibration, a greater number of grains participate in the plastic deformation process, leading to a more even distribution of plastic deformation across different grains. The above-mentioned difference indicates that ultrasonic vibration can promote grain refinement and grain rotation, thus enhancing the coordination of deformation between grains. In conclusion, ultrasonic vibration promotes coordinated intra- and intergranular deformation of the material, leading to a reduction in deformation resistance and enhanced uniformity, ultimately improving formability.

3.2.2. Dislocation Structure

In order to better analyze the mechanism of the influence of UV on material dislocations, the bright-fields TEM images were acquired, as shown in Figure 8. Figure 8a illustrates that in the absence of ultrasonic vibration, the sample exhibits a high dislocation density with a random and disorderly distribution, resulting in significant dislocation pile-ups. Figure 8b displays the dislocations within the sample under conditions of 30 kHz frequency and 7 μm amplitude. Notably, the dislocation lines are mostly parallel and appear to be straight. This results in a more organized and regular distribution of dislocations in contrast to those observed without UV.

It has been reported that ultrasound energy has a propensity to concentrate on localized dislocations [15]. After the application of UV, these dislocations, which have absorbed energy, become more mobile and tend to distribute themselves in a more organized and even manner. This phenomenon subsequently leads to a reduction in dislocation pile-ups, resulting in a decrease in material strength and an increase in plasticity.

The two primary explanations that can be used to explain the aforementioned phenomena are as follows, as illustrated in Figure 9. The first is the dislocation annihilation brought on by ultrasonic vibrations, which, as the physical model in Figure 9a illustrates, reduces the dislocation density. With a review of the reported investigations [29–31], it is obvious that superimposed ultrasonic vibrations can increase the dislocation mobility. This phenomenon aids dislocations located near grain boundaries in moving further and propagating over longer distances within the grains. This raises the possibility of dislocations with opposing signals coming into contact with one another, which could ultimately lead to dislocation annihilation.

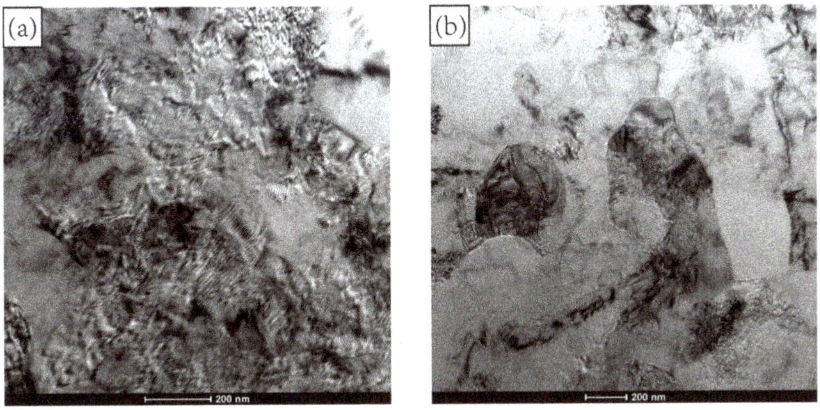

Figure 8. Dislocation distribution of tensile samples without and with ultrasonic vibration: (**a**) without UVA; (**b**) 30 kHz, 7 μm.

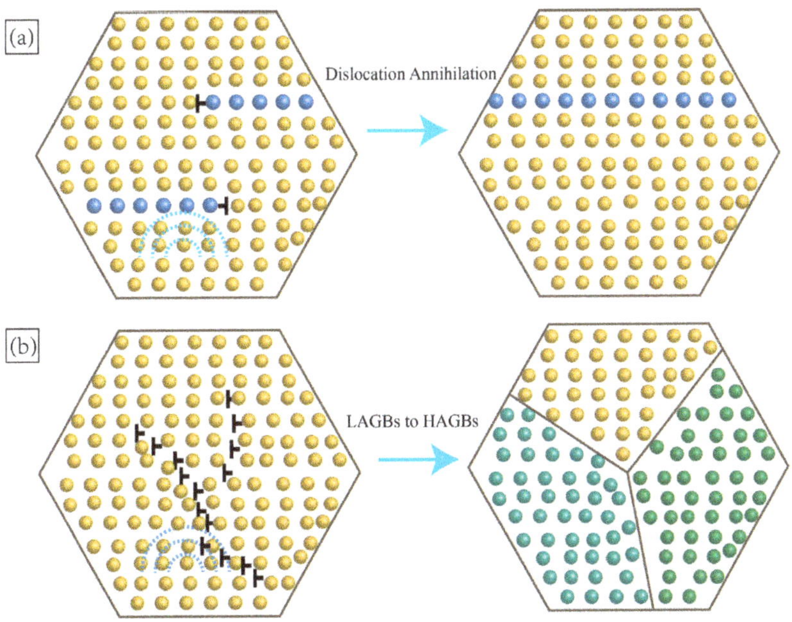

Figure 9. Physical model for micro mechanism in acoustic residual softening: (**a**) Dislocation annihilation; (**b**) LAGBs transformed to HAGBs.

Another reason is the enhancement of lattice rotation brought about by ultrasonic vibrations, which causes the LAGBs to become HAGBs, as depicted in Figure 9b. This was also found in Deng's study [32]. When ultrasonic oscillations are applied, dislocations with the same signal are dispersed and moved until they aggregate along the LAGBs, leading to stronger dislocation walls. In this case, lattice rotation is more likely to occur due to the coupling effect of strain and ultrasonic oscillation stresses. This offers a reasonable explanation for the notable decline in the percentage of LAGBs in the presence of ultrasonic vibration, as shown in Figure 6.

It is reported that during ultrasound-assisted vibratory processing, energy is mainly absorbed by defective locations such as dislocations and vacancies within the material [25].

This absorption results in the contraction of extended dislocations, allowing helical dislocations to cross-slip without the need for thermal activation [33]. In contrast, during thermoforming, thermal energy is uniformly absorbed throughout the entire specimen region. Consequently, the efficiency of acoustic softening is remarkably higher, approximately 107 times more effective than thermal softening [34]. Crystal defects such as dislocations can absorb energy during vibration loading and promote dislocation movement, thus causing a decrease in stress. These microscopic observations offer valuable insights into the mechanisms behind the phenomenon of vibration-induced softening. Nevertheless, it's important to note that the vibration-induced softening effect is temporary and vanishes once the vibration is removed. Additionally, it is challenging to observe the vibration-induced softening phenomenon in situ. This inherent transience and difficulty in real-time observation are the primary reasons why a unified theory explaining the mechanism of vibration-induced softening has remained elusive for an extended period.

3.2.3. Fracture Morphology

The fracture morphology without UV is presented in Figure 10a. This fracture morphology reveals that the wedge-shaped edge is flat and sharp, and the two material parts separate smoothly during the final stages of the fracture. With the application of ultrasonic vibration (UV), as evident in Figure 10a, the fracture surface appears rough and irregular, and there is an absence of a distinct wedge-shaped edge. The fracture surface is mainly composed of some irregularly shaped fiber zone contours, especially when the grain size is larger. In this case, the width of the fiber zone is notably wider. This phenomenon indicates that after loading ultrasonic vibration, the material undergoes a violent deformation at the fracture separation stage. After necking, there is a rapid reduction in the local cross-sectional area of the specimen. This reduction leads to an increase in the energy density of ultrasonic vibration per unit volume in this particular region. The repeated loading and unloading at high intensity prevent the material from undergoing a smooth separation process, ultimately resulting in the formation of complex and irregular fracture morphologies.

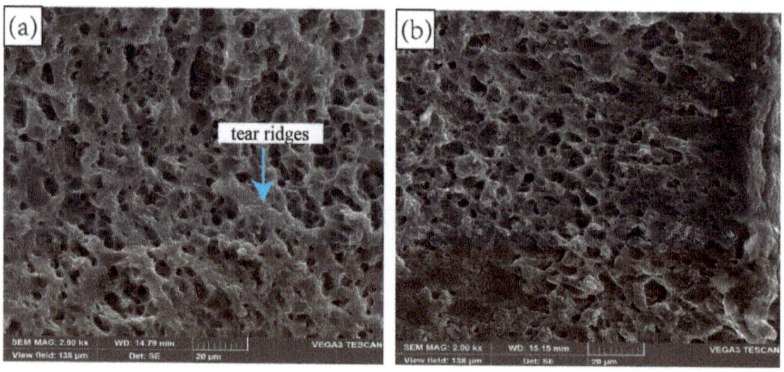

Figure 10. Fracture morphology of tensile samples without and with ultrasonic vibration: (**a**) without UVA; (**b**) 30 kHz, 7 μm.

By comparing the fracture morphology, we can also determine that a large number of dimples appear in the fracture without UV. However, when the UV is applied, the number of dimples decreases. This is mainly due to the fact that when the ultrasonic vibration is applied, the ultrasonic energy impacts defects such as grain boundaries, thereby relieving dislocation blockages. This is less conducive to the initiation and growth of voids, resulting in a reduced number of dimples. It should be noted in particular that, with the application of UV, there is a decrease in the presence of tear ridges on the fracture surface, and the size and depth of the dimples increase. This observation is indicative of the high ductility of the material. The analysis of the fracture morphology aligns well with the statistical results of elongation, confirming the material's enhanced ductility under the influence of ultrasonic vibration.

4. Discussion

The above experimental results describe the influence of parameters, such as vibration frequencies, amplitudes, and strain rates, on the microstructure and mechanical properties of $Mg_{98.5}Zn_{0.5}Y_1$ alloy, giving us a good basis for modeling. In this section, the constitutive equation of $Mg_{98.5}Zn_{0.5}Y_1$ alloy under ultrasonic vibration is established using the Johnson-Cook and its modified model. It's worth noting that the frequency and amplitude parameters were introduced to modify the model. The resulting constitutive equation for $Mg_{98.5}Zn_{0.5}Y_1$ alloy effectively captures the plastic deformation characteristics under ultrasonic vibration. This equation aligns well with the observed plastic deformation behavior of $Mg_{98.5}Zn_{0.5}Y_1$ alloy when subjected to ultrasonic vibration.

4.1. *Johnson–Cook Model of $Mg_{98.5}Zn_{0.5}Y_1$ Alloy without Vibration*

The Johnson-Cook (JC) model is an empirical metal constitutive model proposed by Johnson and Cook in 1983 [35]. Due to its simple formula and high prediction accuracy, the JC model has become one of the most widely used flow stress models. By coupling metal strain hardening, strain rate, and thermal softening, the model can effectively predict the metal stress-strain relationship at different strain rates and temperatures. The original JC model can be described as,

$$\sigma = (A + B\varepsilon^n)\left(1 + Cln\frac{\dot{\varepsilon}}{\dot{\varepsilon}_0}\right)(1 - T^{*m}) \tag{1}$$

where σ is the equivalent flow stress and A is the yield stress at the reference strain rate, indicating the stress resistance to micro plastic deformation. B is the coefficient of strain-hardening, ε is the equivalent plastic strain; and n represents the strain-hardening exponent, describing the strain hardening behavior. C is the coefficient of strain-rate hardening, which reflects the sensitivity of material to strain rate. $\dot{\varepsilon}$ and $\dot{\varepsilon}_0$ are the strain rate and strain rate reference, respectively. T refers to the homogenization temperature; m is the thermal softening index of the material. To enhance the prediction of material properties for magnesium alloys and address the significant deviation in the traditional JC model in describing stress-strain behavior, we incorporated an improved J-C model [23]. This improved model includes a quadratic term of strain and considers the coupling effect of temperature and strain rate on the flow behaviors. With these enhancements, we aim to achieve more accurate and reliable predictions for the material's response under different conditions. Given that the experiment is conducted at room temperature, the temperature component is omitted from the model. The improved J-C model can be expressed as follows:

$$\sigma = \left[A + B_1\varepsilon + B_2\varepsilon^2 + B_3\varepsilon^3\right]\left(1 + Cln\frac{\dot{\varepsilon}}{\dot{\varepsilon}_0}\right) \tag{2}$$

where B_1, B_2, and B_3 are the strain-hardening coefficients. C is the strain rate hardening factor. The remaining parameter definitions are consistent with the original J-C model. The values of the above parameters need to be determined in combination with experimental data. In this study, the reference strain rate is taken as the minimum value of 0.001 s^{-1}, and the improved J-C model can be written as follows:

$$\sigma = \left[A + B_1\varepsilon + B_2\varepsilon^2 + B_3\varepsilon^3\right] \tag{3}$$

By fitting the stress-strain curves under the reference strain rate, as shown in Figure 11, the values of A, B_1, B_2, and B_3 can be obtained as 110.18 MPa, 521.72 MPa, −1659.45 MPa and 1561.44 MPa, respectively. Equation (2) can be transformed into:

$$\sigma = C\cdot\left(A + B_1\varepsilon + B_2\varepsilon^2 + B_3\varepsilon^3\right)\cdot ln\frac{\dot{\varepsilon}}{\dot{\varepsilon}_0} + \left(A + B_1\varepsilon + B_2\varepsilon^2 + B_3\varepsilon^3\right) \quad (4)$$

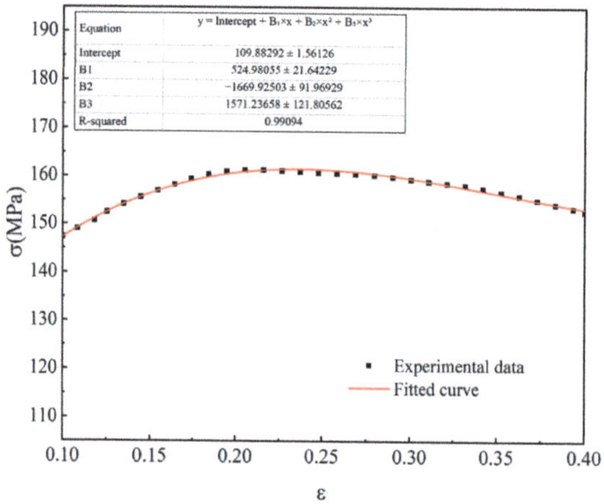

Figure 11. Fitting results of σ and ε for Mg-Zn-Y alloy without UVA tensile test.

Combining formula (4) and experimental data, the values of different strain rates under certain strains can be obtained. In order to determine the value of the strain rate coefficient C, in this experiment, the strain is set in the range of 0.1–0.4, and the strain increment is 0.05. The formula σ–$ln\dot{\varepsilon}/\dot{\varepsilon}_0$ is linearly fitted to obtain the slope of $ln\dot{\varepsilon}/\dot{\varepsilon}_0$, which is C ($A + B_1\varepsilon + B_2\varepsilon^2 + B_3\varepsilon^3$). The fitting results are shown in Figure 12. The constitutive equation of the Mg-Zn-Y alloy without vibration, based on the average value of the seven different C values (calculated to be 0.0287), can be expressed as follows:

$$\sigma = \left[110.18 + 521.72\varepsilon - 1659.45\varepsilon^2 + 1561.44\varepsilon^3\right]\left(1 + 0.0287ln\frac{\dot{\varepsilon}}{\dot{\varepsilon}_0}\right) \quad (5)$$

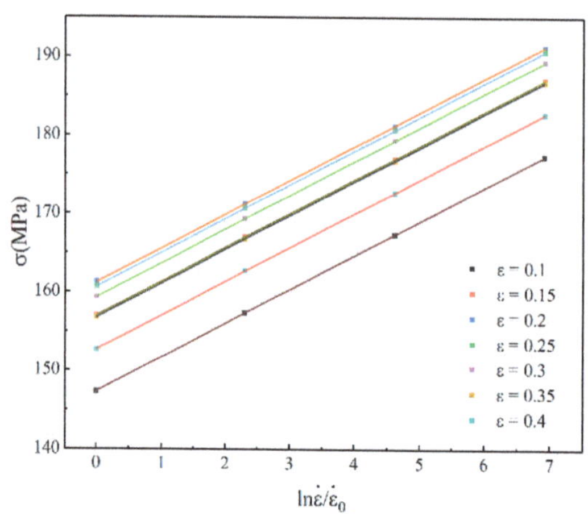

Figure 12. Fitting results of σ and $ln\dot{\varepsilon}/\dot{\varepsilon}_0$ of Mg-Zn-Y alloy without UVA.

4.2. Johnson–Cook Model of $Mg_{98.5}Zn_{0.5}Y_1$ Alloy with Vibration

According to the tensile curves of Mg-Zn-Y alloys under different vibration conditions, the values of parameters under different amplitudes and frequencies were obtained using the experimental method described in Section 4.1. The results are shown in Tables 4–6:

Table 4. Material constants in the modified JC model at 20 kHz.

Amplitude/μm	A/MPa	B1/MPa	B2/MPa	B3/MPa	C
0	110.18	521.72	−1659.45	1561.44	0.0287
3	113.98	624.06	−1764.68	2304.7	0.0296
5	70.22	817.34	−2957.88	3085.99	0.0338
7	60.34	736.83	−2587.12	2721.06	0.0362

Table 5. Material constants in the modified JC model at 25 kHz.

Amplitude/μm	A/MPa	B1/MPa	B2/MPa	B3/MPa	C
0	110.18	521.72	−1659.45	1561.44	0.0287
3	99.94	483.64	−1411.37	1100.37	0.0302
5	71.04	756.17	−2661.68	2678.3	0.0337
7	54.51	805.57	−2931.22	3209.02	0.0371

Table 6. Material constants in the modified JC model at 30 kHz.

Amplitude/μm	A/MPa	B1/MPa	B2/MPa	B3/MPa	C
0	124.43	160.91	−778.08	730.22	0.082
3	87.35	651.5	−2209.28	2576.18	0.031
5	69.99	739.15	−2591.32	2611.52	0.0343
7	59.29	691.36	−2414.21	2437.48	0.0384

In this study, the constitutive model of Mg-Zn-Y alloy under vibration conditions is established according to the material parameter values under different frequency and amplitude conditions. In order to demonstrate the impact of the amplitude on the material constants, the subscripts '0' and 'v' are used to indicate the separation and recording of the material constants with and without oscillation, respectively. Thus, the A_0, B_1, B_2, and B_3 and C_0 represent the parameter values with no vibration. Concurrently, the A_V, B_{1V}, B_{2V}, B_{3V}, and C_V represent the parameter values with vibration. The variation of A_0/A_v, B_{1v}/B_1, B_{2v}/B_2, B_{3v}/B_3, and C_v/C_0 with amplitude can be computed and plotted in Figures 13 and 14.

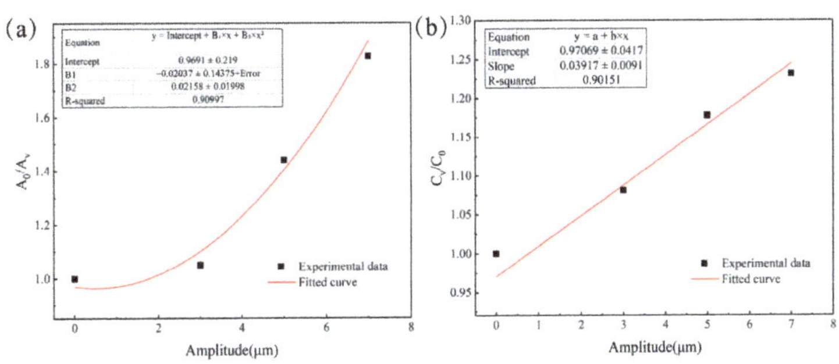

Figure 13. Influence of the amplitude on the parameter ratio at 20 kHz: (**a**) A_0/A_v, (**b**) C_v/C_0.

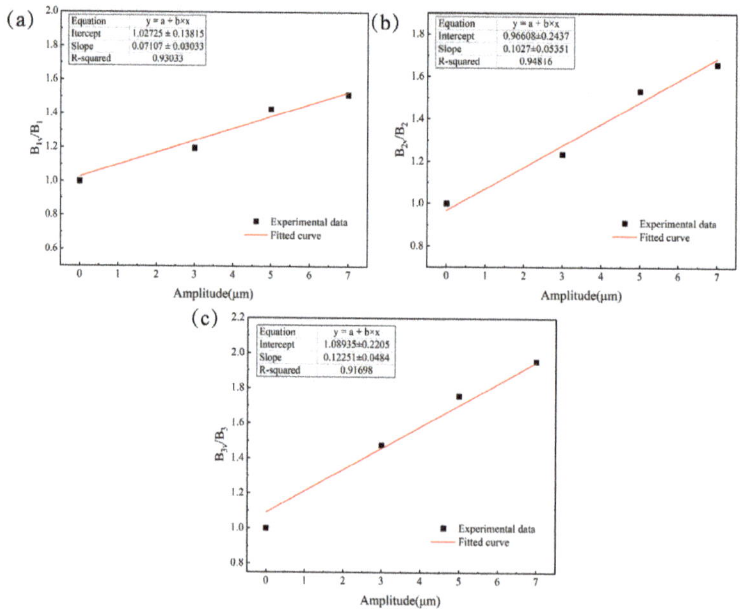

Figure 14. Influence of the amplitude on the parameter ratio at 20 kHz: (**a**) B_{1v}/B_1, (**b**) B_{2v}/B_2, (**c**) B_{3v}/B_3.

Figures 13 and 14 depict the amplitude-based trends for A_0/A_v, C_v/C_0, B_{1v}/B_1, B_{2v}/B_2, and B_{3v}/B_3 in sequential order. Notably, the ratio increases with higher vibration amplitudes. Based on the variation of material parameter values and amplitude shown in the figures, we propose the relations (6)–(10). Here, 'a' represents the amplitude parameter, while the coefficients K, H, and L are derived from the calculation results. The fitting curves in red on Figures 13 and 14 illustrate the outcomes of the fitting process, and the corresponding parameter values are presented in Table 7.

$$A_0/A_v = K_1a^2 + Ha + L_1 \tag{6}$$

$$B_{1v}/B_1 = K_2a + L_2 \tag{7}$$

$$B_{2v}/B_2 = K_3a + L_3 \tag{8}$$

$$B_{3v}/B_3 = K_4a + L_4 \tag{9}$$

$$C_v/C_0 = K_5a + L_5 \tag{10}$$

Table 7. Values of the material parameters at 20, 25 and 30 kHz after fitting.

f/kHz	K_1	K_2	K_3	K_4	K_5	H	L_1	L_2	L_3	L_4	L_5
20	−0.02	0.07	0.10	0.12	0.04	0.02	0.97	1.03	0.97	1.08	0.97
25	−0.03	0.08	0.11	0.14	0.04	0.03	0.98	1.00	0.97	1.05	0.97
30	0.008	0.05	0.07	0.08	0.04	0.07	0.99	1.05	1.07	1.17	0.97

According to the calculation results in Table 7, the relationship equation for the variation of material parameters with amplitude at 20 kHz frequency can be derived as shown in Equation (11):

$$\sigma = \left[\frac{110.18}{-0.02a^2+0.02a+0.97} + (36.52a+537.37)\varepsilon - (165.95a+1609.67)\varepsilon^2\ (187.37a+1686.36)\varepsilon^3\right]\left[1+(0.0011a+0.0278)ln\frac{\dot{\varepsilon}}{\dot{\varepsilon}_0}\right] \quad (11)$$

Using the same method described above, we can derive the vibrational equations of Mg-Zn-Y alloy under the 25 kHz and 30 kHz frequency conditions, as illustrated in Equations (12) and (13).

$$\sigma = \left[\frac{110.18}{-0.08a^2+0.03a+0.98} + (41.74a+521.72)\varepsilon - (182.54a+1609.67)\varepsilon^2 + (218.54a+1639.05)\varepsilon^3\right]\left[1+(0.0011a+0.0278)ln\frac{\dot{\varepsilon}}{\dot{\varepsilon}_0}\right] \quad (12)$$

$$\sigma = \left[\frac{110.18}{0.008a^2+0.07a+0.99} + (26.09a+547.81)\varepsilon - (116.16a+1775.61)\varepsilon^2 + (124.92a+1826.88)\varepsilon^3\right]\left[1+(0.0011a+0.0278)ln\frac{\dot{\varepsilon}}{\dot{\varepsilon}_0}\right] \quad (13)$$

By comparing Equations (11)–(13), it becomes evident that the parameters of the material's vibration constitutive equation undergo changes at different frequencies. Nonetheless, it is a common practice in academic research to concentrate primarily on constitutive equations at a single, specific frequency, with limited exploration of variations in frequency. The calculations presented above clearly demonstrate that changes in vibration frequencies have a substantial impact on the parameters of the vibration equation. Therefore, it becomes imperative to consider the influence of frequency and amplitude on the constitutive equation's parameters for ultrasonic vibration. Notably, the strain rate hardening coefficient remains almost entirely unaffected by the vibration frequency, as evidenced by Equations (11)–(13).

Referring to the literature, it is known that the acoustic energy density in the sample can be calculated using the formula $E = a^2(2\pi f)^2\rho/2$ [36,37]. Additionally, Longenecker's research revealed that the reduction of metal stress under ultrasonic vibration is proportional to the input of sound energy [25]. Building on this understanding, we assume that the material's flow stress reduction under ultrasonic vibration is proportional to the product of vibration amplitude squared and frequency squared, denoted as a^2f^2. Consequently, new parameter relationship Equations (14)–(17) are proposed accordingly. In Equations (14)–(17), a and f represent the amplitude and frequency of ultrasonic vibration, respectively. D, E, G, and I represent the coefficients of the parameter relational equation. By combining this information with the data presented in Tables 4–6, the results of the coupling between frequency and amplitude are depicted in in Table 8.

$$A_0/A_v = D_1a^2f^2 + D_2a + D_3f + D_4 \quad (14)$$

$$B_{1v}/B_1 = E_1a + E_2f + E_3 \quad (15)$$

$$B_{2v}/B_2 = G_1a + G_2f + G_3 \quad (16)$$

$$B_{3v}/B_3 = I_1a + I_2f + I_3 \quad (17)$$

Table 8. Parameter values after fitting the amplitude and frequency.

A_0/A_v		B_{1v}/B_1		B_{2v}/B_2		B_{3v}/B_3	
D1	1.8×10^{-5}	E1	0.07	G1	0.1	I1	0.12
D2	0.083	E2	−0.005	G2	−0.001	I2	−0.007
D3	−0.004	E3	1.14	G3	1.04	I3	1.29
D4	1.03						

Based on the fitting results presented in Table 8, the constitutive equation of the Mg-Zn-Y alloy under ultrasonic vibration is ultimately determined as follows:

$$\sigma = \left[\frac{110.18}{1.8\times10^{-5}a^2f^2+0.083a-0.004f+1.03} + (36.52a - 2.61f + 594.76)\varepsilon - (165.95a - 1.66f + 1725.83)\varepsilon^2 + (187.32a - 10.93f + 2013.69)\varepsilon^3\right]\left[1 + (0.0011a + 8.61 \times 10^{-5}f + 0.025)ln\frac{\dot{\varepsilon}}{\dot{\varepsilon}_0}\right] \quad (18)$$

4.3. *Verification of the Constitutive Equations*

After establishing the ultrasonic vibration constitutive model at the reference temperature, we conducted an evaluation of the material fluidity prediction for the Mg-Zn-Y alloy. The results of this evaluation are depicted in Figure 15a. Evidently, the predicted values from the constitutive model align well with the data obtained from experimental measurements. In order to further compare the accuracy of the model to predict material flow, the maximum error between the predicted stress value σ_p and the measured value σ_m can be expressed as:

$$\text{Error rate} = \frac{|\sigma_p - \sigma_m|}{\sigma_m} \times 100\%$$

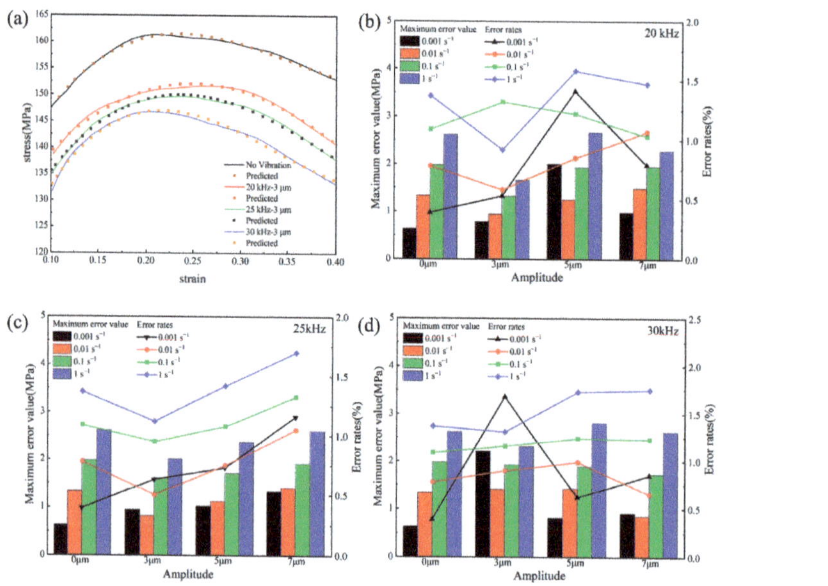

Figure 15. The error between the experimental value and the predicted value: (**a**) 3 μm, (**b**) 20 kHz, (**c**) 25 kHz, (**d**) 30 kHz.

After calculation, the error rate between the improved models σ_p and σ_m are shown in Figure 15b–d. It is evident from the results that the flow stress estimate has a maximum error of only 2.62. In addition, when compared to the original and modified Johnson-Cook models, the improved model offers a more precise estimation of the flow behavior. These results demonstrate that the enhanced Johnson-Cook model presented in this work is appropriate for modeling the Mg-Zn-Y alloys' tensile behavior.

By combining the experimental findings mentioned above, it's apparent that the improved Johnson-Cook's model performs significantly better at predicting the material flow behavior of Mg-Zn-Y alloys under typical tensile conditions. This model exhibits a significant degree of concordance with the experimental results, particularly when it comes to predicting the stress-strain curves of the constitutive equation for Mg-Zn-Y alloys.

5. Finite Element Simulation

5.1. Effect of Ultrasonic Vibration on Stress-Strain Distribution during the Tensile Process

Finite element analysis serves as a potent tool for investigating the plastic forming process of metals. To gain a more profound understanding of the deformation mechanism of Mg-Zn-Y alloy under the influence of ultrasonic vibration, finite element simulations were conducted for both conventional tensile tests and ultrasonic vibration-assisted tensile tests using Abaqus software. In this study, discrete rigid body modeling was employed for the tensile fixture part, and reference points were strategically placed to facilitate the application of boundary conditions. The grid of the tensile specimen part was configured as C3D8R, as depicted in Figure 16. Accounting for the influence of strain rate on stress, the tensile rate was set to 0.001 s^{-1}. Additionally, the frequency was set to 20 kHz and 30 kHz, while the amplitude was varied between 0 μm, 3 μm, 5 μm, and 7 μm.

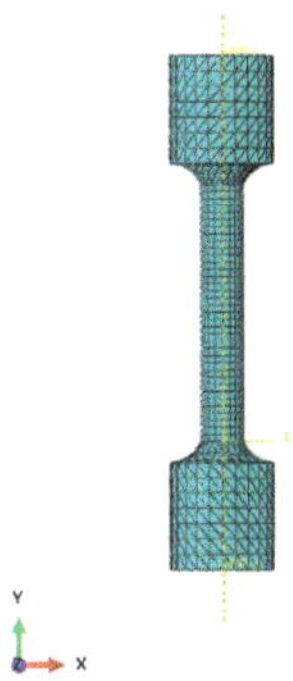

Figure 16. Finite element model of UVA tensile test.

Figures 17 and 18 present a cloud diagram illustrating the stress-strain distribution of the specimen during the stretching process at various amplitude frequencies. In comparison to conventional tensile tests, the introduction of ultrasonic vibration leads to a reduction in the flow stress of the metal material. Importantly, the extent of this reduction is directly correlated with the amplitude of the applied ultrasonic vibration, with larger amplitudes resulting in more substantial reductions in flow stress. Figure 18 displays the magnitude distribution of strain under various amplitude frequency conditions. It is evident that after applying ultrasonic vibration, the maximum strain of the specimen increases with higher frequencies, while variations in amplitude have a relatively smaller impact on strain.

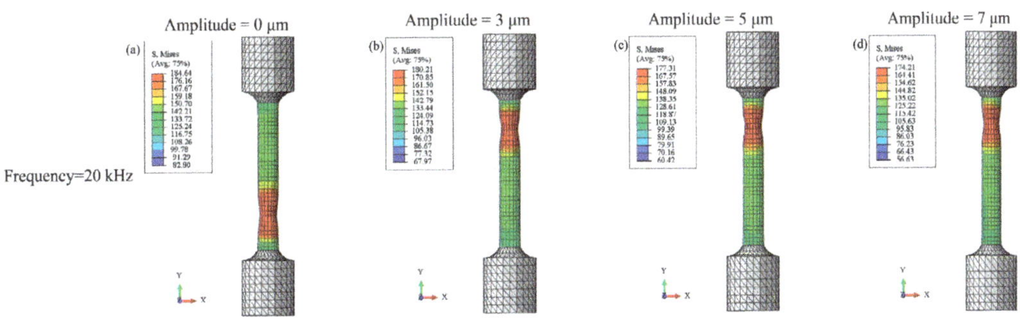

Figure 17. *Cont.*

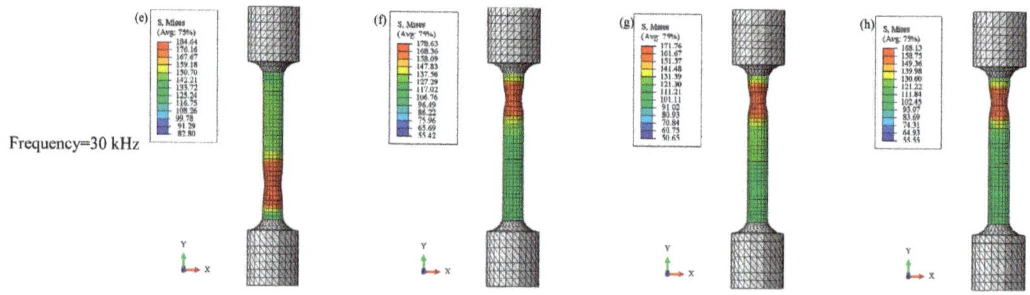

Figure 17. Stress distribution of tensile specimens at different amplitude and frequencies: (**a**–**d**) 20 kHz, (**e**–**h**) 30 kHz.

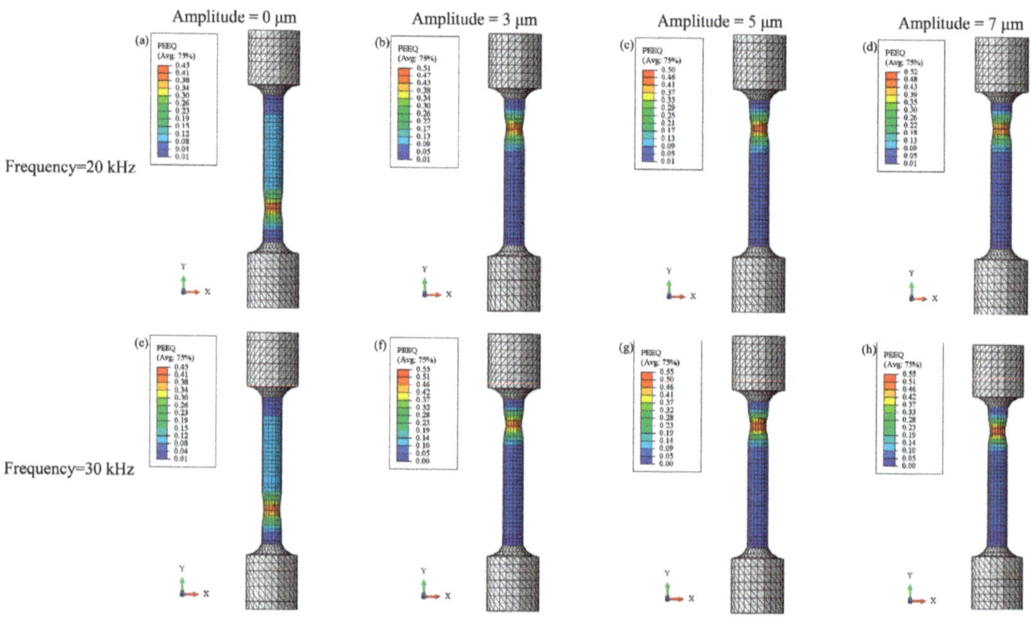

Figure 18. Strain distribution of tensile specimens at different amplitude and frequencies: (**a**–**d**) 20 kHz, (**e**–**h**) 30 kHz.

5.2. Effect of Ultrasonic Vibration on Material Flow Behavior

Figure 19 depict the velocity distribution fields of Mg-Zn-Y alloy specimens under various conditions. Given the immobility of the specimen's bottom during conventional stretching, the velocity at the base remains at zero. However, the material within the stretched portion of the specimen undergoes flow in tandem with the stretching process. After applying ultrasonic vibration, the material's flow velocity undergoes significant transformation. Notably, the flow velocity's direction no longer aligns with the tensile direction; instead, in certain material sections, the flow velocity opposes the tensile direction. By comparing the velocity clouds, it becomes evident that the maximum flow velocity of the material is shifted from the upper part to the main deformed part in the middle of the specimen. Given that vibration primarily demonstrates stress superposition effects within finite element analysis, the vibration frequency exerts a relatively minor influence on stress. However, altering the frequency can induce shifts in material flow velocity. Notably, the figure highlights that an increase in frequency corresponds to an elevation in the material's maximum flow velocity.

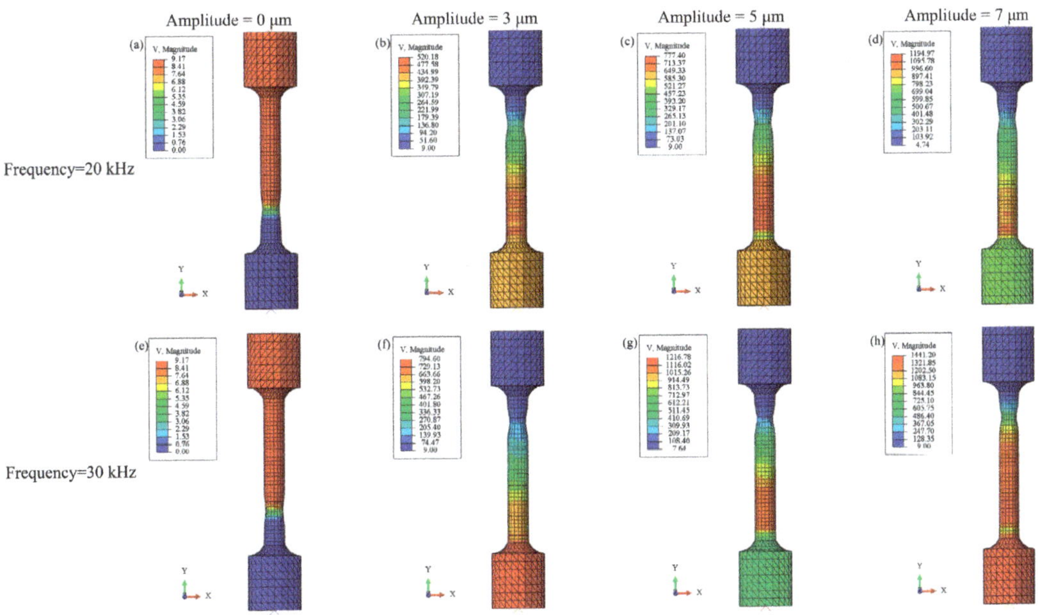

Figure 19. Velocity field distribution of tensile specimen under different conditions: (**a**–**d**) 20 kHz, (**e**–**h**) 30 kHz.

As depicted in Figure 19, with an amplitude of 5 μm, the maximum flow velocity experiences an upsurge as the frequency increases. Remarkably, the material attains a peak flow velocity of 1216 mm·s^{-1} at 30 kHz. Similar to amplitude, frequency incites only an instantaneous rate alteration, resulting in a relatively minor shift in the average flow rate throughout the deformation process.

6. Conclusions

Ultrasonic vibration assisted tension was performed on Mg-Zn-Y alloy. The mechanical behavior and microstructure evolution of the sheet at different amplitudes and frequencies were examined. The constitutive equation of Mg-Zn-Y alloy under tension with vibration is constructed using the modified JC model. Based on the experimental results, the following conclusions can be drawn:

(1) In the ultrasonic vibration stretching process, both the amplitude and frequency of vibration affected the flow behavior of $Mg_{98.5}Zn_{0.5}Y_1$ alloy during the stretching process. The acoustic softening effect caused by ultrasonic vibration is gradually obvious with the increase of amplitude. A comparison of the stress-strain curves at various frequencies and amplitudes leads to the conclusion that amplitude has a more significant effect on the flow stresses of $Mg_{98.5}Zn_{0.5}Y_1$ alloy when compared to vibration frequency.

(2) A hybrid constitutive model coupling the frequency and amplitude of ultrasonic vibration is developed based on the modified Johnson Cook model. Comparing to the measured stress, it is evident that the proposed Johnson-Cook model aligns well with the experimental data, exhibiting a maximum error in the flow stress estimate of only 2.62. That indicates the proposed JC model is very suitable for the flow behavior prediction of $Mg_{98.5}Zn_{0.5}Y_1$ alloy with and without UV.

(3) The finite element analysis results show that ultrasonic vibration has a great influence on the stress-strain distribution in the stretching process. There is great potential in applying vibration-assisted ultrasonic vibration methods to the thermal processing of magnesium alloys. Examples of such applications include vibration-assisted additive manufacturing and vibration-assisted stamping. It can effectively improve the formability of magnesium alloy and improve the quality of formed parts.

Author Contributions: Conceptualization, Z.X., X.G., H.S. and W.Y.; methodology, Z.X. and X.G.; software, W.Y.; validation, W.Y., F.X. and H.Y.; formal analysis, Z.X. and W.Y.; investigation, W.Y., F.X. and H.Y.; resources, Z.X., X.G. and H.S.; data curation, Z.X. and W.Y.; writing—original draft preparation, W.Y.; writing—review and editing, Z.X. and W.Y.; visualization, Z.X. and W.Y.; supervision, Z.X.; project administration, Z.X.; funding acquisition, Z.X., X.G. and H.S. All authors have read and agreed to the published version of the manuscript.

Funding: This research was funded by National Natural Science Foundation of China (52103290), Henan Natural Science Youth Program (232300420302), Henan Postdoctoral Foundation, the Training Program for Young Backbone Teachers in the University of Henan Province, the Fundamental Research Funds for the Universities of Henan Province (NSFRF210333), Natural Science Foundation of Henan Polytechnic University (B2019-42).

Data Availability Statement: Data are contained within the article.

Conflicts of Interest: The authors declare no conflicts of interest.

References

1. Song, J.; She, J.; Chen, D.; Pan, F. Latest research advances on magnesium and magnesium alloys worldwide. *J. Magnes. Alloys* **2020**, *8*, 1–41. [CrossRef]
2. Yin, Z.; Qi, W.; Zeng, R.; Chen, X.; Gu, C.; Guan, S.; Zheng, Y. Advances in coatings on biodegradable magnesium alloys. *J. Magnes. Alloys* **2020**, *8*, 42–65. [CrossRef]
3. Doege, E.; Dröder, K. Sheet metal forming of magnesium wrought alloys -formability and process technology. *J. Mater. Process. Technol.* **2001**, *115*, 14–19. [CrossRef]
4. Wang, L.; Qiao, Q.; Liu, Y.; Song, X. Formability of AZ31 Mg alloy sheets within medium temperatures. *J. Magnes. Alloys* **2013**, *1*, 312–317. [CrossRef]
5. Liu, N.; Lai, Z.; Cao, Q.; Han, X.; Huang, Y.; Li, X.; Chen, M.; Li, L. Effects of air on metallic sheet deformation by electromagnetic forming. *Int. J. Adv. Manuf. Technol.* **2019**, *103*, 311–324. [CrossRef]
6. Fauzi, E.R.I.; Jamil, M.S.C.; Samad, Z.; Sheikh, M.A.; Najib, A.M. Influence of non-conventional beam profile on edge effects in laser forming of AISI 304 stainless steel plate. *Int. J. Adv. Manuf. Technol.* **2019**, *104*, 1593–1601. [CrossRef]
7. Langenecker, B.; Frandsen, W.; Kennedy, A. Effects of ultrasound on deformation characteristics of structural metals Part 2. Acoustic heating and zone melting. *Tech. Prog. Rep.* **1965**, *371*, 14.
8. Wen, T.; Wei, L.; Chen, X.; Pei, C. Effects of ultrasonic vibration on plastic deformation of AZ31 during the tensile process. *Int. Miner. Metal. Mater.* **2011**, *18*, 70–76. [CrossRef]
9. Hung, J.; Hung, C. The influence of ultrasonic-vibration on hot upsetting of aluminum alloy. *Ultrasonics* **2005**, *43*, 692–698. [CrossRef]
10. Siegert, K.; Ulmer, J. Superimposing ultrasonic waves on the dies in tube and wire drawing. *J. Eng. Mater. Technol.* **2001**, *123*, 517–523. [CrossRef]
11. Mousavi, S.A.A.A.; Feizi, H.; Madoliat, R. Investigations on the effects of ultrasonic vibrations in the extrusion process. *J. Mater. Process. Technol.* **2007**, *187–188*, 657–661. [CrossRef]
12. Aziz, S.A.; Lucas, M. Characterising the acoustoplastic effect in an ultrasonically assisted metal forming process. *IOP Conf. Ser. Mater. Sci. Eng.* **2012**, *42*, 12014–12017. [CrossRef]
13. Cao, M.; Li, J.; Yuan, Y.; Zhao, C. Flexible die drawing of magnesium alloy sheet by superimposing ultrasonic vibration. *Trans. Nonferrous Met. Soc. China* **2017**, *27*, 163–171. [CrossRef]
14. Bagherzadeh, S.; Abrinia, K.; Han, Q. Ultrasonic assisted equal channel angular extrusion (UAE) as a novel hybrid method for continuous production of ultra-fine grained metals. *Mater. Lett.* **2016**, *169*, 90–94. [CrossRef]
15. Liao, J.; Zhang, L.; Xiang, H.; Xue, X. Mechanical behavior and microstructure evolution of AZ31 magnesium alloy sheet in an ultrasonic vibration-assisted hot tensile test. *J. Alloys Compd.* **2022**, *895*, 162575. [CrossRef]
16. Yao, Z.; Kim, G.; Faidley, L.; Zou, Q.; Mei, D.; Chen, Z. Effects of superimposed high-frequency vibration on deformation of aluminum in micro/meso-scale upsetting. *J. Mater. Process. Technol.* **2012**, *212*, 640–646. [CrossRef]
17. Hu, J.; Shimizu, T.; Yang, M. Investigation on dynamic impact effect of ultrasonic-assisted compression test. *Mater. Sci. For.* **2018**, *920*, 102–107. [CrossRef]

18. Storck, H.; Littmann, W.; Wallaschek, J.; Mracek, M. The effect of friction reduction in presence of ultrasonic vibrations and its relevance to travelling wave ultrasonic motors. *Ultrasonics* **2002**, *40*, 379–383. [CrossRef]
19. Yao, Z.; Kim, G.; Wang, Z.; Faidley, L.; Zou, Q.; Mei, D.; Chen, Z. Acoustic softening and residual hardening in aluminum: Modeling and experiments. *Int. J. Plast.* **2012**, *39*, 75–87. [CrossRef]
20. Prabhakar, A.; Verma, G.C.; Krishnasamy, H.; Pandey, P.M.; Lee, M.G.; Suwas, S. Dislocation density based constitutive model for ultrasonic assisted deformation. *Mech. Res. Commun.* **2017**, *85*, 76–80. [CrossRef]
21. Sedaghat, H.; Xu, W.; Zhang, L. Ultrasonic vibration-assisted metal forming: Constitutive modelling of acoustoplasticity and applications. *J. Mater. Process. Technol.* **2019**, *265*, 122–129. [CrossRef]
22. Meng, B.; Cao, B.N.; Wan, M.; Wang, C.J.; Shan, D.B. Constitutive behavior and microstructural evolution in ultrasonic vibration assisted deformation of ultrathin superalloy sheet. *Int. J. Mech. Sci.* **2019**, *157–158*, 609–618. [CrossRef]
23. Lin, J.; Li, J.; Liu, T.; Xie, Z.; Zhu, L.; Wang, Y.; Guan, Y. Investigation on ultrasonic vibration effects on plastic flow behavior of pure titanium: Constitutive modeling. *J. Mater. Res. Technol.* **2020**, *9*, 4978–4993. [CrossRef]
24. Cao, M.Y.; Hu, H.; Jia, X.D.; Tian, S.J.; Zhao, C.C.; Han, X.B. Mechanism of ultrasonic vibration assisted upsetting of 6061 aluminum alloy. *J. Manuf. Process.* **2020**, *59*, 690–697. [CrossRef]
25. Langenecker, B. Effects of ultrasound on deformation characteristics of Metals. *IEEE Trans. Sonics Ultrason.* **1966**, *13*, 1–8. [CrossRef]
26. Liu, P.; Wang, R.; Liu, X.; Ren, R. Effect of surface ultrasonic rolling on evolution of surface microstructure of EA4T axle steel. *J. Mater. Eng. Perform.* **2021**, *30*, 1270–1279. [CrossRef]
27. Zhou, H.; Cui, H.; Qin, Q.; Wang, H.; Shen, Y. A comparative study of mechanical and microstructural characteristics of aluminium and titanium undergoing ultrasonic assisted compression testing. *Mater. Sci. Eng. A* **2017**, *682*, 376–388. [CrossRef]
28. Kamaya, M. Assessment of local deformation using EBSD: Quantification of local damage at grain boundaries. *Mater. Charact.* **2012**, *66*, 56–67. [CrossRef]
29. Liu, Y.; Wang, C.; Bi, R. Acoustic residual softening and microstructure evolution of T2 copper foil in ultrasonic vibration assisted micro-tension. *Mater. Sci. Eng. A* **2022**, *841*, 143044. [CrossRef]
30. Wang, X.; Wang, C.; Liu, Y.; Liu, C.; Wang, Z.; Guo, B.; Shan, D. An energy based modeling for the acoustic softening effect on the Hall-Petch behavior of pure titanium in ultrasonic vibration assisted micro-tension. *Int. J. Plast.* **2021**, *136*, 102879. [CrossRef]
31. Meng, D.A.; Zhao, X.; Li, J.; Zhao, S.; Han, Q. Mechanical behavior and microstructure of low-carbon steel undergoing low-frequency vibration-assisted tensile deformation. *J. Mater. Res.* **2017**, *32*, 3885–3893. [CrossRef]
32. Deng, T.; Liu, H. A Study of Mechanical Characteristics and Microstructural Evolution of Copper-Nickel Alloy Sheet Undergoing Ultrasonic Vibration Assisted Uniaxial Tension. *Mater. Sci. Eng. A* **2023**, *885*, 145608. [CrossRef]
33. Rusinko, A. Analytical description of ultrasonic hardening and softening. *Ultrasonics* **2011**, *51*, 709–714. [CrossRef] [PubMed]
34. Siddiq, A.; El Sayed, T. Ultrasonic-assisted manufacturing processes: Variational model and numerical simulations. *Ultrasonics* **2012**, *52*, 521–529. [CrossRef]
35. Johnson, G.R.; Cook, W.H. A constitutive model and data for metals subjected to large strains, high strain rates and high temperatures. *Eng. Fract. Mech.* **1983**, *21*, 541–548.
36. Frederick, J.R. *Ultrasonic Engineering*; Wiley: New York, NY, USA, 1965; p. 362.
37. Pierce, A.D. Basic Linear Acoustics. In *Springer Handbook of Acoustics*; Thomas, D.R., Ed.; Springer: New York, NY, USA, 2014; pp. 29–115.

MDPI AG
Grosspeteranlage 5
4052 Basel
Switzerland
Tel.: +41 61 683 77 34

Crystals Editorial Office
E-mail: crystals@mdpi.com
www.mdpi.com/journal/crystals

www.ingramcontent.com/pod-product-compliance
Lightning Source LLC
Chambersburg PA
CBHW061841210926
51850CB00006B/939